Nanotechnology for Electronic Materials and Devices

Nanostructure Science and Technology

Series Editor: David J. Lockwood, FRSC
National Research Council of Canada
Ottawa, Ontario, Canada

Current volumes in this series:

Alternative Lithography: Unleashing the Potentials of Nanotechnology
Edited by Clivia M. Sotomayor Torres

Interfacial Nanochemistry: Molecular Science and Engineering at Liquid–Liquid Interfaces
Edited by Hiroshi Watarai, Norio Teramae, and Tsuguo Sawada

Nanoparticles: Building Blocks for Nanotechnology
Edited by Vincent Rotello

Nanoscale Assembly: Chemical Techniques
Edited by Wilhelm T.S. Huck

Nanostructured Catalysts
Edited by Susannah L. Scott, Cathleen M. Crudden, and Christopher W. Jones

Nanotechnology for Electronic Materials and Devices
Edited by Anatoli Korkin, Jan Labanowski, Evgeni Gusev, and Serge Luryi

Nanotechnology in Catalysis, Volumes 1 and 2
Edited by Bing Zhou, Sophie Hermans, and Gabor A. Somorjai

Nanotechnology in Catalysis, Volume 3
Edited by Bing Zhou, Scott Han, Robert Raja, and Gabor A. Somorjai

Ordered Porous Nanostructures and Applications
Edited by Ralf Wehrspohn

Polyoxometalate Chemistry for Nano-Composite Design
Edited by Toshihiro Yamase and Michael T. Pope

Self-Assembled Nanostructures
Jin Z. Zhang, Zhong-lin Wang, Jun Liu, Shaowei Chen, and Gang-yu Liu

Self-Organized Nanoscale Materials
Edited by Motonari Adachi and David J. Lockwood

Semiconductor Nanocrystals: From Basic Principles to Applications
Edited by Alexander L. Efros, David J. Lockwood, and Leonid Tsybeskov

Surface Effects in Magnetic Nanoparticles
Edited by Dino Fiorani

A Continuation Order Plan is available for this series. A continuation order will bring delivery of each new volume immediately upon publication. Volumes are billed only upon actual shipment. For further information please contact the publisher.

Anatoli Korkin Jan Labanowski
Evgeni Gusev Serge Luryi

Editors

Nanotechnology for Electronic Materials and Devices

Springer

Anatoli Korkin
Nano & Giga Solutions
1683 E. Spur Street
Gilbert, AZ 85296
USA
korkin@nanoandgiga.com

Jan Labanowski
204A Nieuwland Science Hall
University of Notre Dame
Notre Dame, IN 46556
USA
jlabanow@nd.edu

Evgeni Gusev
IBM T. J. Watson Research Center
P.O. Box 218
Yorktown Heights, NY 10598
USA
gusev@us.ibm.com

Serge Luryi
Department of Electrical
and Computer Engineering
State University of New York at Stony Brook
Stony Brook, NY 11794
USA
serge.luryi@stonybrook.edu

Series Editor:
David J. Lockwood
National Research Council of Canada
Ottawa, Ontario
Canada

Cover illustration: The fabrication procedure for nanoscale molecular-switch devices by imprint lithography. A monolayer of switchable molecules is deposited over a Pt nanowire bottom electrode made by imprint lithography. A blanket of Ti contact layer is deposited on top of the molecular layer, which also protects the molecules from damage during the fabrication process. Finally, reactive ion etching is used to remove the blanket Ti protective layer, leaving a crossbar device with the molecular monolayer sandwiched between two metal nanowires.
Permission for front cover art granted by Hewlett-Packard Co. (authors: Yong Chen and R. Stanley Williams)

ISBN-13: 978-1-4419-3612-7   e-ISBN-13: 978-0-387-49965-9

Printed on acid-free paper.

9 8 7 6 5 4 3 2 1

springer.com

Preface

The high level of attention and interest of the global community to NANO science and technology to a large extent is linked to the GIGAntic challenges for the continuing growth of information technology, which sparked an unprecedented level of interdisciplinary and international cooperation among industrial and academic researchers, companies, IT market rivals, and countries, including former political and military rivals. Microelectronics technologies have reached a new stage in their development: The latest miniaturization of electronic devices is approaching atomic dimensions, interconnect bottlenecks are limiting circuit speeds, new materials are being introduced into microelectronics manufacture at an unprecedented rate, and alternative technologies to mainstream complementary metal–oxide semiconductors (CMOSs) are being considered.

The very dynamic stage of science and technology related to the advanced and future electronics and photonics creates a growing gap between the large number of rapid publications and nanotechnology highlights in media on one side and fundamental understanding of underlying phenomena and an adequate evaluation of scientific discoveries and technological innovations on the other side. Writing a tutorial book on fundamentals of science and technology for electronics at this time is almost the same level of challenge as writing a history book during a revolution.

This book, published in the series Nanostructure Science and Technology, is primarily based on the lecture notes presented at the Summer School of the second Nano & Giga Forum in Krakow, Poland. We made an effort to enhance the tutorial component and to provide material complementary to the book which published the school lectures from the first meeting in Moscow (*Nano and Giga Challenges in Microelectronics*, Elsevier, 2003). The book is designed as an introduction for graduate students, engineers, and researchers wishing to obtain a fundamental knowledge and a snapshot in time of the cutting edge in electronics technology research. As a natural consequence, it is also designated to be an essential reference for the "gurus" wishing to keep abreast of the latest directions and challenges in microelectronic technology development and future trends. The combination of viewpoints presented in the book can help to foster further research and cross-disciplinary interaction needed to surmount the barriers facing future generations of technology design.

The first chapter (by Cerofolini and Mascolo) provides a detailed and critical analysis of the possible routes from advanced microelectronics toward future nanoelectronics and molecular electronics. The second chapter (by Cristoloveanu) offers a comprehensive tutorial on state-of-the-art silicon-on-insulator (SOI) technologies. Basics on the operation and strategies for design of semiconductor lasers, which are in many ways second only to transistors on their impact on today's high-tech industries, are described in Chapter 3 (by Mao). Written by the authors from the Freescale Semiconductor research group, the fourth chapter (by Rao, Sadd, Steimle, Swift, Gasquet, and Stoker) describes new technologies for nanocrystal memories. The introduction of new high-permittivity (*high-k*) dielectrics, which is crucial for future generations of CMOS devices, is described in Chapter 5 (by Lee, Korkin and Huff). Tutorials on one of the most important analytical tools in advanced electronics and other nanotechnology areas, scanning force microscopy, are presented in Chapter 6 (by Such, Krok, and Szymonski). Chapter 7 (by Asenov, Brown, Cheng, Watling, Roy, and Alexander) provides an overview of the problems and simulation technique for nano-CMOS devices bridging the atomic-scale device structure with the device performance and system architecture. The final chapter (by Alexandrov) presents a comprehensive description of the idea and modeling results for polaron-based switches made of nanowires and quantum dots.

It is a great pleasure and honor for the editors to present these collected chapters and we thank our distinguished authors for sharing their insights and expertise as well as the sponsors of NGCM2004 for their gracious support. We invite our readers to join our next Nano & Giga forum in Arizona, March 12–16, 2007 (http://asdn.net/ngc2007/), and future meetings in this series and contribute in the living and developing legacy of science and technology for advanced and future electronics and photonics.

Evgeni Gusev, IBM, Yorktown Heights, NY
Anatoli Korkin, Nano & Giga Solutions, Phoenix, AZ
Jan Labanowski, Notre Dame University, Notre Dame, IN
Serge Luryi, Stony Brook University, New York, NY

Contents

1

A Hybrid Route from CMOS to Nano and Molecular Electronics

G. F. Cerofolini and D. Mascolo

1.1. INTRODUCTION

The exponential increase of complexity of integrated circuits (ICs) has already (2005) allowed the production of approximately 10 μmol ($\approx 6 \times 10^{18}$) of transistors; if the current trend toward higher and higher integration continues, an Avogadro number of transistors will be manufactured in the next 20 years—in a way, microelectronics is already "molecular" electronics, if not for the transistor size for the number of transistors at least. Hence, the following question: *Is there indeed a need of genuine molecular electronics?*

This chapter is essentially devoted to discuss this question. Since predicting the future is notoriously a risky activity, we will greatly constrain the scenario assuming the following hypotheses:

- No technological breakthrough will occur in fields (like spintronics, photonics, etc.), which are potential competitors to microelectronics.
- The concerted research and development activity outlined by the road map will guarantee that the evolution of microelectronics will continue in the next 10 years as described by Moore's laws.

1.2. MICROELECTRONICS TOWARD THE NANO ERA

Most of ICs are formed by metal–oxide–semiconductor (MOS) field-effect transistors (FETs) connected in a way that allows assigned electrical functions to be performed. The more sophisticated the functions, the more complex the circuit.

STMicroelectronics, Post-Silicon Technology, I 20041 Agrate MI, Italy gianfranco.cerofolini@st.com; danilo.mascolo@st.com

The early IC market was slow to develop and its exponential development was triggered by the Fairchild decision to sell IC chips at a price lower than the sum of the prices of the individual components necessary to make an equivalent circuit [1]. The cost reduction was possible thanks to the possibility of scaling down in a relatively easy way the size of the elementary device (the transistor) forming the circuit.

Most ICs are indeed produced with the MOS technology and are addressed to data handling. In turn, MOS ICs may be classified in two families, *logics* and *memories*, according to the fact that they operate in a combinatorial or sequential way. Memories, in turn, may be regarded as *ROMs* (read only memories, containing fixed information) or *RAMs* (random access memories, where stored data can be changed quickly). Arranged in a same device, a ROM (storing the algorithm of the machine), a RAM (where input–output data are stored), and the logic (working as an arithmetic–logic unit) form a *microprocessor.* The distinction between RAM and ROM is complicated by the fact that some ROMs may be reprogrammed; according to the speed of this process, they may be regarded as ROMs (in which case, one speaks of *EPROMs*, electrically programmable ROMs) or RAMs (in which case, one speaks of *NVRAMs*, nonvolatile RAMs). RAMs where the information is lost, unless refreshed, even in the presence of an applied power supply are known as *DRAMs* (dynamic RAMs).

1.2.1. Moore's Laws and MOSFET Paradigm

The development of microelectronics in the last 25 years is described by a set of statements collectively known as Moore's laws (from the name of the person[1] who first called the attention on them). The first law is a statement describing the increase with time of circuit complexity. Its original formulation [2],

> The complexity for minimum component cost has increased at roughly a factor of 2 per year. Certainly over the short term this rate can be expected to continue, if not increase.

has substantially maintained true over approximately 40 years, except for a modest correction to the characteristic doubling time [3] and its specialization for different ICs.

The Moore law is now formulated as follows:

> **First Moore law** *The number of transistors on integrated circuits doubles every 18 months for memories or 24 months for microprocessors.*

Figure 1.1 confirms the validity of these trends.

---

[1] The person who noticed the exponential development of the IC complexity is Gordon E. Moore. One of the inventors of the silicon-gate technology (a technology for the self-alignment—an alignment not requiring that a mask is aligned on a previous one with minimum uncertainty—of the MOSFET gate with respect to source and drain), Moore is also famous as a co-founder of Intel, of which he is currently president.

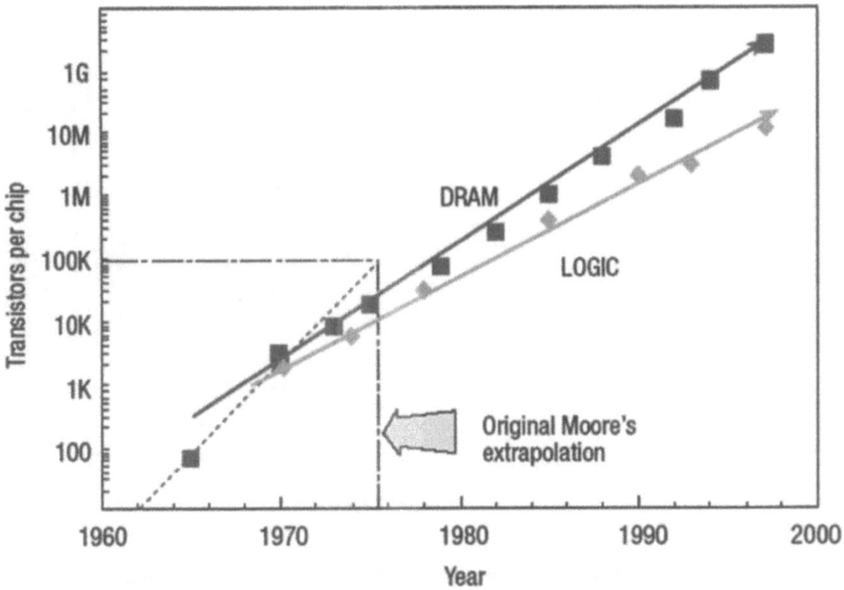

FIGURE 1.1. The first Moore law: time variation of the number of transistors in ICs.

Preparing denser and denser ICs is, however, a difficult job that has been done by building progressively more and more expensive fabrication facilities (fabs). The fab cost has increased at a lower rate than the bit integration:

Second Moore law *The average cost of fabs for manufacturing ICs increases by a factor of 2 every 3 years.*

Figure 1.2 (taken from the projections and survey data of Refs. 4 and 5, respectively) describes pictorially this trend.

The combination of the first and second Moore laws has resulted in a progressive decrease of the average cost per transistor, dropped to one-millionth of a cent compared to about a penny per transistor in the late 1970s. (Even though economic considerations may seem awkward in a technical–scientific book, they are, however, given here because we accept the idea that *the production cost of a good in a market economy is in ultimate analysis related to the overall thermodynamic yield of the process required for the production of that good.*) The evolution of the price per bit is sketched in Fig. 1.3, taken from Ref. 6: It sustains that the microelectronic industry was able to reduce the average price of MOSFETs by one order of magnitude in an average period of 5 years. Actually, the engine of the information revolution has just been this decrease, and that revolution will continue to be fueled until microelectronics continues to respect Moore's laws.

As already stated, the basic constituent of any IC is the FET. The shrinking of FETs has been doomed by the idea of scaling: Imagine that one wants to scale down by a factor of α ($\alpha < 1$) an FET with certain characteristics. When all of the voltages and dimensions are reduced by the scaling factor α and the doping and

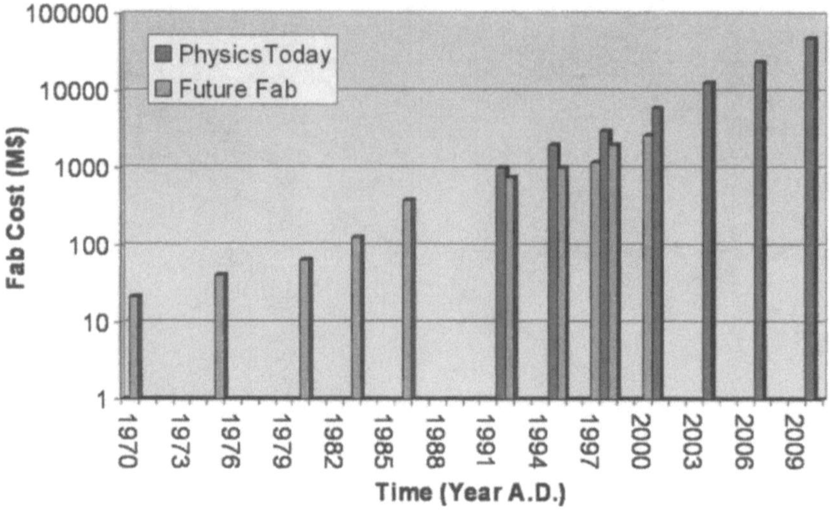

FIGURE 1.2. The second Moore law: time variation of the cost of fabrication facilities for IC manufacturing. Upper data are projections from Ref. 4, lower data are a survey from Ref. 5.

charge densities are increased by the inverse factor, α^{-1}, the electric field inside the FET remains almost the same as it was in the original unscaled FET (approach known as constant-field scaling and shown in Table 1.1. To some extent, this allows the scaled device to work similarly to the original one, gaining in performances.

The process of shrinking has brought the MOSFET from its original feature size ($\simeq$ 50 μm) to the present one (0.09 μm, at the forefront of the technology),

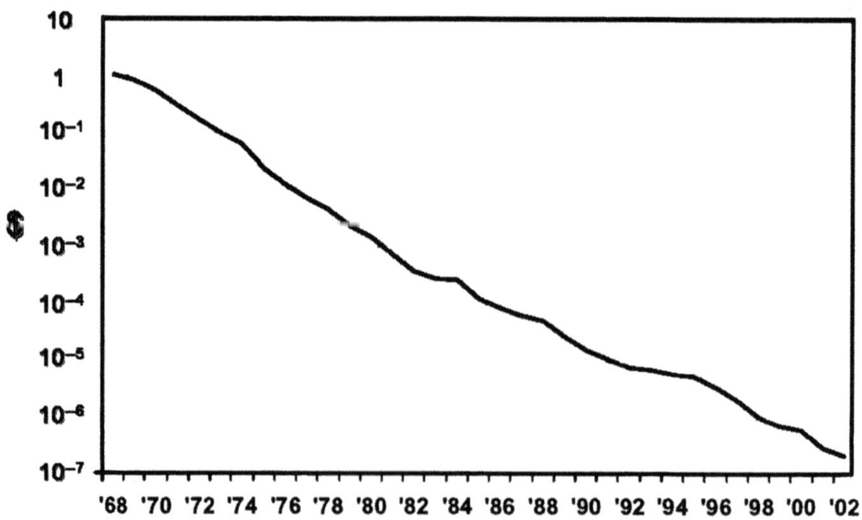

FIGURE 1.3. Time evolution of the bit price (taken from Ref. 6).

TABLE 1.1. Scaling rules for both constant-field scaling and generalized scaling. The factor α is the dimensional scaling factor and ϵ is the electric field scaling parameter.

	Scaling factor	
Physical quantity	Constant electric field	Scaled electric field
Channel length	α	α
Channel width	α	α
Insulator thickness	α	α
Electric field	1	ϵ
Supply voltage	α	$\epsilon\alpha^{-1}$
ON current (per device)	α	$\epsilon\alpha^{-1}$
Dopant concentration	α^{-1}	$\epsilon\alpha^{-1}$
Device area	α^2	α^2
Gate capacitance	α	α
Gate delay	α	α
Power dissipation (per device)	α^2	$(\epsilon\alpha)^2$
Power density	1	ϵ^2

corresponding to an increase of bit density by a factor of about 2×10^5 (the remaining increase being due to improved design).

Figure 1.4 compares a MOSFET of the mid-eighties with a MOSFET at the forefront technology; the comparison is made by either arranging the length scales so as to have the transistors with the same channel length (upper comparison) or using the same length scale for both transistors (lower comparison). It is noted that while the scaled transistors (upper comparison) have approximately the same shape for their inner parts (channel, source, and drain), they are, instead, completely different for the outer parts, whose scaling behaviors are not described by Table 1.1.

The process of shrinking has never been trivial and several seemingly insurmountable bottlenecks have been overcome. This impressive (r)evolution has been possible thanks to the setting up of the silicon technology. The silicon technology is essentially based on the combination and repetition of a few basic technologies, in turn requiring the use of relatively few materials (primarily Si, SiO_2, S_3N_4, Al, $TiSi_2$, TiN, and W, in addition to sacrificial materials like photosensitive polymers and gas or liquid etchants).

1.2.2. The Menu of the Silicon Technology

The fabrication of ICs is based on the silicon technology, whose pivotal areas are photolithography, oxidation, doping, etching, and chemical and physical vapor deposition.

1.2.2.1. Photolithography

Photolithography plays a central role in the process of shrinking, which has characterized the evolution of the IC technology. This technology is essentially a set

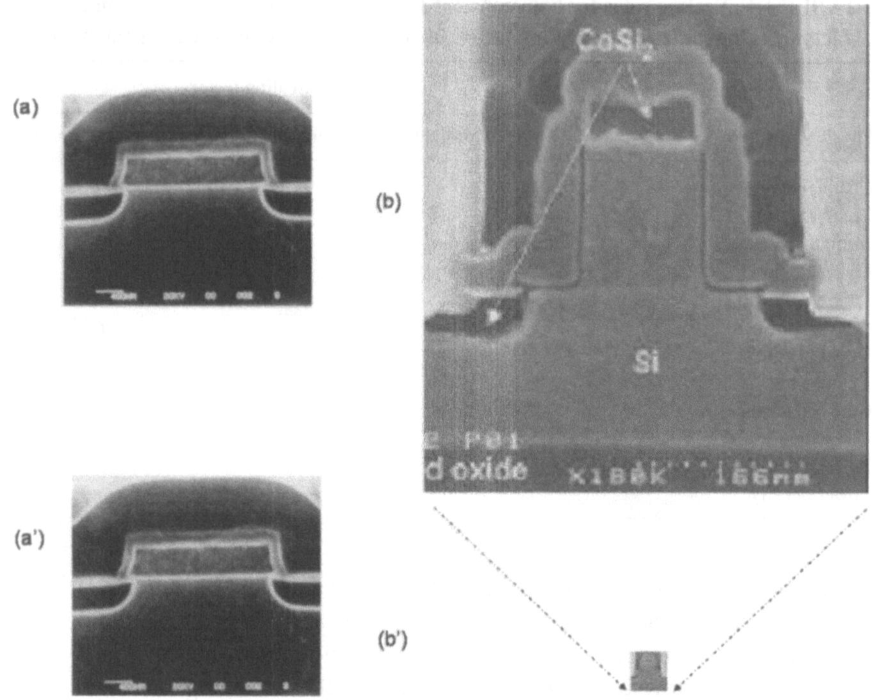

FIGURE 1.4. Comparison of the cross sections of two MOSFETs prepared with a technology of the mid-eighties (left) or with a current technology (2005) at the forefront (right). In the upper part, the length scales have been arranged so as to have the transistors (a) and (b) with the same channel length, whereas in the lower part the same length scale was used for both transistors (a′) and (b′).

of methods by which patterns are defined over a substrate. Generally speaking, a lithographic system is essentially constituted by a radiation source, a sample coated with a photosensitive material ("resist"), and an image-control system, regulating the part of the substrate being illuminated. Illumination modifies the chemical properties of the resist and, depending on the type of resist, the exposed (positive process) or the unexposed (negative process) part of the resist may be removed selectively by a development etching process. The pattern, so defined into the resist, can be transferred over the substrate by a subsequent selective chemical etching.[2] When visible light is used as radiation, the process is usually referred to as optical lithography. Depending on the wavelength used, a distinction is made between ultraviolet (UV, 365–436 nm), deep UV (DUV, 157–250 nm), extreme UV (EUV, 11–14 nm), and X-ray (<10 nm) lithography. Moreover, the use of electrons or ions instead of light brings us to electron-beam or ion-beam lithography.

[2] The name "resist" stems from this step: that part of the substrate covered by the resist is protected by the etchant agent and the resist material must be resistant to the etch process

1.2.2.2. Oxidation

Oxidation of the silicon surface remains one of the fundamental processes in the silicon technology (perhaps this fact is the answer to the question of why electronics is based on silicon—see, however, the discussion of Section 1.2.3). It is an effective route to provide silicon with a good quality, trap-free, insulating film whose thickness can be controlled down to the monolayer level. Oxidation is usually performed by heating the silicon surface to a temperature ranging from 900°C to 1100°C in an atmosphere containing oxygen or water vapor ("dry" or "wet" oxidation, respectively). Other interesting properties of thermally grown SiO_2 are as follows:

- SiO_2 is highly stable for temperatures over 900°C.
- SiO_2 can be removed by a selective chemical etching (aqueous solution of HF, HF_{aq}), leaving the underlying silicon almost unaffected.
- SiO_2 is a barrier against the diffusion of dopants (B, As, P) and other impurities.
- SiO_2 has a high dielectric breakdown field (1 V nm^{-1}).
- SiO_2 has good insulating properties (with resistivity ρ higher than 10^{16} Ω m) and wide band gap (with energy gap E_g of about 8 eV).

1.2.2.3. Doping

Doping is usually carried out by inserting a given amount of impurities in a shallow region close to the surface ("predeposition") and producing their migration into the bulk to attain the desired concentration profile ("diffusion").

Ion Implantation

Ion implantation is now the preferred method for putting dopants into a semiconductor. In simple terms, an ion gun is used to shoot dopants into the semiconductor. Some important features of ion implantation are briefly listed in the following:

- Dopants can be placed with much greater precision than with other chemical techniques, especially when small dopant amounts are required.
- As a counterpart, the ion-implantation apparatus is much more expensive than predeposition reactors, particularly when working with large batches of wafers.
- Ion implantation is done at room temperature, so there are more masking materials that can be used, including the photoresist.
- If needed, dopants can be shot clear through a masking layer into the underlying semiconductor.

Diffusion

Once a given dopant amount is put into the substrate, it must be redistributed in the silicon to achieve the desired concentration profile. This is usually carried out via thermally activated diffusion. Whenever a high concentration of some species exists in one place, the natural tendency for particles constituting the species is to diffuse toward regions of lower concentration. The whole process is a consequence of the random motion at the microscopic level.

The general approach for dopant diffusion is based on introducing a large amount of the dopant material at the surface of a wafer (thus generating therein a strong concentration gradient) and then raising the temperature. As the temperature increases, the diffusion coefficient reaches a sufficiently high value to pull dopants toward the deeper silicon zone.

1.2.2.4. Etching

Etching processes usually refer to steps in which material is removed from the substrate (subtractive method). Depending on the etch process, the removal occurs in both horizontal and vertical directions: The anisotropy of an etch process is given by complementing to 1 the ratio between the horizontal and the vertical etch rates.

Etching processes can be classified into physical or chemical etching, according to the mechanism involved in the removal procedure.

Chemical Etching

Chemical methods are usually performed by means of chemical reactions between reactive species, which can be either in liquid solution (wet etching) or in gas atmosphere (dry etching) and part of the substrate uncovered by the mask. Chemical reactions are usually activated by temperature (but radio-frequency or microwave excitation can be used too—in which case, one speaks of plasma etching) and can be carried out so as to select a material with respect to others because of chemical specificity. Wet etching, being characterized by almost equal horizontal and vertical etch rates, is usually isotropic. Table 1.2 lists etchants for a few materials of interest in semiconductor processing, together with the species that are not attacked. Dry etching, on the contrary, when assisted by the electromagnetic fields inside the reactor is often quite anisotropic.

Physical Etching

Physical etching does not make distinctions of the chemical species being removed and is usually performed by ion bombardment (the technique is usually referred to as sputtering). This technique relies on momentum transfer from particles hitting and eroding the surface. The process is generally characterized by

TABLE 1.2. Selectivity and wet etching of the basic materials of the silicon technology.

Material	Etchant	Selectivity
Si	$(HNO_3 + HF)_{aq}$	Si_3N_4
Si	KOH_{aq}	Si_3N_4, SiO_2
SiO_2	HF_{aq}	Si, Si_3N_4
Si_3N_4	H_3PO_4	Si, SiO_2

a vertical etch rate greater than the horizontal one by several orders of magnitude (thus being strongly anisotropic) and the material is removed only in one preferred direction.

1.2.2.5. Deposition

Especially for the preparation of interconnects, the deposition of additional materials onto the substrate is necessary. This is achieved via physical vapor deposition (PVD) or via chemical vapor deposition (CVD).

Physical Vapor Deposition

Physical vapor deposition is fundamentally a coating technique, involving the transfer of material brought in atomic form by vaporization of a solid or liquid target. PVD processes are carried out under vacuum conditions and incorporates the following processes:

Evaporation A target consisting of the material to be deposited is bombarded by high-energy electrons (in which case, one speaks of electron-beam evaporation) or ions (in which case, one speaks of sputtering). The energy released in this process dislodges atoms from the target surface, producing their evaporation.

Transport It consists of the movement of evaporated atoms from the target to the substrate to be coated.

Reaction In some cases, coatings will consist of metal oxides, nitrides, carbides, or other such materials. In these cases, the target will consist of the metal. The atoms of metal will then react with the appropriate gas during the transport stage. For the above examples, the reactive gases may be oxygen, nitrogen, and methane.

Deposition It is the process of coating buildup on the substrate surface. Depending on the actual process, some reactions between target materials and the reactive gases may also take place at the substrate surface simultaneously with the deposition process.

Chemical Vapor Deposition

Whereas the materials undergoing PVD are in solid or liquid form, precursors for CVD are introduced to the reaction chamber in the gaseous phase. In a typical CVD process, the substrate is exposed to one or more volatile precursors, which react or decompose either in the gas phase or on the substrate surface to produce the desired deposit while the volatile by-products are removed by gas flow through the reaction chamber. CVD is widely used in the semiconductor industry; it is mostly used to deposit polycrystalline, amorphous, and epitaxial silicon, SiO_2, Si_3N_4, SiO_xN_y, silicon–germanium alloys, tungsten, and various insulators with a high dielectric constant κ. CVD processes are usually classified according to the specific thermodynamic conditions as follows:

Atomic layer CVD Two complementary precursors are alternatively introduced into the reaction chamber. Usually one of the precursors is adsorbed onto the substrate saturating its surface. Moreover, it cannot completely decompose without the second precursor. Thus, the film thickness is controlled by the number of precursors cycles rather than the deposition time, as it is the case for conventional CVD processes.

Low-pressure or ultrahigh-vacuum CVD Environments at subatmospheric or very low pressure (typically of the order 10 Pa) are used to reduce unwanted gas-phase reactions and improve film uniformity across the wafer.

Plasma-enhanced CVD A plasma is used to enhance chemical reaction rates of the precursors. This allows deposition at lower temperatures, which is often critical in the manufacture of semiconductors.

1.2.3. The Historical Evolution of Microelectronics

Combining the shrinking of FETs with the development of silicon technology, the evolution of microelectronics has been dominated by the MOSFET paradigm:

MOSFET paradigm *Families of progressively denser devices are built scaling down silicon-based FETs.*

where "MOSFET" is used here and in the following with the restrictive meaning of "metal–oxide–*silicon* FET."

It has become common usage to refer to the the process of scaling down large circuits to produce similar, still working, smaller devices as the "top-down" route.

That the evolution of the IC technology has been dominated by the ability to increase the complexity of devices rather than by intrinsic properties of silicon or of the MOS transistor is supported by the following analysis, which is taken, with minor changes, from a review going back almost 15 years [7,8]: The current integration limit has been obtained through a complex evolution in which families of technologies are classified in relation to their ability to achieve a certain class of integration (see Table 1.3, taken almost *verbatim* from a work published in 1984 [9]). For instance, the small-scale integration (SSI, characterized by a number of bits per device in the interval 2–128) was obtained by means of the bipolar

TABLE 1.3. Complexity of semiconductor integrated circuits.

			Size		
Year	Class	Bit	Transistor (μm)	Chip (mm^2)	Wafer (mm)
1960–1968	SSI	2–128	>10	1–15	25
1965–1975	MSI	64–4 K	3–10	10–25	50
1972–1983	LSI	2 K–128 K	1.5–4	15–50	50–100
1980–1988	VLSI	64 K–4 M	0.75–2	25–75	100–125
1985–1993	ULSI	2 M–128 M	0.5–1	50–200	125–200
1990–1999	GSI	64 M–4 G	<0.5	100–400	≥200

The acronyms SSI, MSI, LSI, VLSI, ULSI, and GSI mean small-, medium-, large-, very large-, ultralarge-, and giga-scale integration, respectively.

TABLE 1.4. Change of technology and physical properties associated with the change of device.

Device	Technology	Physical property
Tube → transistor	Vacuum → solid state	v_{sat}, $3 \times 10^{10} \rightarrow 10^7$
Ge → Si	Mesa → planar	v_{sat}, unchanged; $\mu_{Si} < \mu_{Ge}$
Bipolar → pMOS		l, 0.1–1 → 10–100
pMOS	Al gate → Si gate	ρ, $10^{-6} \rightarrow 10^{-3}$
pMOS → nMOS	Ion implantation + LOCOS (111) → (100)	μ, 300(p) → 1000(n)
nMOS → CMOS		Channel: P, 300 → 10–100; μ, 1000(n) → 300(p) Interconnections: Al → Cu SiO_2 → low-κ Gate:
	Oxidation → CVD	SiO_2 → high-κ

Symbols: v_{sat} [cm/s], saturation velocity; μ [cm^2/V s], mobility; l [μm], device length (base in bipolar transistor, gate in MOS transistor); ρ [Ω cm], resistivity; P [μW/gate], dissipated power; low-κ, dielectric with low dielectric constant ($\kappa < 2$); high-κ, dielectric with high dielectric constant ($\kappa \gg 10$).

technology; the medium-scale integration (MSI) was possible either by the bipolar technology or by the aluminum gate, *p*-channel, MOS technology; the large-scale integration (LSI) was characterized by the silicon gate, local-oxidation-of-silicon (LOCOS), *n*-channel, MOS technology (nMOS). There is a general consensus that the MOS technology will continue to be the leading technology of future microelectronics.

Although the evolution of the microelectronic technology was not without branching, it is not an excessive overidealization to sketch the major technological changes that occurred after the invention of the transistor, as reported in Table 1.4. The thorough inspection of this table leads to the following conclusions:

1. Only in one case [replacement of the p-channel MOS (pMOS, in short) technology by the nMOS technology], a change of technology involved an improvement of intrinsic properties.
2. Only in one case [nMOS → CMOS (complementary MOS)], the change of technology was not associated with an improvement of design rules; this change was, however, dictated by the impossibility of dissipating the heat produced during functioning.
3. A technology or material continues to be used even when the reasons that motivated its success are no longer valid (e.g., SiO_2, which was introduced because of its excellent masking properties and of the existence of an etchant, HF_{aq}, which etches SiO_2 selectively with respect to silicon; the latter property is no longer interesting in plasma etching).

TABLE 1.5. Comparison of switching times of elementary devices built with different technologies.

Device	Switching time (s)
Electronic	10^{-11}–10^{-10}
Superconductor	10^{-12}–10^{-11}
Electro-optic	10^{-15}–10^{-12}
Photonic	10^{-15}–10^{-14}

4. A technology or device wins when it makes up for lower intrinsic properties by means of greater circuit complexity at lower cost.

The last item is the key for understanding the evolution of microelectronics. For instance, when switching times of existing or hypothetical devices are compared (as done in Table 1.5) to deduce the future replacement of silicon by GaAs or of silicon-based microelectronics by photonics (etc.) for volume applications, one neglects that in most cases it is possible to make up for lower intrinsic properties by a larger circuit complexity.

Remembering that at the present level of complexity silicon microelectronics has a cost of the order of 10^{-7} \$/bit, one is forced to conclude that *microelectronics in the next decade will be silicon based.*

Whether or not this prediction can be extended further will be discussed in Section 1.2.4. Assuming that the near future will be characterized by the absence of technological discontinuities, the top-down development of the IC technology within the frame of the MOSFET paradigm will, however, require important additions to the list of the basic materials. High-κ dielectrics as gate insulators [10,11] as well as low-κ dielectrics [12,13] and copper (or gold) for interconnections [14,15] are already in the reach of the silicon technology.

1.2.4. Trying to Sustain the Validity of the First Moore Law

Of course, the Moore law cannot hold true indefinitely—sooner or later physical, technological, or economical constraints will limit the effort toward higher integration.

The International Technology Roadmap for Semiconductors (ITRS, the "Roadmap") is essentially a Gantt chart listing the major technical improvements that are expected to be necessary to maintain the validity of Moore's law in the near future [16]. Its last version (2005) has predicted the introduction of the 45-nm-node MOS devices into mass production by 2010. This will require the technological capability of integrating transistors with pitch shorter than 25 nm.

On a length scale below 20 nm, the conventional MOSFET architecture is expected to fail since the control over short channel effect is lost as well as its current-driving capability. New inspired-to-MOSFET devices have been proposed (for a recent review, see Ref. 17) that might theoretically be useful for even harder geometries (in the range of 10 nm), leaving open the problem of which new factors (like thermal fluctuations, quantum effects, and power dissipation) will ultimately limit the transistor performances.

There are only feasibility indications beyond the limits of the Roadmap, and no corresponding technology has hitherto been hypothesized for practical productions. Even limiting the attention to the goals of the Roadmap, it is not clear whether or not they can actually be achieved in an economically sustainable way.

The top-down route of the silicon technology has indeed been relatively easy to run until its basic step (optical lithography) has met its first physical limits (minimum feature size around the light wavelength). Further improvements have been possible only because of masterly combinations of reticle enhancement techniques (optical proximity correction, phase-shifting masks, and off-axis illumination), which have allowed [in combination with sublithographic patterning techniques, such as the spacer patterning technology (SPT) or the resist-ashing and oxide-trimming process steps] the definition of geometries with a size lower than the wavelength by a factor of 2–3 (see Fig. 1.5).

By means of SPT (which is, according to Ref. 18, the most attractive way for overcoming the limits of conventional lithographic techniques—including electron-beam lithography—in terms of pattern fidelity, critical-dimension variation, and pattern density), Intel has already been able to produce generations of MOSFETs with gate lengths of 15 nm [19].

The development of the IC technology is potentially hindered by two factors that are expected to limit the exponential growth of microelectronics:

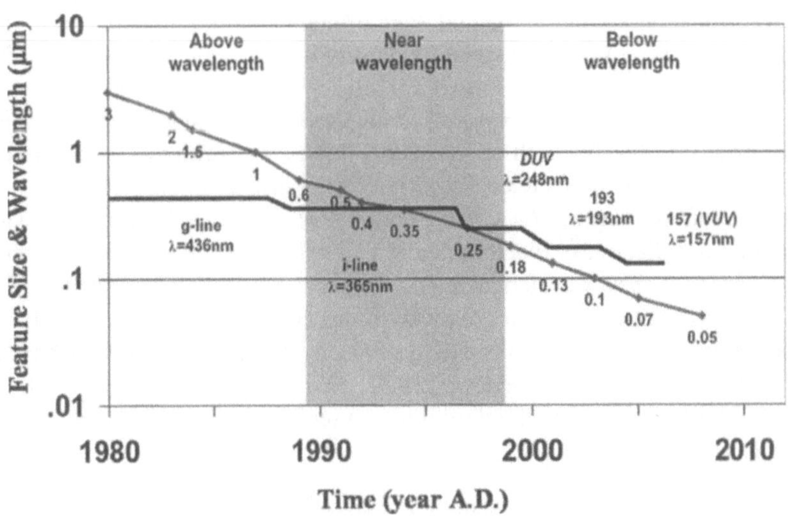

FIGURE 1.5. Time evolution of the technologically attainable feature size and the wavelength allowing its definition.

- The *physical limit*: the nonscalability of the MOS transistor below a critical size L_{phys}.
- The *lithographic limit*: the impossibility of batch defining, using proximity masks, feature sizes below a critical one L_{litho}.

By "physical limit," we denote here not only the failure of the transistor model on which the MOSFET paradigm is based but also the fact that the main MOSFET electrical features (like the infinite channel-gate impedance, the negligible current in OFF condition, and the saturation current in ON condition, features that have allowed the integration of more and more functions in a unique circuit like a microprocessor) are lost below a critical size. Of course, these failures do not necessarily imply that it is not possible to construct electronic circuits with such elements (this use seems conceptually possible down to a size scale where nonlinear current–voltage characteristics are preserved [20]), but rather that the circuit will progressively lose its digital behavior to assume an analogic character—which notoriously makes LSI difficult. Designing complex circuits with linear rather than digital devices is notoriously a major difficulty, which has eventually led to the almost complete occupation of the IC market with digital circuits.

The comparison of the shrinking rate of MOS devices with that of new functions added to the IC shows that the first rate is much higher than the second one. In fact, the increase of design complexity with the number of constituting devices has led to progressively less efficient design solutions, favoring the design time rather than the bit area. In the Moore diagram, the effect of this choice is reflected in the different growth rates for repetitive and simple structures (like memories) and for complex circuits (like microprocessors). The *design gap*, between the number of technologically integrable devices and the number of those practically integrated, is remarkable (compare in Fig. 1.1 the evolution of memories and microprocessors) and a major progress in the methodology of automatic design could increase by one order of magnitude the integration density with no new technological efforts [21].

The mainstream of IC research is engaged on one side in the attempt of reducing L_{phys}, working both on transistors (high-κ dielectrics as gate insulators) and interconnects (low-κ dielectrics as intermetal insulators), and on another side in the attempt of reducing L_{litho} using new exposure sources (extreme ultraviolet, synchrotron radiation, or electron beam) and extending the reticle enhancement techniques to the new wavelength region. Limits to the scaling down of MOSFETs and to their integration in progressively more complex ICs have recently been discussed in Refs. 22–25; limits and the expected future of lithography have also been subjected to extended reviews [26,27]. Which between the above (physical or technological) factors will indeed limit the scaling down of ICs is not clear; nor it is clear if

- The *economic limit*: the climbing investment cost for fabs

(predicted, hitherto correctly, by the second Moore's law and projected to rise to about 10 B$ in 2010—see Fig. 1.2) will eventually limit the minimum feature size to a value L_{eco}. On historical bases, the fab cost has essentially increased because the cost of photolithographic equipments does not progressively reduce with the maturity of the technology—and the prospected lithographies for next-generation ICs seem to run in this state of affairs.

Let L^* denote the maximum among L_{phys}, L_{litho}, and L_{eco}. At present, it is difficult to make a sensible prediction about the value of L^*. An upper limit is suggested by the Roadmap: $L^* \lesssim 30$ nm. Assuming L_{phys} as the lower limit, there remains to find the ultimate physical limits of computation. Even though extreme projections seem to indicate that they are eventually determined by space-time properties at the Planck length scale [28], we will conservatively assume that such limits are encountered on the nanometer length scale (NLS).[3] This is just the molecular scale length—and this coincidence is particularly interesting because certain molecules have already proved to be able to perform logical operations.

In view of the above considerations, L^* is thus in the interval 1–30 nm. The closer L^* is to the upper limit of this interval, the sooner microelectronics will become a mature industry. On another side, the closer L^* is to the upper limit, the wider is the room for the development of new (at present, hypothetical) ICs in which the data-handling elements are single molecules.

In molecular electronics, the preparation of single devices forming the circuit is left to the chemist, and it involves steps with which he or she is familiar: molecular design and modelling, synthesis, purification, etc. The IC preparation requires the arrangement and interconnection of these molecules to achieve an assigned electrical function—which requires the development of a new *bottom-up* technology.

Assume that there are the conditions for the development of a bottom-up technology; for $L^* \simeq 30$ nm (a feature size that, stipulating the constancy of the current logarithmic shrinking rate of the MOSFET, should be reached by about 2014), hypothetical ICs built via the bottom-up technology might have bit densities three orders higher than the ultimate circuits that can be built via the top-down technology.

1.2.5. Theoretical Limits of Computation

The forthcoming considerations, based on the use of the uncertainty principle, are essentially qualitative. Factors of the order of unity (related to the detailed form of the wavefunction in the considered state) are therefore lost.

The energy dispersion ΔE of a state cannot be larger than the energy ΔE_{bit} required to switch from the corresponding logical state to another, $\Delta E < \Delta E_{bit}$. The minimum time required by the system to switch to a distinguishable (thus orthogonal state) is given by the Heisenberg relation [29]:

$$\Delta t \geq \hbar/2\Delta E \geq \hbar/2\Delta E_{bit},$$

[3] This assumption will be discussed in Section 1.2.5.

where $\hbar$ is the reduced Planck constant, $\hbar = h/2\pi$. The minimum switching time τ_{min} is thus given by

$$\tau_{\text{min}} = \Delta t \geq \hbar/2\Delta E_{\text{bit}}.$$

Assuming a duty cycle of 50%, the maximum switching frequency ν_{max} of a single logical gate is given by $\nu_{\text{max}} = 1/2\tau_{\text{min}}$; that is,

$$\nu_{\text{max}} = \Delta E_{\text{bit}}/\hbar. \tag{1.1}$$

A similar argument applies to the minimum size of the logical device. Assume that the different logical states are associated with the occupation or not of a certain region of size Δq by an electron [30]. Localizing the electron within Δq imparts to it a momentum p that cannot be lower than the uncertainty Δp allowed by the position–momentum uncertainty principle:

$$p \geq \Delta p \geq \hbar/\Delta q.$$

Thus, localization imparts to the electron a kinetic energy $E_{\text{kin}} = p^2/2m_0$, where m_0 is the electron mass. In the ground state, $p = \Delta p$ so that the kinetic energy resulting from localization is given by

$$E_{\text{kin}} = \hbar^2/2m_0(\Delta q)^2. \tag{1.2}$$

In this scheme, ΔE_{bit} represents the minimum energy required by the electron to escape from its state, and the condition for localization to be possible is

$$E_{\text{kin}} < \Delta E_{\text{bit}}.$$

Combining this inequality with Eq. (1.2), one has

$$\Delta q \geq \hbar/\sqrt{2m_0\Delta E_{\text{bit}}},$$

which immediately gives the minimum size Δq_{min} allowing computation:

$$\Delta q_{\text{min}} = \hbar/\sqrt{2m_0\Delta E_{\text{bit}}}. \tag{1.3}$$

In a planar arrangement, the maximum bit density ρ_{max} per unit area is thus given by $\rho_{\text{max}} = K/(\Delta q_{\text{min}})^2$; that is,

$$\rho_{\text{max}} = 2m_0 K \Delta E_{\text{bit}}/\hbar^2, \tag{1.4}$$

where K is a factor related to the fact that addressing and sensing lines forbid a full occupation of the plane by active elements (for a simple square arrangement with pitch of $2\Delta q_{\text{min}}$, this factor is given by $K = \frac{1}{4}$).

The figure of merit in performing calculations of a circuit is given by the product of the device density per unit area, ρ, times the switching frequency, ν. This figure of merit for computation rate has thus an upper limit given by

$$\rho_{\max}\nu_{\max} = 2m_0K(\Delta E_{\text{bit}})^2/\hbar^3. \tag{1.5}$$

Equation (1.5) shows that the calculation rate can in principle be as high as wanted provided that the energy ΔE_{bit} required for switching is adequately increased. However, the power dissipation P involved in this process, $P = \rho_{\max}\nu_{\max}\Delta E_{\text{bit}}$, increases with the third power of ΔE_{bit}:

$$P = 2m_0K(\Delta E_{\text{bit}})^3/\hbar^3. \tag{1.6}$$

Of course, since the removal of the dissipated power is among the major problems of integrated circuits (if not the major one) even at the current level of integration, one would like to maximize $\rho_{\max}\nu_{\max}$ and minimize P. Equations (1.5) and (1.6) show, however, that this goal is manifestly impossible. Still interesting is the condition of optimum figure of merit (computation rate) when the switching energy ΔE_{bit} is kept to its minimum.

Of course, ΔE_{bit} depends on the particular device used to manipulate the information. The analysis of a few devices showed, however, that ΔE_{bit} is of the order of k_BT [31]. This value, in turn, is consistent with the minimum energy $\Delta E_{\min}$ required for computation as results from the Shannon–Von Neumann–Landauer analysis:

$$\Delta E_{\min} = k_BT\ln 2, \tag{1.7}$$

with k_B being the Boltzmann constant and T the operating temperature [32].

Inserting Eq. (1.7) into Eqs. (1.1), (1.3), (1.4), (1.5) and (1.6), one eventually gets the optimum frequency, localization, density, figure of merit, and dissipated power, respectively:

$$\nu_{\max} = k_BT\ln 2/\hbar, \tag{1.8}$$

$$\Delta q_{\min} = \hbar/\sqrt{2m_0k_BT\ln 2}, \tag{1.9}$$

$$\rho_{\max} = K\hbar^2/2m_0k_BT\ln 2, \tag{1.10}$$

$$\rho_{\max}\nu_{\max} = 2m_0K(k_BT\ln 2)^2/\hbar^3, \tag{1.11}$$

$$P = 2m_0K(k_BT\ln 2)^3/\hbar^3. \tag{1.12}$$

For operation at room temperature ($k_BT = 25$ meV), Eqs. (1.8) and (1.9) give

$$\begin{aligned}\nu_{\max} &= 25\times 10^{12}\ \text{s}^{-1},\\ \Delta q_{\min} &= 1.49\ \text{nm},\end{aligned} \tag{1.13}$$

independent of K. All other quantities depend on K. Assuming $K = 0.25$ (extremely aggressive, but consistent with the nature of the above arguments), one has

$$\rho_{\max} = 1.1 \times 10^{13}\ \text{cm}^{-2},$$
$$\rho_{\max}\nu_{\max} = 27.5 \times 10^{25}\ \text{cm}^{-2}\ \text{s}^{-1},$$
$$P = 7.48 \times 10^{5}\ \text{W cm}^{-2}$$

The maximum dissipated power is manifestly inconsistent with the properties of ordinary matter (for which P is of the order of 1 W cm^{-2}). The dissipated power can only be kept to acceptable limits only limiting the number of memory cells simultaneously switched. More consistent with the properties of matter is, instead $\Delta q_{\min}$. In fact, the estimate of Eq. (1.13) shows that, irrespective of geometric factors, *the optimum localization for computation is in the nanometer length scale—the molecular scale length.*

1.3. MOLECULAR ELECTRONICS

Molecular electronics is the stage of electronics in which the data-handling device coincides with the constituting material (the molecule). Even assuming the existence of molecules able to perform logic operations by themselves (discussed in Section 1.3.1), their availability does not guarantee *per se* the possibility of molecular electronics, unless one is able to arrange them in an ordered and accessible way and to probe their states.

The problem of interconnecting molecular devices in such a way that they may perform complex operations and of connecting the resulting network to the external world in such a way that the operations are performed on demand is the central problem of molecular electronics—the spectrum of opinions about its possible solution ranges from moderate optimism [33] to complete optimism [34]. Several techniques have been developed to make samples of a few, or even one, molecules accessible to electrical probes: break junctions, mercury-drop junctions, scanning probe microscope, and so forth [35]. At the moment, however, there are no clear ideas how the molecular devices (wires, switches, memories, etc.) can be arranged in an assigned order and can be coupled with the external world [36].

1.3.1. Molecules of Potential Interest for Molecular Electronics

There are plenty of prospected solutions for molecular devices. Among them, the ones that have attracted the most attention are reconfigurable molecules, characterized by two states with largely different conductances. States with different conductances are produced by modifying the structure of the frontier orbitals—the highest occupied molecular orbital (HOMO) and the lowest unoccupied molecular

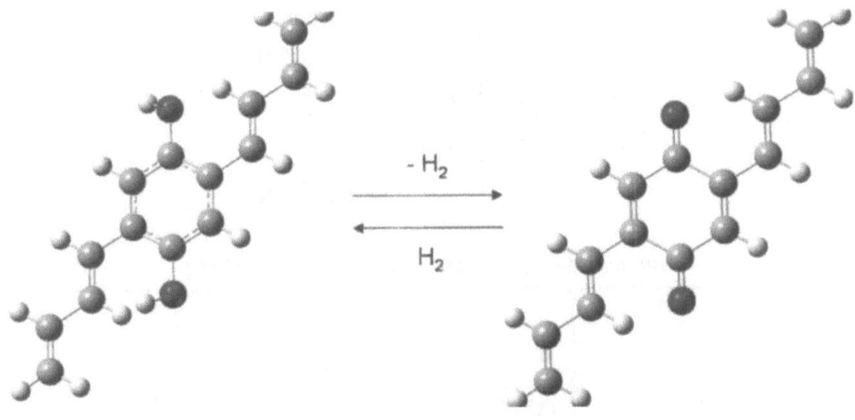

FIGURE 1.6. How the conjugation in a π-conjugated molecular wire may be destroyed by oxidation: left, two poly-acetylene chains linked by a hydroquinone group form a semiconducting π-conjugated wire; right, the quinone group, resulting from dehydrogenation of the hydroquinone group, destroys the π conjugation.

orbital (LUMO). This may be achieved, for instance, by destroying the conjugation in a π-conjugated molecular wire (see Fig. 1.6) or by producing a partial occupation of the frontier molecular orbital (an example will be shown in Fig. 1.16).

In general, the two conductance states are produced via one or the other of the following methods:

1. *Internal redox*: the molecule contains a donor-acceptor pair separated by a bridge D ⋈ A and is brought to a metastable ionized state $D^+ \bowtie A^-$ via internal redox reaction by the application of a potential difference.
2. *External redox*: the molecule contains a donor (acceptor) site that is brought to a metastable ionized state D^+ (A^-) via external redox reaction by electron exchange with an external electrode.

In both cases, the molecule is brought in one or the other state by the application of a relatively large potential difference;its state is instead probed measuring the current flowing under a small potential difference.

1.3.1.1. Molecules Involving Internal Redox

The idea of molecular electronics rests on the proposal of Aviram and Ratner of using asymmetric molecular tunnel junctions as rectifiers [37]. The visionary molecular device they proposed was formed by a molecule between a pair of electrodes performing the function of a rectifier. The Aviram–Ratner molecule is a structural analogue of a silicon-based p-n junction (diode).

The working principle is based on the difference in energy of the frontier orbitals of two separated donor and acceptor π-systems: A spacer between these two groups preserves the energy differences of the frontier orbitals, allowing, to some extent, electronic transport, as electronic conduction shows a preferential

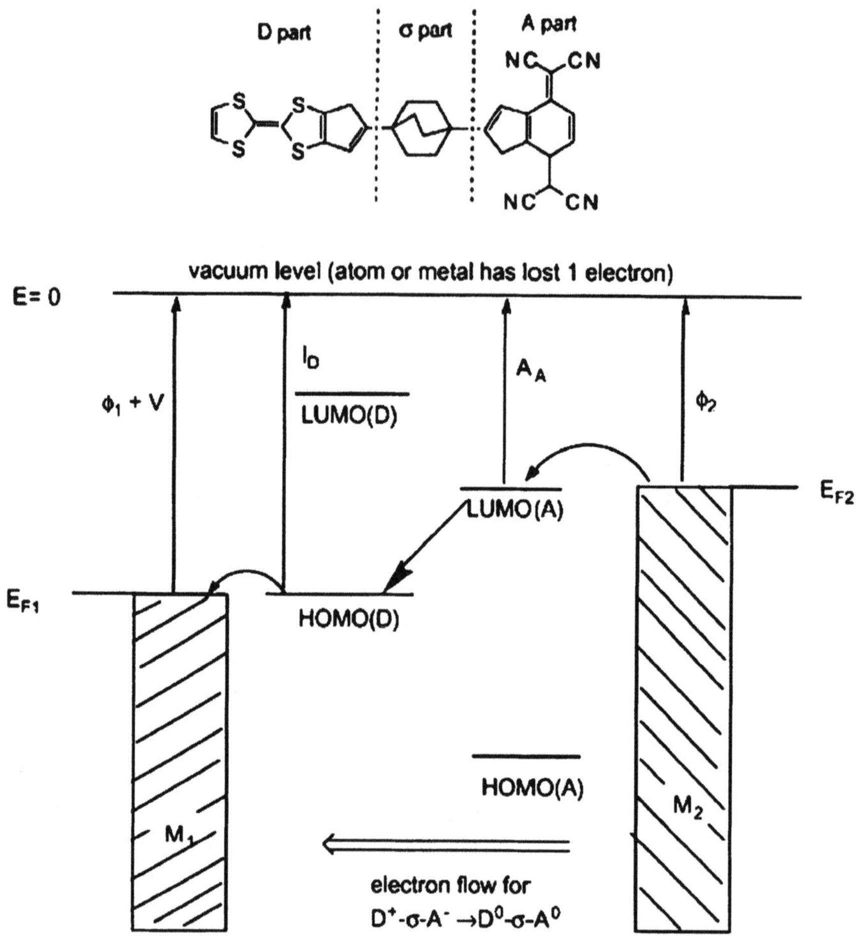

FIGURE 1.7. Aviram–Ratner molecular rectifier and the mechanism allowing electron conduction in the ON state. Reprinted from Synthetic Metals, 124, Robert M. Metzger, "Rectification by a single molecule", 107–112, Copyright (2001), with permission from Elsevier.

direction. Termination D (tetrathiafulvalene) is an organic electron donor, σ is a covalent saturated bond ('sigma bridge') working as a spacer, and A (tetracyanoquinodimethene) is an organic electron acceptor (see Fig. 1.7). The whole molecular structure, placed between two metal electrodes M_1 and M_2, forms the coupled junctions M_1-D and A-M_2 linked by a σ bridge ⋈: An easy electron transfer from M_2 to M_1 is facilitated by the "downhill" through-molecule tunneling from the electronically excited state $D^+ \bowtie A^-$ to the ground state $D^0 \bowtie A^0$, whereas electron flow is inhibited in the opposite direction.

1.3.1.2. Molecules Involving External Redox

The most studied reconfigurable molecules involving external redox are rotaxanes. A rotaxane is a supramolecular system formed by a cyclic molecule ("ring")

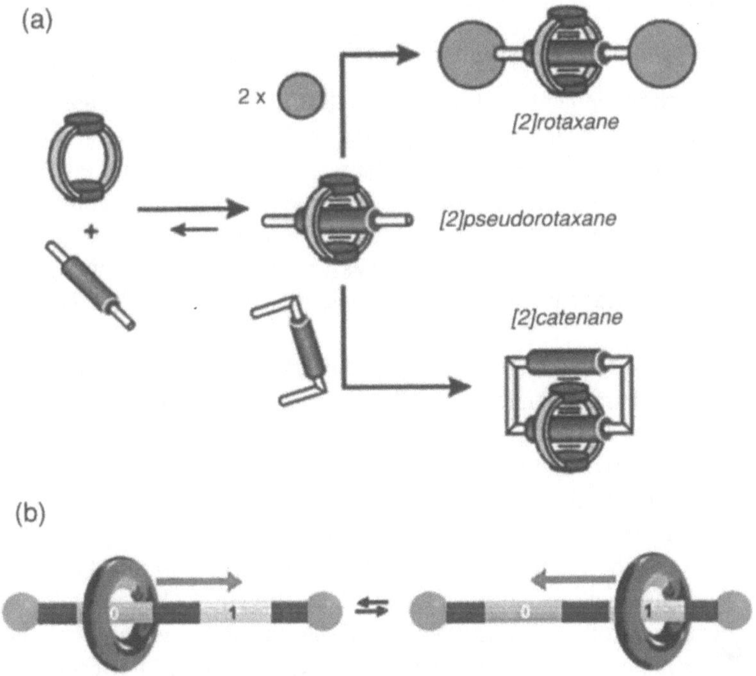

FIGURE 1.8. Pictorial views of (a) a rotaxane and catenane and (b) how the two states of a [2]-rotaxane can be used to map the 0 and 1 logic states chemically.

threaded on a linear molecule ("axle") with bulky caps on the heads to prevent the ring from falling off. Chemically speaking, the ring is usually a rigid, nearly flat arene (a famous example being the "paraquat"—two doubly–charged bipyridinium units linked by two benzene rings, with the net positive charge balanced by four PF_6^- anions), whereas the axle is formed by *n* sites (for instance, benzene units) typically connected by chains of ethoxy groups. The *n* sites behave as preferred positions for the ring, and the rotaxane is correspondingly denoted as [*n*]-rotaxane. Linking the heads of the rotaxane, one gets the simplest catenane—a [2]-catenane (see Fig. 1.8). For the topochemistry of such species, see Refs. 38–41.

Asymmetric [2]-rotaxane may be designed to behave as reconfigurable two-terminal devices exhibiting a hysteretic response to the applied voltage. The supramolecular systems, in fact, can be arranged endowing it with two recognition sites that provide the desired bistability. Cycles of oxidation, reduction, and Coulomb repulsion operate to mechanically change the states.

Thus, the conductivity of the rotaxane can be controlled by the position of the ring along the axle and interesting switching behavior has been first experimentally demonstrated in the solution phase [42,43].

1.3.2. Molecular Electronics Ex Novo

1.3.2.1. Supramolecular Systems as Simple Solid-State Devices

After the first implementations of Aviram–Ratner ideas [44], intense experimentations have started, aiming at integrating molecular monolayers in solid-state devices. A large variety of molecules have been considered and tested for their electrical properties. It is not a goal of this chapter to review this rapidly increasing field; for our purposes, it is sufficient to direct the reader to the recent review of Joachim et al. [45], mentioning molecules with the behavior of wires, switches, rectifiers, and storage elements, and quoting 106 references—and the list should be updated.

In most cases, the study was limited to synthesis, grafting to nano-electrodes, and electrical characterization. The efforts stopped just before the implementation of the molecular device as a functional element of an IC, with one remarkable exception due to a collaboration between groups from Hewlett–Packard (HP) and the University of California at Los Angeles (UCLA).

It is difficult to provide in a few pages a rationale of the major results of the HP–UCLA collaboration. Nonetheless, the following, necessarily abridged, sketch is perhaps sufficiently balanced and up to date. The HP–UCLA scientists focused their attention on rotaxanes and catenanes. Exploiting their switching behavior after embedding molecules into solid-state electronic devices has, however, revealed a difficult task.

The first attempt at integrating rotaxane monolayers as electronically configurable tunnel junctions in simple solid-state devices was done by embedding the rotaxane monolayer between two perpendicularly oriented electrodes (Fig. 1.9). The fabrication process comprised the steps of deposition of a rotaxane monolayer as a Langmuir–Blodgett (LB) film[4] over photolithographically patterned Al wires constituting the bottom electrode. The evaporation of a metallic top electrode completes the device fabrication. Moreover, a passivating layer of Al_2O_3 is formed over the bottom Al electrode. It was found that the molecular switching (Fig. 1.9) is strictly dependent on the metals used for the upper electrode: The combined use of Ti | Al, Ti | Au, or Cr | Al metallizations showed some success, whereas the use of stand-alone Al destroyed the switching behavior of the junction. Whether or not some reaction between the metallic upper electrode and the molecular layer had occurred was, however, hard to establish, as well as the molecular integrity at the end of process [46].

A similar approach with similar results has been successively developed [47], where a [2]-catenane-based molecular layer was sandwiched between an n-type poly-Si bottom electrode and a top electrode formed by Ti(5 nm) | Al(100 nm) film (see Fig. 1.10).

---

[4] The Langmuir–Blodgett technique is a technique for the deposition of ordered and compact layers of amphiphilic organic molecules (characterized by being nearly linear and with one hydrophilic head and the other hydrophobic) via their transfer from a polar liquid (where the molecule form a monolayer film with the hydrophobic head up) to a nonpolar target.

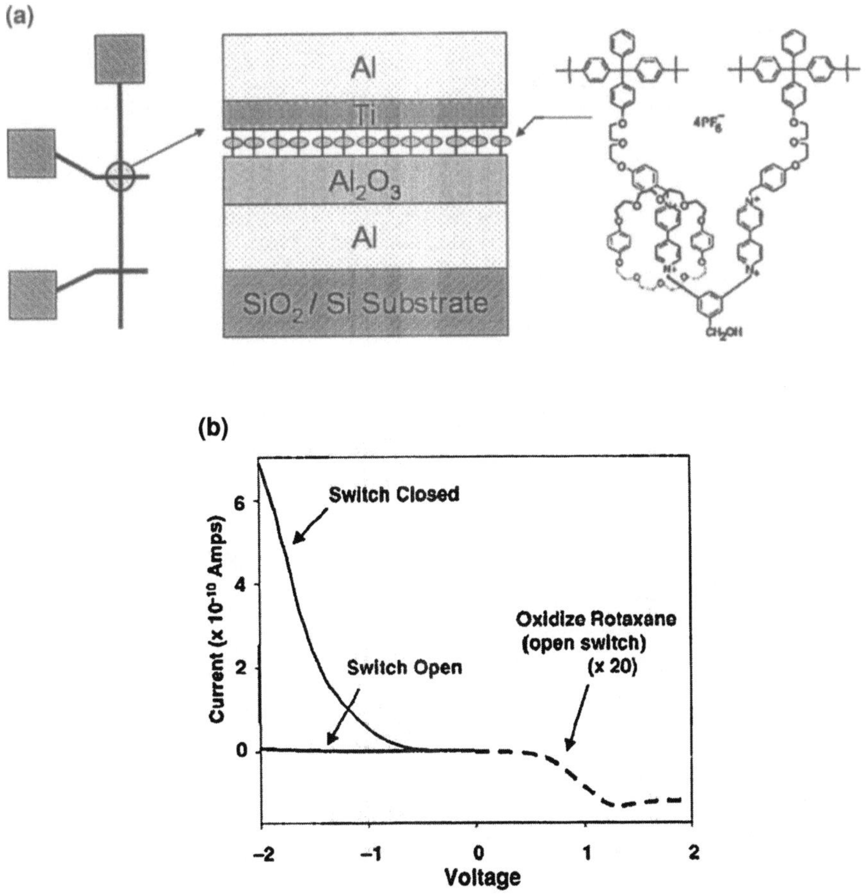

FIGURE 1.9. HP–UCLA first solid-state molecular device. (a) Top view of a linear array of molecular devices (left), side view cross section of a single device junction (middle), and the structural formula of the used molecule (right). Each device contained several million molecules sandwiched between two perpendicularly oriented macroscopically accessible electrodes. (b) Current–voltage characteristics showing the operation of the devices. Reprinted with permission from *Science* **285**, 391–394, C.P. Collier, E.W. Wong, M. Belohradsky, F.M. Raymo, J.F. Stoddart, P.J. Kuekes, R.S. Williams, and J.R. Heath, "Electronically configurable molecular-based logic gates", Copyright (1999), AAAS. Readers may view, browse, and/or download material for temporary copying purpose only, provide these users are for noncommercial personal purpose. Except as provided by law, this material may not be further reproduced, distributed, transmitted, modified, adapted, performed, displayed, published, or sold in whole or in part, without prior written permission from the publisher.

Further steps toward hybrid nano-ICs have implied more complex arrangements. Among them, the one that seems easier to implement is the crossbar architecture, consisting typically of two arrays of conductive wires crossing each other on different planes and separated from one another by the active molecular film [48].

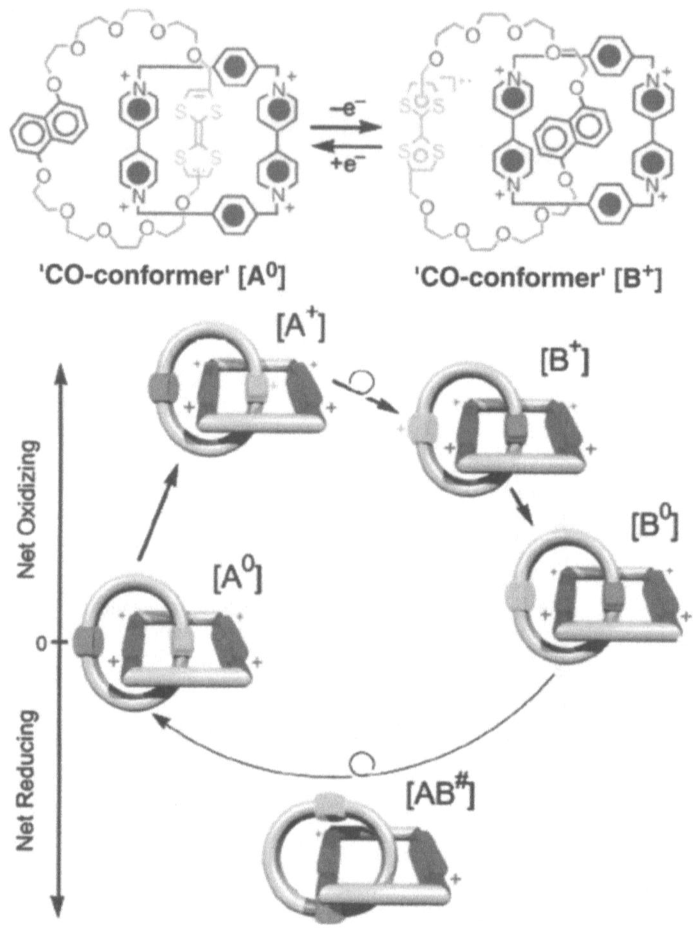

FIGURE 1.10. Switching mechanism in [2]-catenane. Reprinted with permission from *Science* **289**, 1172–1175, C.P. Collier, G. Mattersteing, E. W. Wong, Y. Luo, K. Beverly, J. Sampaio, F.M. Raymo, J.F. Stoddart, and J.R. Heath, "A [2]Catenane-based solid state electronically reconfigurable switch", Copyright (2000), AAAS. Readers may view, browse, and/or download material for temporary copying purpose only, provide these users are for noncommercial personal purpose. Except as provided by law, this material may not be further reproduced, distributed, transmitted, modified, adapted, performed, displayed, published, or sold in whole or in part, without prior written permission from the publisher.

1.3.2.2. A Molecular Random Access Memory

The first hybrid nano-IC was developed by the HP–UCLA collaboration (however, totally ignoring the complex addressing and sensing circuitry) using a [2]-rotaxane molecule as the memory element [49]. Even though the conceptually simplest architecture for a memory is that of an array of 2^n rows crossing 2^n columns with 2^{2n} memory elements at the crossing points, the silicon technology has been unable to

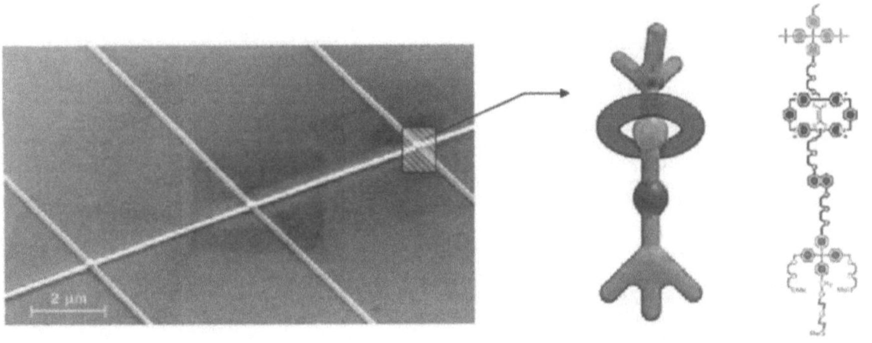

FIGURE 1.11. Two-dimensional hybrid circuit with [2]-rotaxanes as memory elements. Reprinted from *Chem. Phys. Chem.*, **3**, Y. Luo, C.P. Collier, J.O. Jeppesen, K.A. Nielsen, E. Delonno, G. Ho, J. Perkins, H.-R. Tseng, T. Yamamoto, J.F. Stoddart, and J.R. Heath, "Two dimensional molecular electronics circuits", 519–525, Copyright (2002), with permission from Wiley-VCH Verlag GmbH & Co KG".

produce circuits with this architecture. In practice, memories are organized with the memory elements at the crossing of 2^N bit lines with 2^k word lines (with $N \gg k$). The absence of an integrated addressing, sensing, and writing circuitry makes the HP–UCLA memory manifestly useless for practical applications. Nonetheless, such a structure is useful for familiarizing oneself with the problems that are presumably faced in the development of a nano-IC technology. The HP-UCLA collaboration prepared a crossbar structure with a process resembling the one described in Ref. 47, but scaling the wire width down to 70 nm by means of electron-beam lithography. In this way, the crossing bars defined junction areas (down to 5×10^{-3} μm^2) containing as few as approximately 3.5×10^3 molecules (Fig. 1.11).

The circuit may be operated to work as a memory: Storing of binary information is achieved via the application of suitable potential differences (in the interval of $\pm$ 2 V) to the corresponding electrodes; sensing is obtained by measuring at low voltage (about 0.2 V) the current flowing through the molecule; and bit discrimination is achieved by means of two different current levels associated with the two different equilibrium positions of the ring along the axle.

Further improvement in scaling down the crossbar architecture has involved a novel nano-fabrication technique based on the imprint lithography. In this method, electron-beam lithography is used to produce an imprint mold [50]. The use of a mold, in turn, avoids direct writing over a titanium-covered LB molecular layer. The crossbar fabrication technique comprised the following steps (see Fig. 1.12):

(1) Defining imprinted trenches by transferring the pattern from the imprint mold to a polymer coating a SiO_2 layer on a silicon substrate.
(2) Depositing sequentially 5-nm Ti and 10-nm Pt metal films by evaporation onto the substrate.

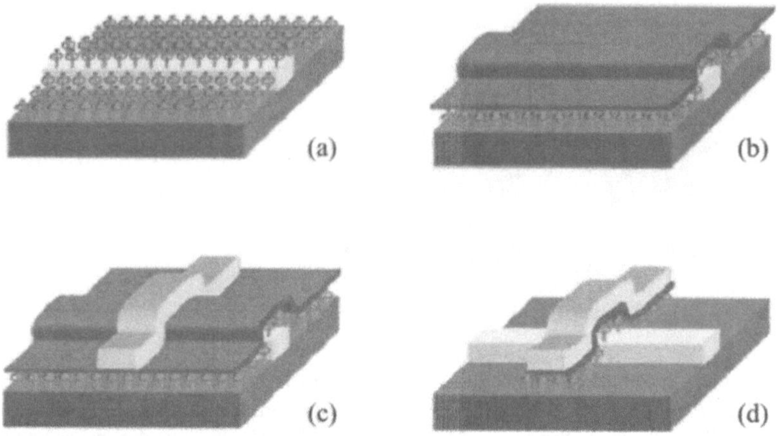

FIGURE 1.12. Crossbar architecture fabrication steps. Reprinted with permission from *Appl. Phys. Lett.* **82**, 1610–1612, Y. Chen, D.A.A. Ohlberg, X. Li, D.R. Stewart, J.O. Jeppesen, K.A. Nielsen, J.F. Stoddart, D.L. Olynick, and E. Anderson, "Nanoscale molecular-switch devices fabricated by imprint lithography", Copyright (2003), American Institute of Physics.

(3) Defining the Ti | Pt wires (with width of 40 nm) by the lift-off removal process of the unpatterned field (bottom electrode).
(4) Transferring an LB film of the bistable [2]-rotaxane covering the imprinted bottom metal electrodes.
(5) Coating the organic monolayer with a *protective* 7.5-nm-thick Ti film deposited by evaporation.
(6) Depositing 5-nm Ti and 10-nm Pt metal films by evaporation onto the substrate.
(7) Defining the Ti | Pt top electrode by the said imprint technique, aligning the mold orthogonally to the bottom electrodes so as to ensure the crossing between wires.
(8) Removing the unmasked film by an anisotropic reactive-ion etching [51,52].

The final crossbar architecture has been able to host approximately 1100 molecules embedded between a bottom Pt electrode and a top Ti electrode. The basic addressing unit is therefore a metallic–organic–metallic junction with an active area of about 1600 nm^2 (see Fig. 1.13).

Even though the demonstration that the circuit does really work as intended is presumably invalidated by artifacts (see Refs. 53 and 54 and the comments in Ref. 55), the basic idea is certainly interesting, perhaps revolutionary, and deserves to be reformulated to overcome the limits of the actual implementation.

The actual implementation manifestly suffers from several limitations. Without pretending completeness, the following ones are immediately recognized:

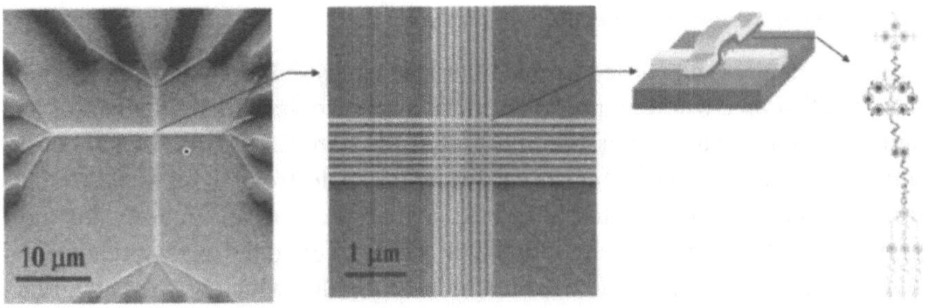

FIGURE 1.13. Crossbar architecture. Reprinted with permission from *Nanotechnology*, **14**, Y. Chen, G.Y. Jung, D.A.A. Ohlberg, X. Li, D.R. Stewart, J.O. Jeppesen, K.A. Nielsen, J.F. Stoddart, and R.S. Williams, "Nanoscale molecular-switch crossbar circuits", 462–468, Copyright (2003), Institute of Physics Publishing.

(i) the limited performances of the rotaxane in the writing phase; (ii) the high reactivity of metals depositing in atomic form; (iii) the process sequence that forces the organic compound to feel the severe counterelectrode deposition conditions; (iv) the limits of imprint lithography; and (v) the weak bonding linking molecules to bottom and top electrodes. The following are brief comments on the limitations:

(i) The limited performances are intrinsically associated with the mechanism through which the conductance of the [2]-rotaxane is changed. The switching from one conduction state of the axle to the other requires the motion of the ring, whose mass is higher than the electron mass by a factor of at least 3×10^5. By itself, this higher mass implies an increase (with respect to the equivalent process obtained with electron motion) of the time constant by a factor given by the square root of the mass ratio—almost 10^3. The resulting time constant is consistent with EPROMs only.

(ii) The high reactivity of metal atoms impinging the organic layer is likely the factor responsible for the mentioned artifact [53–55].

(iii) The sequence of crossbar preparation, in which the counterelectrode is deposited only after having transferred the LB film, forces the organic monolayer to tolerate an intense radiation dose that, if not completely destroying the original molecules, produces random damages in each memory element.

(iv) Imprint lithography is a contact (rather than proximity) lithographic method, which makes its application potential difficult to predict.

(v) The LB technique involves an organic layer having hydrophilic and hydrophobic terminations that control the link to the metallic contacts. Such links, however, are very weak and render the overall configuration structurally unstable.

1.3.3. Hybrid Routes to Molecular Electronics

The fundamental limits posed by quantum mechanics suggest that the connection of the microscopic world of molecules to the macroscopic world of humans or machines will require a stage of amplification of the signal at the mesoscopic level.[5] Since single devices of present or near-future microelectronic ICs are mesoscopic in nature and may be operated to behave as measurement apparatuses, the most promising architecture is the hybrid one, in which a very dense array of molecular devices is hosted in a silicon-based microelectronic IC, which provides the functions for addressing, reading, and writing, in addition to the input–output power stage and a sensing circuitry [56]. The hybrid approach, in which microelectronic circuits host molecular devices grafted to the silicon surface, seems the most realistic one for future electronics [57].

This state of affairs suggests that molecular electronics will necessarily involve a combination of micro- and nano-technologies. *Self-assembly* (i.e., the self-organized arrangement, usually driven by Van der Waals forces, in certain regions of the microelectronic device of suitable molecules in such a way as to perform a given function) and *grafting* (i.e., the stabilization of such structures via the selective chemical bonding of certain groups of the molecules to electrodes in the microelectronic regions) seem the key problems for what concerns the "bottom" side of the hypothetical bottom-up technology, whereas the definition (via the silicon technology) of structures able to host molecular devices is probably the major problem for the "up" side of the bottom-up technology.

1.3.3.1. An Opportunistic Approach

The basic idea underlying this approach started from the observation that the MOS structure offers a parallel-plate geometry where objects of molecular size can be attached—the region separating the gate electrode from source or drain ones [56]. Figure 1.14 shows two examples taken from current MOS production.

Attaching any molecule whose conduction state can be changed from OFF to ON by an external potential in the region between the single-crystalline silicon surface and the gate electrode would allow the limits of the MOSFET paradigm to be violated. The device on which we focus our attention, sketched in Fig. 1.15, is a MOSFET in which the drain region is connected to the gate electrode by a molecular Schmitt trigger controlled by an external potential. This structure allows not only all functions of microelectronics to be reproduced (it suffices that the external potential is fixed to a value keeping the molecule in the OFF state) but also new functions to be obtained.

Although in the following the attention will be limited to molecules connecting the drain and gate electrodes, this restriction is dictated by simplicity of discussion. Actually, simple process modifications would result in molecules connecting the source to the gate, the source and drain to the gate, and even the channel to the gate.

[5] This matter is discussed in Section 1.6.3.

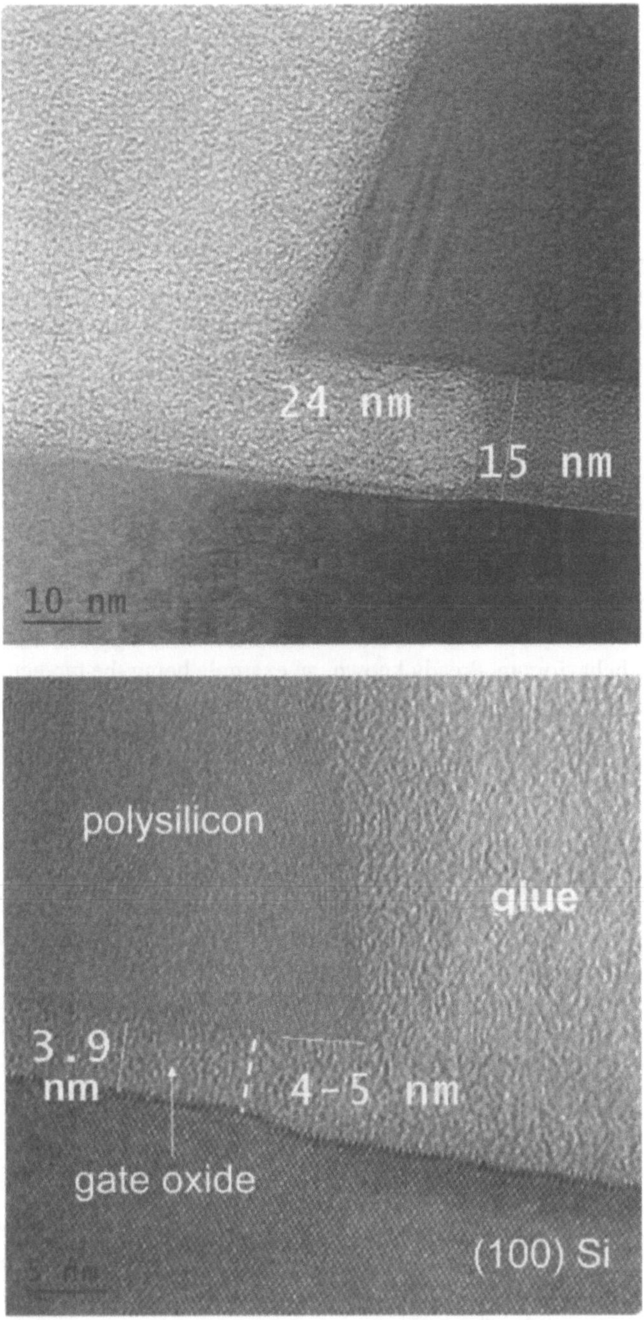

FIGURE 1.14. Two recessed regions obtained after etching with HF_{aq} the oxide not protected by the polysilicon gate. The separation between the upper (gate) and lower (source or drain) electrodes is controlled by the thickness of the gate oxide.

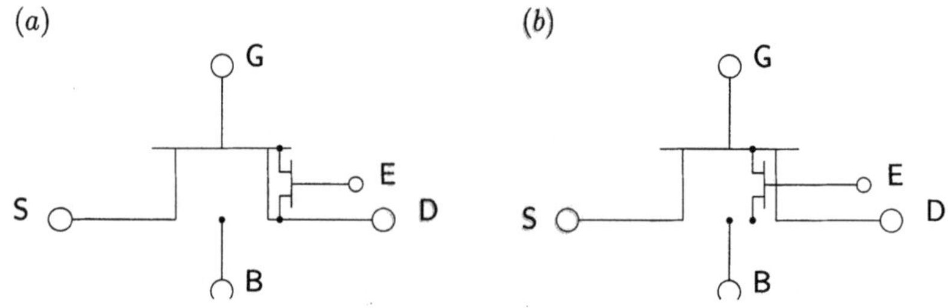

FIGURE 1.15. The considered hybrid devices: (a) the device in which the controlled conduction path links the drain and gate and (b) the device in which the controlled conduction path links the channel and gate. Symbols for electrodes are as follows: S, source; D, drain; G, gate; E, external control; B, body.

In the last case, when the control electrode forces to OFF the conduction state of the molecule, the device is nothing but a MOS transistor; when the control electrode is at a potential switching to ON the conduction state of the molecule, the device behaves like a bipolar device. Hence, the possibility of using such molecules for writing and erasing nonvolatile memories follows.

The reason why we mentioned a molecular Schmitt trigger is that molecules with such a behavior are already known, an example being the molecule (showed in Fig. 1.16) formed by a bipyridimium core, two alkane chains, and two thiol terminations [33].

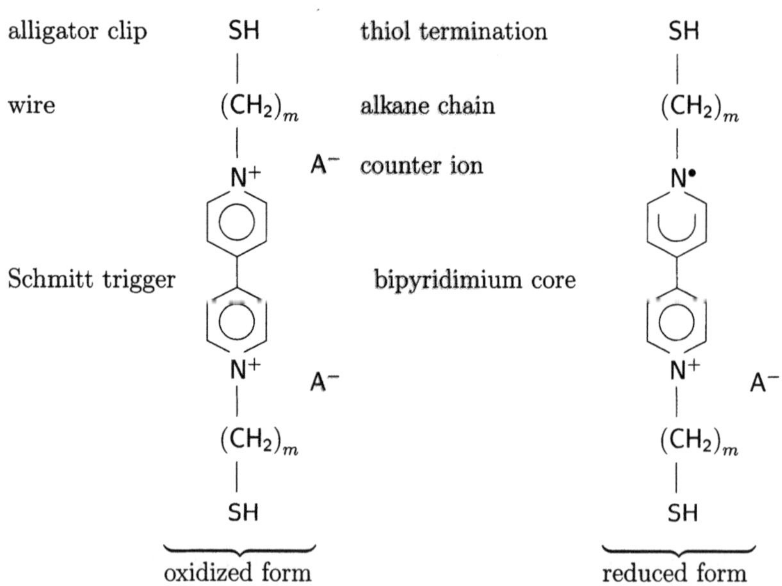

FIGURE 1.16. Attribution of electrical functions to the segments forming the molecule.

1.3.3.2. A Dedicated Approach

In this approach, the hypothetical bottom-up route is imagined as a kind of host–guest technology, in which the host is committed to microelectronics while the guest is committed to chemistry. Combining host and guest is possible only if the hosting site is adapted to the guest size (that requires a control of geometries on the sub-NLS) and if the molecular linkage between the hosting site and the guest is sufficiently strong to ensure long-term operations.

Denoting by "nanodevice" any device with size in one dimension at least on the NLS, because of the gate-oxide thickness the MOSFET is a nanodevice since the nineties. In fact, the thickness of the SiO_2 gate dielectric is around 3 nm for logics and 5 nm for nonvolatile memories. However, since the factors limiting the IC integration are horizontal sizes, in electronics one properly speaks of nanodevice when its size in one *horizontal* dimension at least is on the NLS.[6]

The preparation of features on the NLS, manifestly impossible via optical lithography, can be done using X-ray or electron-beam lithography. These techniques, however, are very expensive and still in their infancy, at least for what concerns their exploitation in the industrial practice.

Geometries on the NLS can, however, be produced with relative ease without the use of such lithographic techniques by means of the SPT [i.e., transforming vertical features (like film thickness) in the vicinity of a step of a sacrificial layer into horizontal features]. The ultimate length producible in this way is controlled by the following:

- The steepness of the step defining the sacrificial layer.
- The uniformity of the deposited or grown films.
- The anisotropy of its etching.

Silicon nanowires with a width below 10 nm and a pitch half the one achievable via electron-beam lithography have been prepared with this technique [59].

Although useful for the preparation of a few devices with special needs, the above trick does not allow, by itself, the development of a nanotechnology, where *each layer* useful for defining the FET should be on the NLS and aligned on the underlying geometries with tolerances on the NLS. Setting up such a nanotechnology is a major problem that will involve the IC industry in the post-Roadmap era.

Regardless of the detailed structure of the basic constituents of nano-ICs (molecules, supramolecular structures, clusters, etc.), any nano-IC can hardly be prepared without the ability to produce arrays of conductive strips with a pitch on the NLS. The following section is devoted to describing a scheme (essentially based on the existing silicon technology) for their production without the use of advanced lithography and their use to host molecular devices.

---

6 The US Patent and Trademark Office supports this view: "For an invention to qualify, at least one of its dimensions should be between 1 and 100 nanometres in size, and the tiny size of the device must be essential to its function. This means that some ingredients in sunscreen may well be nanotech, for example, but a computer chip may not be" [58].

1.4. IMPLEMENTATION OF THE DEDICATED HYBRID ROUTE

One can make up for both the high reactivity of metal strips resulting from the top-electrode deposition steps (likely responsible for the mentioned artifacts) and the limits of imprint lithography using polysilicon strips as address lines and defining them via a generalization of SPT—the *multispacer patterning technology*, hereafter referred to as S^nPT.

In the following, we discuss a strategy for the production of crossbar memories in which the matrix crossbar structure is obtained using the S^nPT and single bits have a molecular nature. The considered example is essentially based on the silicon technology; since it requires only familiar materials (like polycrystalline silicon, SiO_2 and Si_3N_4) and exploits the etching selectivity of these materials, it may be seen as a *conservative extension of the silicon technology to a nano-IC technology* [60,61].

1.4.1. Preparing the Host

Which materials, geometries, and alignments are required for the preparation of nano-ICs is manifestly largely unpredictable. Nevertheless, the assignment of technological goals (able to clarify which difficulties will be met in the long road to nanoelectronics and which technologies should be developed for that) is certainly useful.

1.4.1.1. The Spacer Patterning Technology

As already mentioned, the MOSFET paradigm has supported the miniaturization of denser geometries (particularly in the MOS gate length L and width w) and faster circuits. Conventional photolithography limits the critical dimension to hundreds of nanometers. It is only thanks to nonphotolithographic methods that minimum feature sizes in the scale of few tens of nanometers have been achieved. Among these methods, the most attractive one is certainly the SPT, which was first applied in the early 1980s for submicron transistors [62] and in grid fabrication [63], and eventually used for the definition of the polysilicon gate in NLS MOS processing [64,65].

The SPT allows the definition of geometries on the NLS by transforming vertical features (like film thickness) in the vicinity of a step of a sacrificial layer into horizontal features. The ultimate length producible in this way is controlled by the steepness of the step defining the sacrificial layer and the uniformity of the deposited or grown films.

The SPT exploits a combination of conventional photolithography, anisotropic etchings, and the excellent homogeneity and reproducibility of conformal CVD processes. It comprises the following steps (Fig. 1.17):

(1) The deposition of a sacrificial layer.
(2) The definition of a vertical step by means of lithography and anisotropic etching.
(3) The deposition of a conformal layer.
(4) A final anisotropic etching.

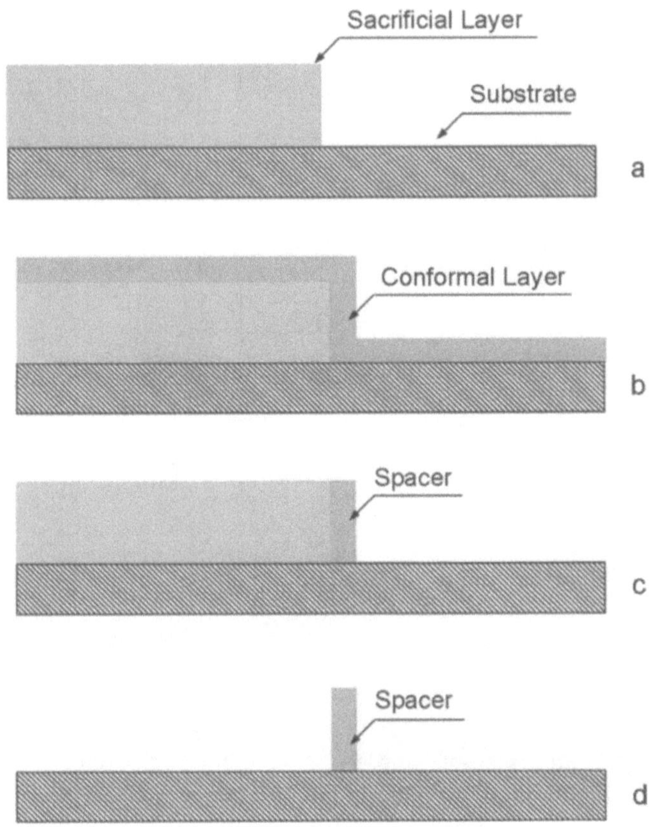

FIGURE 1.17. A way for obtaining sublithographic features via the SPT.

This process results in a well-defined spacer as a side wall on the sacrificial layer. The spacer can finally act as a hard-mask layer with NLS feature size, after removing the sacrificial layer by selective chemical etching. In this way, the lithographic step is necessary to define the spacer position only, whereas the deposition process and the anisotropic etching are critical for NLS definition on the lateral direction of the top of the substrate.

After a score of years of experimentations, a number of materials typical of the standard silicon technology has been successfully deposited as single spacers at the nanometer scale, aiming to overcome specific issues. In this way, the SPT has become a kind of process scheme for the definition of sublithographic features.

1.4.1.2. The Multispacer Patterning Technology

The S^nPT is a technology that exploits the possibility of reiterating the SPT. In fact, a spacer sequence of different materials can be easily deposited reiterating the process steps of conformal deposition and anisotropic etching as many times as the number of desired layers one needs to form. In such a way, it is relatively easy

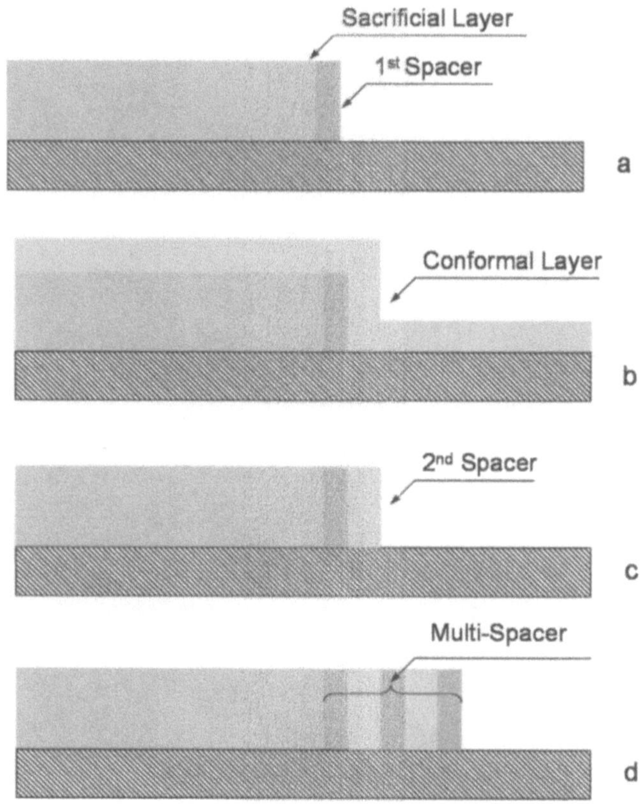

FIGURE 1.18. Obtaining a spacer array via the S^nPT.

to build an NLS structure in which a set of conductive (heavily doped polysilicon) and insulating (SiO_2 or Si_3N_4) spacer pairs follow each other (Fig. 1.18).

The development of a S^nPT, with process steps taken from the conventional silicon technology, might play a key role in the fabrication of NLS crossbarlike architectures on industrial scale. A possible scheme involves the following steps (see Fig. 1.19):

(a) Preparation of a first-floor array of double spacers (poly-Si|SiO_2).
(b) Thermal growth of a thin SiO_2 film, whose thickness is tuned on the molecules being hosted.
(c) Preparation of a second-floor array of double spacers crossing the first-floor array, being careful to preserve the original interlayer oxide or to regrow another oxide of the same thickness after its complete etch at each stage of the second-floor S^nPT.
(d) Selective chemical etching of the interlayer SiO_2.

The first three steps are necessary to built a two-floor structure characterized by conductive spacers separated by an insulating layer and crossing each other

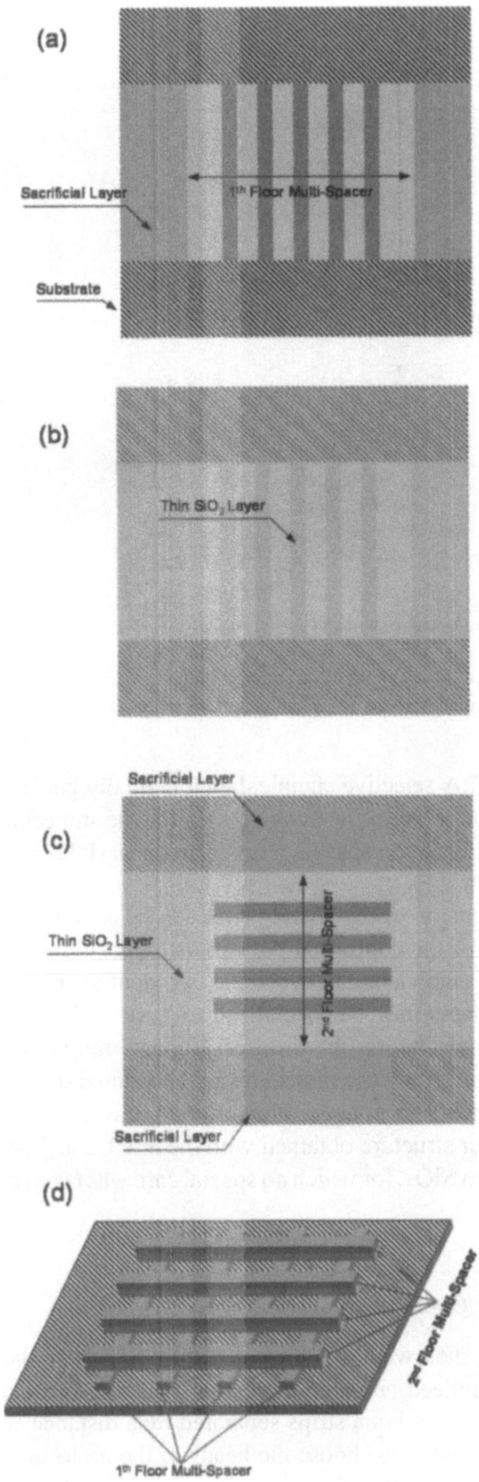

FIGURE 1.19. Crossbar fabrication steps.

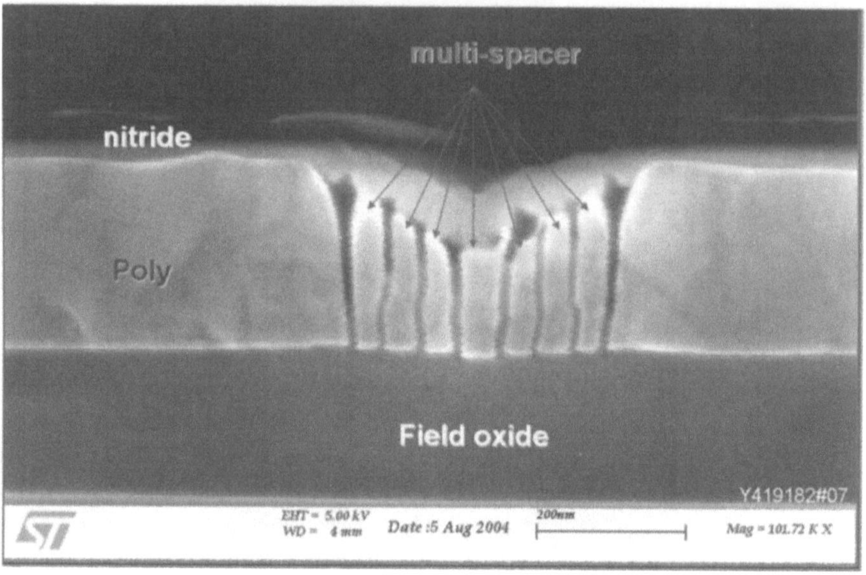

FIGURE 1.20. Image at the scanning electron microscope of a cross section of a crossbar structure with seven spacers obtained with the S^nPT using undoped polysilicon and thermally grown SiO_2.

on different planes. A selective chemical etch is finally performed to remove the interlayer oxide and to form the hosting sites for the molecular guests, whereas the second-floor polysilicon spacers work as hard-mask layers for the underlying double-spacer layer.

The final dimension of the hosting sites can be adapted to the molecule size with the thickness of the conformal polysilicon depositions and the thickness of the insulator layer sandwiched between the two double-spacer floors. It is worth noting that the process may be specialized to have different (p or n) doping of polysilicon in the two floors; the different dopants might be used to allow the self-assembly of nonsymmetric molecules in the wanted direction.

Figure 1.20 shows an image at the scanning electron micoscope of a cross section of a crossbar structure obtained with the S^nPT using undoped polysilicon and thermally grown SiO_2, for which no special care was taken to maintain a planar structure.

1.4.2. Inserting the Guest

Assume that in our shelf we have molecules able to perform the wanted electrical function and have indeed prepared a crossbar structure formed by two mutually perpendicular arrays of silicon strips separated by a distance equal to the molecular length. There remain to choose the heads of the molecule and to modify the terminations of silicon to allow the spontaneous grafting of the molecules between the upper and lower strips.

In the last few years, the derivatization of the silicon surface with functional organic molecules has become a hot research area because of its impact on fundamental chemistry, surface science, and molecular electronics [56,57,66]. Several modes for the addition of organic molecules to silicon are known:

1. The *cycloaddition* of molecules with alkene or alkyne functions to the clean, 2 × 1 reconstructed, (1 0 0)-oriented surfaces [67,68].
2. The *condensation* of molecules containing alcoholic groups at hydroxyl-terminated silicon surfaces [69].
3. The *silanization* of hydroxyl-terminated surfaces [70] as well as of hydrogen-terminated surfaces [71] via reaction with chlorosilane-terminated molecules and elimination of HCl.
4. The *silanization* of hydrogen-terminated surfaces via reaction with azanes and elimination of NH_3 [71].
5. The *arylation* of hydrogen-terminated silicon via its reaction with diazonium salts and N_2 elimination [72].
6. The *hydrosilation* of alkenes at hydrogen-terminated surfaces [67,68,73].

These processes can also be classified according to silicon terminations:

(i) Although very elegant and of large conceptual interest, cycloaddition is not considered of practical interest because it is limited to the extremely reactive, *clean, 2 × 1 reconstructed, (1 0 0) surface of single-crystalline silicon* (which, chemically speaking, may be considered as a two-dimensional ordered layer of silene groups with extremely bulky side chains).
(ii) Since no process is known for the controlled preparation of *hydroxyl-terminated silicon surfaces*, condensation and silanization of these surfaces seem difficult to implement.
(iii) On the contrary, device quality, relatively stable, *hydrogen-terminated silicon surfaces* may be prepared by etching with aqueous solution of HF a sacrificial film of SiO_2 thermally grown on the surface [74].

All of the pathways considered above are sketched in Fig. 1.21. This classification would limit the scope of candidate reactions to modes 3–6. Of them, the one involving the least aggressive pathway (and thus *a priori* the most compatible one with the numerous layers, especially the metal ones, characterizing integrated circuits) is certainly hydrosilation. Having in mind a hybrid route to molecular electronics, the most serious candidate process for silicon functionalization is thus the hydrosilation of unsaturated hydrocarbons at the hydrogen-terminated silicon surface.

As sketched in Fig. 1.22, hydrosilation involves the interaction of a π bond with a silanic bond. This process may mechanistically be described as a hydrogen shift from silicon to a carbon atom involved in the π bond and binding of the silicon atom to the other carbon atom (see Fig. 1.22). Although typically carried out for 1-alkenes [75–81], $C{\equiv}C$, $C{=}O$, $C{=}S$, and $C{=}N$ bonds can be hydrosilated too, in addition to the $C{=}C$ bond [82]. Even though the grafting of short molecules is

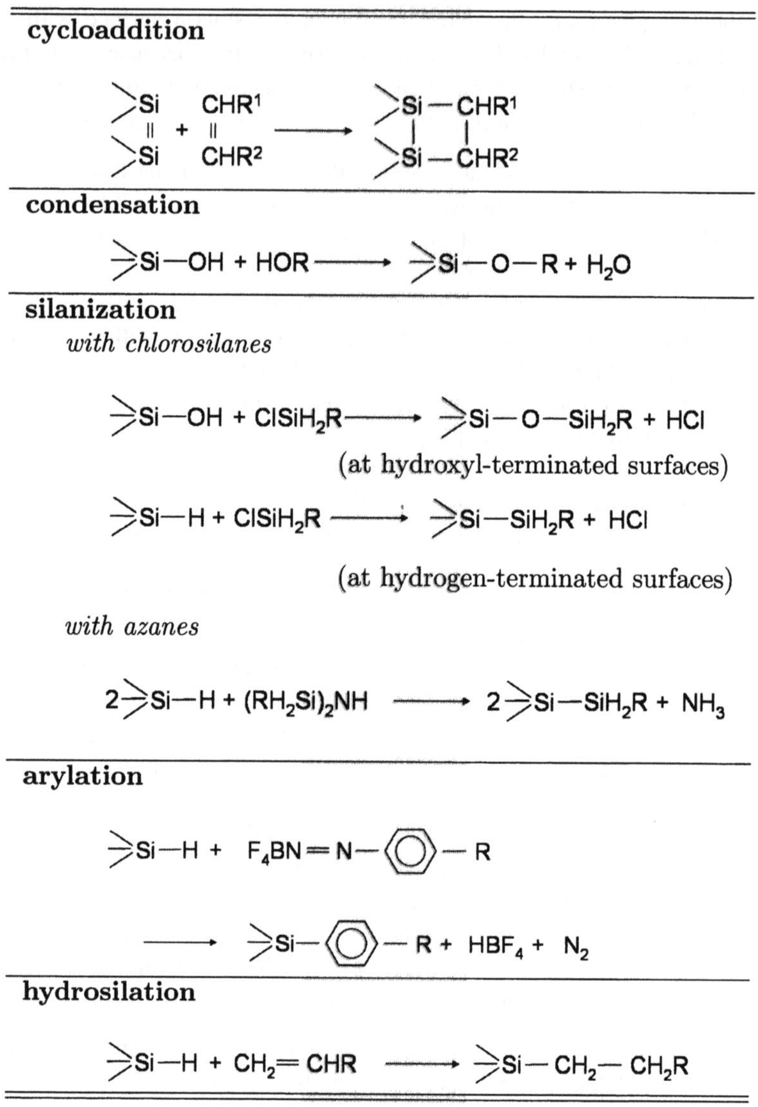

FIGURE 1.21. The considered pathways for silicon functionalization.

said to require radical activators or a platinum catalyst [68] (scarcely compatible with IC processing), it has been shown that the reaction proceeds spontaneously in the laboratory time scale at temperature around 200°C [83–87]. Figure 1.23 shows the C 1*s* spectra obtained via X-ray photoemission spectroscopy (XPS) from a hydrogen-terminated (100) surface and from the same surface after derivatization with 1-octyne at 170°C (the spectrum from the pristine surface is due to adventitious carbon physically or chemically adsorbed from the atmosphere). Since hydrosilation involves one silicon atom only (see Fig. 1.22), this reaction is expected not to be dominated by the crystalline face of silicon in the strips.

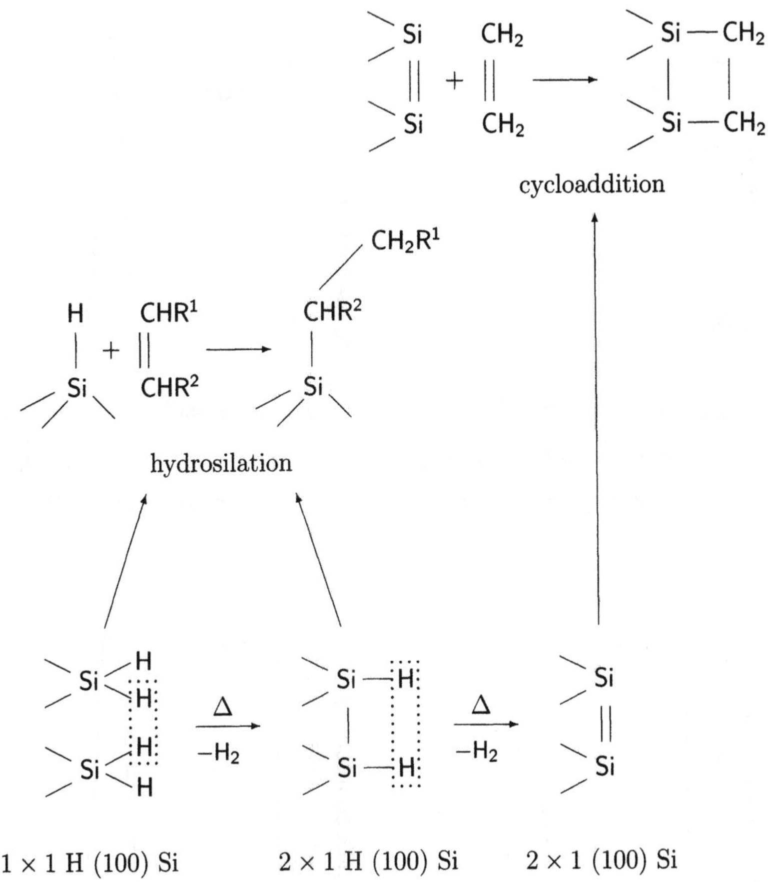

FIGURE 1.22. Reactions of unsaturated hydrocarbons with silicon: cycloaddition at the clean 2 × 1 (100) surface and hydrosilation at hydrogen-terminated surfaces.

Hydrosilation with alkenes or alkynes results in environmentally stable superficial Si–C bonds. Figure 1.23 shows that the exposure of a pristine hydrogen-terminated surface produces the adsorption of an amount of adventitious carbon that progressively undergoes oxidation (as demonstrated by the shoulders at C 1s binding energy higher than 286 eV). On the contrary, the spectrum of the alkyne-derivatized surface remains substantially unchanged even after exposure to air for 9 months.

1.4.3. Addressable Nanowires and the Nano-to-Litho Link

Even admitting the technological capability of realizing parallel conductive wires on the NLS, the problem on how to contact them to the rest of the device architecture is still challenging. The development of strategies able to address ultrahigh dense architecture on the NLS is under systematic investigation [88–92]. Of course, linking the highly repetitive pattern producible via the S^{n}PT

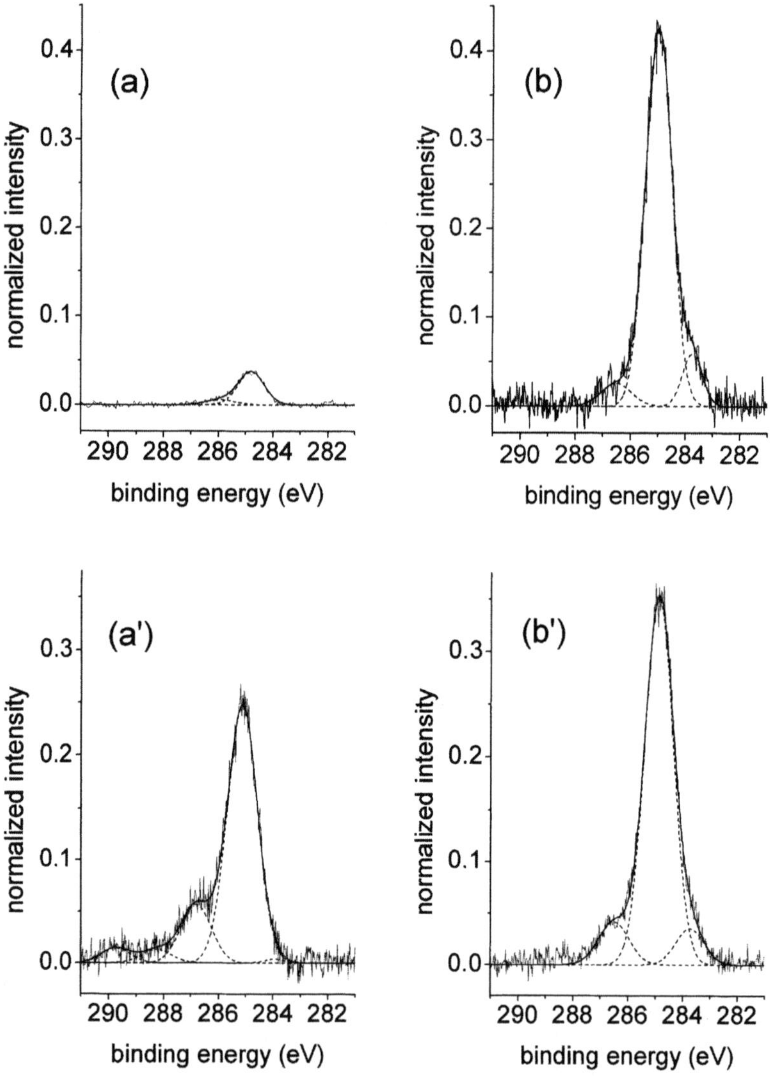

FIGURE 1.23. C 1*s* XPS spectra from (a) a hydrogen-terminated (100) silicon surface, (b the same surface after derivatization with 1-octyne, (a′) a surface such as (a) after prolonged (2.6×10^7 s) to air, and (b′) a surface such as (b) after prolonged exposure (2.6×10^7 s) to air.

or advanced imprint lithography to the hosting microelectronic circuits requires some form of "hardware demultiplexing," which can hardly be achieved without using electron-beam lithography. However, although electron-beam lithography has resolution compatible with the geometries achievable with S^nPT or advanced imprint lithography, it is unable to match their ultimate pitches (60 nm vs. 20 nm).

One major advantage of S^nPT is that it can be arranged to generate automatically a hardware demultiplexing simply via an appropriate design of the mask

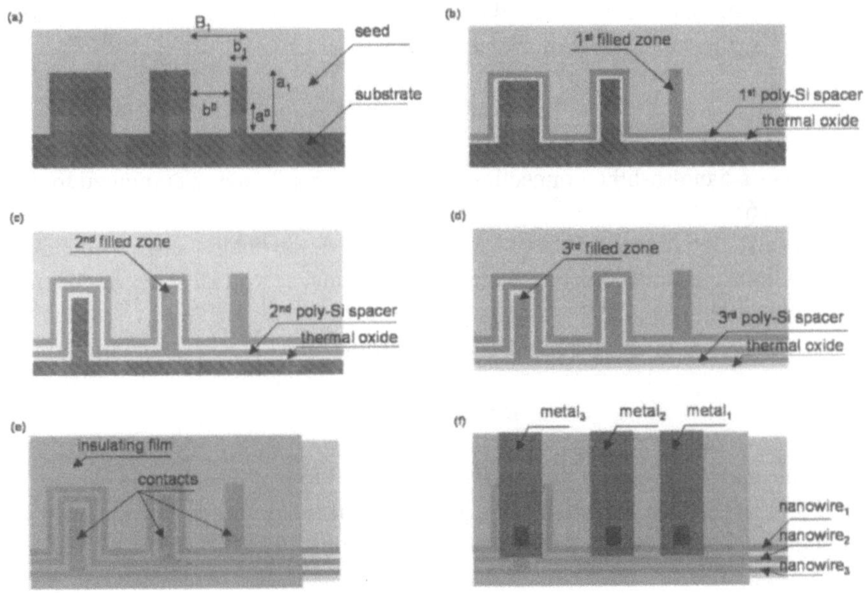

FIGURE 1.24. Plan view of the mask defining the wires and the trick adopted to allow the separate contact of each wire using electron-beam lithography.

defining the sacrificial layer. The basic idea underlying this application is described in the following.

Imagine that the mask defining the sacrificial layer has the shape shown in Fig. 1.24a, where the kth indentation extends along, and perpendicularly to, the wire by lengths b_k (with pitch B_k) and a_k, respectively. These quantities satisfy the following conditions:

$$b_k = 2t_{\mathrm{Si}} + 2(k-1)t_{\mathrm{sp}}, \quad (1.14)$$

$$B_k = b^0 + b_k, \quad (1.15)$$

$$a_k = (n-1)t_{\mathrm{sp}} + a^0 = a \text{ (independent of } k), \quad (1.16)$$

where t_{Si} is the thickness of the deposited polysilicon (transformed by SPT in polysilicon width), t_{sp} is the width of each double spacer, b^0 and a^0 are sizes that allow the alignment of contacts (defined by electron-beam lithography) in the zone where a single polysilicon deposition fills the underlying region. Figures 1.24b–1.24f show the evolution of the structure after the deposition of 1, 2, 3, etc. spacers provided that Eqs. (1.14)–(1.16) are really satisfied, and they show how it is possible to contact each wire individually with resolution afforded by electron-beam lithography.

It is noted that the above conditions are not the ones that minimize the area occupied by the contact region. For instance,

$$a_k = (k-1)t_{sp} + a^0$$

still allows a nano-to-litho connection, but is more space-saving compared to that of Eq. (1.16).

It is also noted that the above trick can be used "vertically" also, by defining with a unique mask, in correspondence to the original sacrificial layer, a sequence of n rectangular masks with side b_k along the wires given by Eq. (1.14), pitch B_k given by Eq. (1.15), and extension a perpendicularly to the wire given by

$$a = nt_{sp} + a^0,$$

where a^0 and b^0 preserve the same meaning as above. The central spacer in Fig. 1.20, larger than the other ones, is actually formed via the "fusion" of two neighboring layers deposited in a restricted geometry.

1.4.4. Circuit and Process Architecture

It is difficult to make sensible predictions on circuit and process architecture in the absence of a clear-cut definition of the addressing and sensing elements and the nature of the guest memory element.

Nonetheless, in part in analogy with what is done on current ICs and in part with the expected organic nature of the memory element, a few (likely not totally meaningless) hypotheses can be made and the first work in this field have appeared [90,91].

In general, any IC may be viewed as the union of nonintersecting zones: the active zone, hosting transistors and capacitors, and the field, which provides electrical insulation between adjacent active devices. Active devices are interconnected via metal conductors and polysilicon resistors flowing from one contact to another through active zones and fields and separated from them by interlayer dielectrics. The process architecture is organized in two segments: the front end (ion implantation, diffusion, oxidation, etc.) and the back end (metal and dielectric deposition, alloying, etc.). In turn, the heat treatments are hierarchically organized in such a way that the subsequent heat treatment does not appreciably modify the overall dopant distribution resulting from all of the previous treatments. In practice, the first treatments are carried out at temperatures around 1000°C (well diffusion and field oxidation) and the front end is completed with the segregation annealing of the gettering process at approximately 700°C. The back end is, instead, characterized by several heat treatments at temperatures around 400–500°C [93].

Assuming, as done in this work, that the crossbar structure is formed by polysilicon strips (thus producible at temperature of 600–700°C) and the active elements are instead organic molecules (thus typically destroyed at temperature of, say, 300°C), the following circuit and process architectures seem reasonable.

The circuit is formed by two nonoverlapping zones: the *nanozone*, hosting the molecular devices, and the *lithozone* (in turn divided in active zone and field) hosting the silicon devices. Nanozones and lithozones are interconnected by the strips that define rows and columns of the crossbar memory.

The process for the production of the nano-IC should comprise a *zeroth mask* protecting with Si_3N_4 or SiO_2 the nanozone from the *front-end initial stage*, in which stage transistors, capacitors, and short-range polysilicon interconnections are defined. At a certain point of this stage, the protection of the nanozone is removed and the polysilicon seed is allowed to cross the nano–litho frontier. After protecting the active zone (again with Si_3N_4 or SiO_2), the *strip-formation stage* is started. After its conclusion, the nanozone is protected (once again with Si_3N_4 or SiO_2) and the *front-end final stage* is carried out. The process goes to completion as in the standard IC process. After that, the dielectric layer protecting the nanozone is selectively removed and the *grafting* of the active elements to the polysilicon strips is achieved by the simple exposure to the organic molecules: The molecules are addressed to the wanted regions simply by the appropriate distribution of chemical potentials. In turn, that requires the ability to control the chemical terminations of the surfaces of the most important IC films. The passivation with organic resins of the overall nano-IC completes the process.

1.5. READING AND WRITING MOLECULES AS QUANTUM PROCESSES

The description of electrical properties of microelectronic passive elements on the NLS is expected to require major changes with respect to the current one. How phenomena like the mobility in restricted geometry, failure of continuum models of the solid, quantum size, or Casimir effects may impact the electrical performance of silicon wires is still a matter of discussion, but will not be consiudered here. Rather, we intend to focus our attention on the possible quantum effects that are expected to occur when a few molecules (and eventually only one) will be the active circuital device.

At the current level of technological development, the area of each crosspoint is of the order of 10^3 nm^2. Assuming a cross section of 1 nm^2 for functional molecules, each memory element may be formed by approximately 10^3 closely packed molecules. This number is presumably large enough to allow a statistical description of the system. This number will progressively be reduced with the progress of the technology and currently may be further reduced, limiting the grafting to the crosspoint perimeter (this situation is exactly the same as sketched in Fig. 1.19).

In this way, hybrid molecular electronics could soon face the situation in which few molecules or even *one single molecule* only will be the memory element. This situation is critical for two reasons:

- The first, statistical in nature, is associated with the fact that, due to the process of molecular insertion, the number of molecules forming the memory element will be a random variable (presumably with a Poisson distribution).

- The second, of a quantum nature, is associated with the fact that, due to the microscopic nature of molecules, writing and reading of the memory must be seen as preparation and measurement quantum processes.

The first point requires a circuit architecture that is both intrinsically statistical (to manage nominally indistinguishable elementary devices with a random distribution of electrical characteristics) and fault tolerant (to make up for the number of crosspoints with no inserted molecules). This is a hot research area that will probably take advantage of the current progress in the design of system architectures.

The possibility of testing macroscopically the conduction state of single molecules introduces conceptual problems (in addition to potential applications) [94] and is intrinsically associated with an emerging research area—single-molecule detection and spectroscopy ([95–98, and references quoted therein]. Its discussion requires a better specification of the functional element and of its operation mode. In this work we rest at what may be considered the simplest configuration: a reconfigurable molecule with two electronic states (one stable and the other metastable) with sharply different static current–voltage, I–V, characteristics.

Since the circuit considered by the HP–UCLA collaborations was just of this kind, one could imagine that circuit as the material model of application of theoretical arguments. Since the conductance of, for instance, a [2]-rotaxane is determined by its unoccupied molecular orbitals, a macroscopic measurement at low voltage is, in ultimate analysis, able to determine the *energy* structure of the molecule. On the other side, since the [2]-rotaxane conductance is uniquely controlled by the *position* of the ring along the axle, the situation is expectedly controlled by the noncommutativity of energy and position operators. However, since the *electron* conduction along the [2]-rotaxane is controlled by the motion of the ring *molecule*, the quantum nature of the problem is almost immediately lost. In fact, the assumption that a measurement of current determines both the *energy* of the electronic levels and the *position* of the ring does not pose problems with respect to the measurement of noncommuting observables because most molecules in most situations are adequately described by the Born–Oppenheimer approximation,[7] for which *nuclear positions are parameters rather than dynamical variables* with respect to the electronic observables.

The situation becomes more complex when the electronic and nuclear states are not so intrinsically linked, as it happens for the [2]-rotaxanes considered by the

[7] Since most molecules appear in position states, it has been hypothesized that the decoherence of quantum systems eventually responsible for their classical behavior is effective just at the molecular level [99, Sect. 3.2.4]. In this case, the Born–Oppenheimer approximation would not simply be an (certainly accurate) *approximation*, but, rather, a *fundamental law* reflecting (paraphrasing Kundera) "the unbearable heaviness of being molecules."

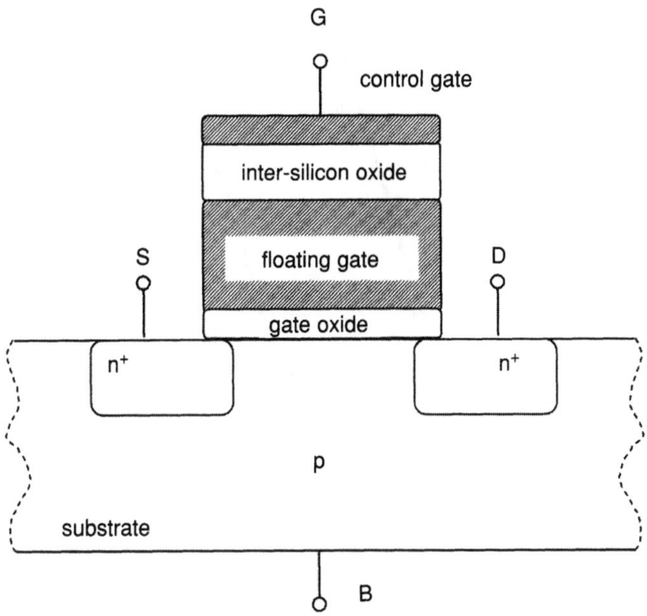

FIGURE 1.25. Cross section of a flash memory device.

HP–UCLA collaboration. This situation will be considered later, but its analysis requires a preliminary description of the simplest example of memory—the flash memory.

1.5.1. Conventional Flash Memory Devices

A memory cell is the element of an electronic system in which information can be stored and possibly modified. A flash device bases its storage ability on the modulation of the threshold voltage of a MOSFET with a modified gate stack structure. A floating gate is placed between a topmost control gate and the MOSFET channel. Both gates are usually built in polysilicon and are embedded in insulating oxides (the oxide separating the channel from the floating gate is usually referred to as the gate oxide, whereas the other, between the floating and control gates, is known as the interpoly oxide). A schematic cross section of a generic floating-gate flash memory is shown in Fig. 1.25.

The functionality of a flash cell can easily be understood by determining the relationship between the floating-gate potential (controlling the channel conductivity) and the control-gate potential. As any other MOSFET, the flash cell operates as a switch with a threshold voltage that can be tuned according to the amount of charge placed in the floating gate. This allows the cell to behave as a memory. It is possible to define two different states by choosing a suitable threshold shift in the turn-on transistor voltage [100,101]: The higher the charge placed in the floating gate, the higher the MOSFET threshold voltage.

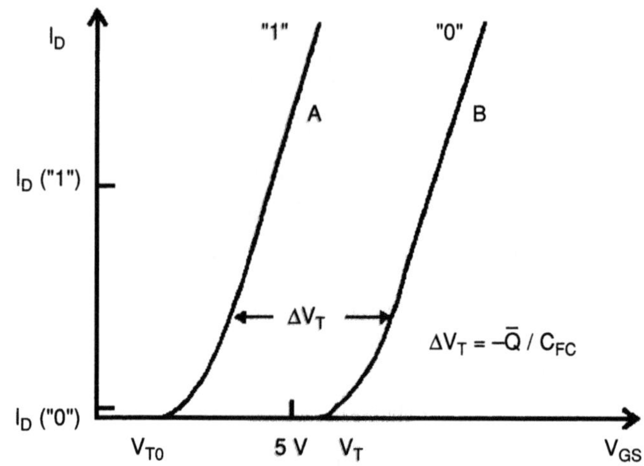

FIGURE 1.26. Current–voltage characteristics of a flash device for two different values of charge stored within the floating gate respectively denoting two different states: $Q = 0$, state A, and $Q < 0$, state B.

In this scheme, the cell is written by confining electrons in the floating gate. In such a situation, their negative charge obstructs the channel formation by screening the underlying silicon from the control gate. This effect is manifested in terms of a higher turn-on control-gate voltage. On the other hand, the memory is erased by removing the stored charge and restoring the voltage threshold to its original value. The threshold-voltage shift resulting from charging of the floating gate is known as the "memory window" and the memory state is sensed by measuring the current in the MOSFET when a control-gate bias is applied within the memory windows. Figure 1.26 shows the I–V characteristics of a flash device before (state A) and after (state B) the injection of charge in the floating gate, with the relative variation in the threshold voltage.

The transport mechanisms of the electrons in and out of the floating gate involve tunneling phenomena during both write and erase operations.

Write Under typical programming conditions, large positive drain and control-gate voltage are applied to produce a strong electric field. This field sets the conditions for a current to flow from the source to the drain. A fraction of these electrons, accelerated by the strong field in the channel near the drain region, gains a sufficient kinetic energy to overcome the SiO_2 barrier of 3.2 eV and reaches the floating gate. This programming method is known as channel hot-electron injection; the hot electrons refer to the highly energetic electrons produced by the strong electric field in the channel region close to the drain. Upon surmounting the barrier, the electrons thermalize rapidly, losing their kinetic energy through interactions with the atomic lattice vibrations, neutral or ionized dopants, and the other electrons. They trickle down and are collected in the floating gate, thus modulating the threshold voltage of the FET.

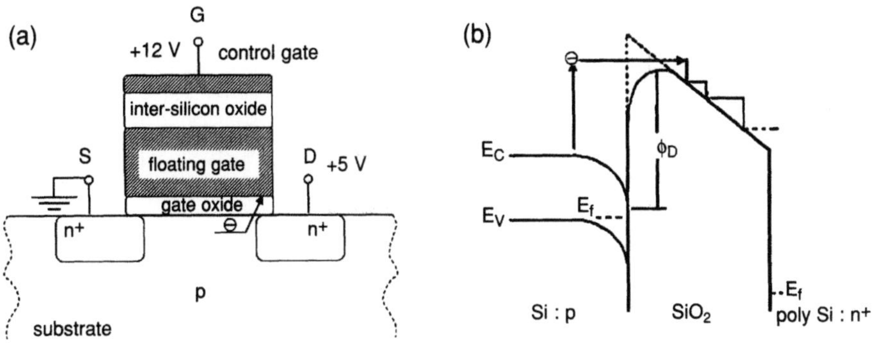

FIGURE 1.27. Energy-band diagram of a flash memory channel in the "write" operation mode.

Erase Under erase conditions, a large negative control-gate voltage is applied with respect to the source contact. The electric field established between the floating gate and the source region modulates the shape of the potential barrier felt by the electrons in the floating gate. At a high field, the electrons may thus escape from the floating gate, tunneling through the oxide to the silicon, where they thermalize through interactions with the source electrons and the crystal lattice. This mechanism, of high-field-assisted tunneling, is known as Fowler–Nordheim tunneling.

Figures 1.27 and 1.28 are schematics of the energy band in the two different working modes as well as the transport of electrons into or from the floating gate toward the substrate.

1.5.2. A Possible Molecular Flash Device

One advantage of the flash structure and functioning is that they may be reproduced almost exactly using molecules. The following example shows this analogy (with the remarkable exception of carrier sign, holes, for the reason discussed later).

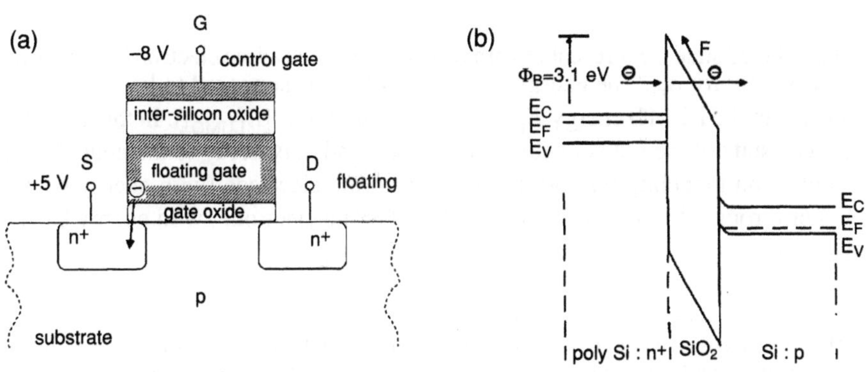

FIGURE 1.28. Energy-band diagram of a flash memory channel in "erase" operation mode.

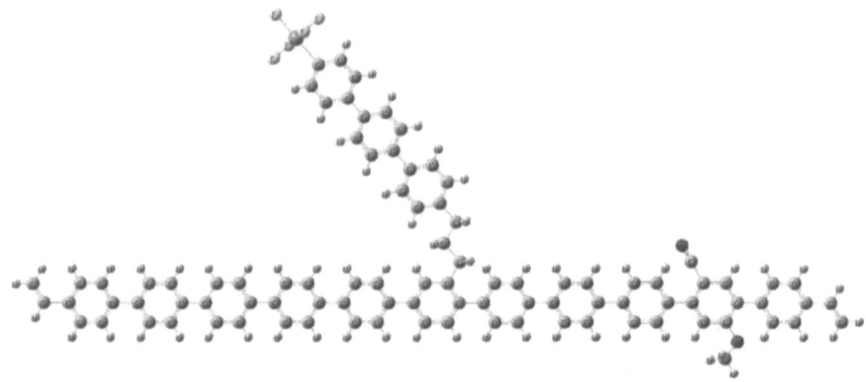

FIGURE 1.29. A candidate molecule for mimicking a flash memory cell, where the wire is formed by a π-conjugated poly-*para*phenylene chain, the arm is a rigid chain linked to the wire by a group discontinuing the π conjugation, and the redox center, SbF_5, is a group with high electron affinity. The chain contains two groups (a donor CN and acceptor OCH_3 group on the same benzene ring) behaving as rectifiers.

Consider a molecule connecting two heavily doped *p*-type polysilicon electrodes. The molecule is formed by three moieties: a *wire* (for instance, a π-conjugated chain), covalently bonded at its extremes to the silicon electrodes; a side *redox center* R, whose state charge may be varied via electron transfer from, or to, the neighboring electrodes; and an *arm*, linking materially the redox center to the wire. The arm is supposed to decouple the redox center from the wire, in such a way that the state of the latter is affected by the state of the former only electrostatically.

Figure 1.29 sketches a hypothetical molecule with the above characteristics. Of course, such a molecule has been designed to satisfy the electronic requirement listed in the following; without modifications of the benzene-ring terminations, in practice it would be not processable.

The redox center admits two states: neutral R^0 or reduced R^-,

$$R^0 + e^- \longrightarrow R^-.$$

The reduced state is more stable with respect to R^0 plus a free electron by a positive electron affinity A_{el}. The energy level $-A_{el}$ of R^- is supposed to be appreciably lower than the LUMO energy E^a_{LUMO} of the arm ($A_{el} > |E^a_{LUMO}|$). In this case, the reduced state of the redox center is metastable and may decay to the neutral state via electron tunneling through the vacuum to the electrode (with energy barrier A_{el}) or through the arm to the wire and then to the electrode (with energy barrier $A_{el} - |E^a_{LUMO}|$).[8]

[8] The decay has been supposed to be not thermally activated. If the process were thermally activated, the lifetime should be given by $\tau_0^- \exp((A_{el} - |E^a_{LUMO}|)/k_B T)$, where τ_0^- is the period outer-electron oscillation in R^-.

In the absence of negative charge on the redox center, the wire is supposed to be an insulator, which means (for hole conduction) that the HOMO energy E^{w}_{HOMO} of the wire is lower than the valence band edge E_{VB} ($|E_{VB}| < |E^{w}_{HOMO}|$). When a negative charge is injected into the redox center, however, its effect is to increase (for purely Coulombic interaction) the HOMO energy of the wire. If the arm length is short enough and the wire is suitably chosen, it may thus happen that E^{w}_{HOMO} will become higher than the valence band edge ($|E_{VB}| > |E^{w}_{HOMO}|$). This situation allows the injection of holes from the electrode into the wire and, in ultimate analysis, brings it in a conductive state. Thus, according to the $0/-$ charge state of the redox center, the wire will be in different insulating/conductive states—which allows the use of the described molecular arrangement to code a bit and to sense it by the application of a weak electric field between the polysilicon contacts.

It remains to be explained how the redox center may be modified. Assuming the neutral state as the equilibrium one, the reduced state may be obtained via the application between the contacts of a potential difference so strong as to allow an avalanche injection of electrons into the region containing R^0. Once an electron is captured by the center, the current will then flow mainly through the wire that will allow an external circuitry to switch off the writing potential. Erasing the R^- state may be achieved via tunnelling of the trapped electron into the writing electrode, sustained by the application of a large reverse potential difference between the contacts (sufficiently large to increase the tunneling rate to a polysilicon conduction band but not not so large to produce an avalanche injection). This can be achieved without an unwanted flow of current through the wire via the insertion along it of a rectifying moiety (like the one hypothesized by Aviram and Ratner [37]) or doping the polysilicon wires with different (*p* or *n*) dopants.

The active molecule is almost necessarily asymmetric—which opens the problem of the correct grafting to the two sides of the polysilicon electrodes. For solving this problem, one can either accept that part of the grafted molecules (expectedly around 50%) do not participate in the writing–erasing–sensing memory activity or modify process and molecule to have asymmetric host and guest. An example is given in the following:

- Nitridation (for instance, by exposure to NO) of the Si | SiO_2 interface of the lower electrode immediately before the S^nPT construction.
- Attack with aqueous solution of HF of the interlayer oxide, leaving the lower electrodes protected by a monolayer of nitrogen-terminated silicon.
- Exposure of the structure to functional molecules where only one side has alkene $CH_2 = CH-$ or alkyne $CH \equiv C-$ termination and the other is terminated with a Lewis acid –R.
- Grafting of the molecule to the upper electrode via hydrosilation and to the lower electrode via acid–base reaction.

The reason why we have considered hole conduction through the wire (rather than electron conduction, thus destroying the complete analogy with conventional flash memories) is that the transition $R^0 + e^- \longrightarrow R^-$ usually involves an energy

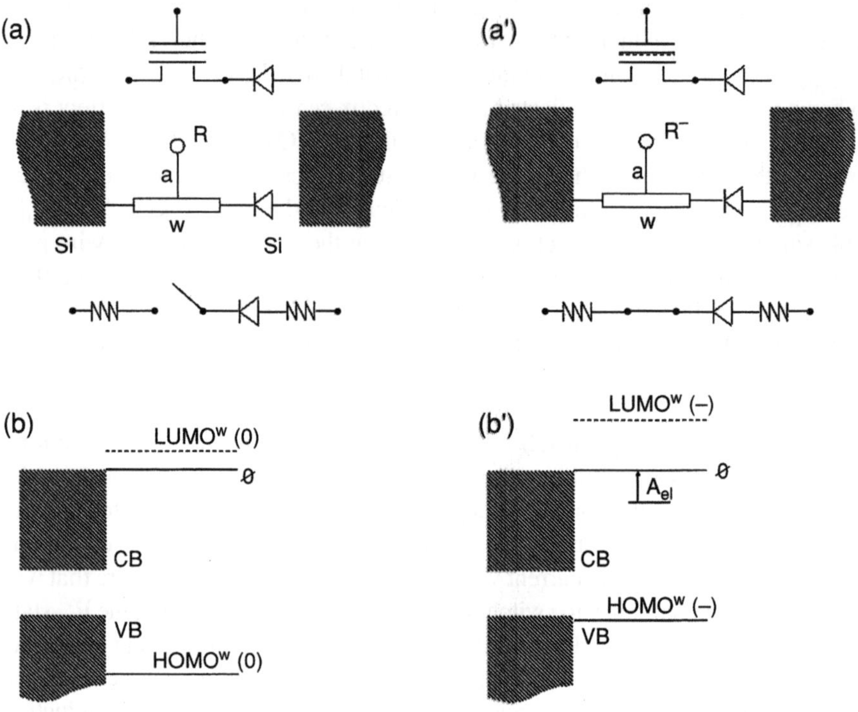

FIGURE 1.30. Energy configurations for the molecule in the neutral and charged states, and their analogy with the flash memory cell.

in the electron-volt energy scale, whereas the transition $R^0 \longrightarrow R^+ + e^-$ involves an energy in the energy scale of 10 eV.

Figure 1.30 shows the electronic configurations of the molecule in the considered states. Let **0** and **1** denote the neutral and negatively states, respectively, of the redox center. As discussed above, the **0** state imparts to the wire a low conductivity, whereas the **1** state imparts a high conductivity, which allows the determination of the charge state of the redox center by a measurement of the wire conductivity under static or quasi-static conditions—a quantity that seems accessible to currently available apparatuses. The physical meaning of such a measurement for single molecules is discussed in the following subsection.

1.5.3. What Do We Measure Measuring Static $I - V$ Characteristics of Single Molecules?

The description of charge conduction along a molecule has been an extremely hot research area in last years. For instance, several works have been devoted to the study of the influence of thermal reservoir [102], relaxation phenomena [103], temperature rise [104], and electron–vibron interaction and electron–electron correlation [105] on the current flowing through a molecular junction connecting two

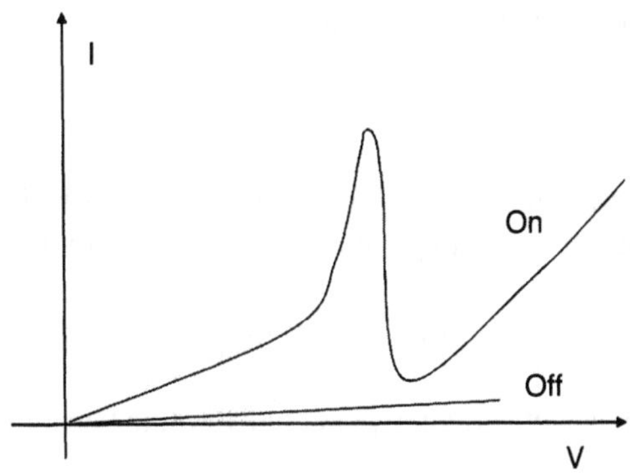

FIGURE 1.31. Different conduction behaviors of a hypothetical molecule in the ON or OFF state.

metallic contacts (see Ref. 106 for a recent review of this topic). Although they are useful for predicting quantitatively molecular conductances, the complexity of the treatments hides the fundamental problems associated with the processes of writing and reading a bistable single molecule. To avoid this difficulty, we describe the current flow through the wire simply as a tunneling through a given (static) potential barrier, eventually described by a certain nonlinear characteristic $I^{(0)}$–V (the upper index is used to denote the charge state of the redox center). At present, it is difficult not only specifying the exact functional relationship $I^{(0)}(V)$ but even predicting the order of magnitude. What can be predicted is that the conduction will be resonant around a few voltages when the energy spectrum is discrete (see Fig. 1.31), whereas the molecule will switch from insulator to conductor at a certain voltage when the energy levels form a band.[9]

The ionized state, instead, is metastable and is characterized by two decay modes. The first mode is associated with the fact that since the localized electronic state is embedded in a continuous spectrum of extended states, the wavefunction giving the electronic probability distribution will eventually spread to the entire electrode region. This evolution is the unitary evolution of quantum mechanics described by the time-dependent Schrödinger equation. The second mode, on the contrary, is essentially statistical in character and is associated with the fact that the electron, being embedded in a thermal reservoir at temperature T, eventually acquires an energy ΔE^* sufficient to escape from the redox site to the wire (in which case $\Delta E^* = A_{\rm el} - |E^{\rm a}_{\rm LUMO}|$) or to one of the electrodes (in which case, $\Delta E^* = A_{\rm el}$).

The first process is characterized by an evolution with time constant Δt related to the portion of the continuous energy spectrum ΔE required to represent the

[9] For the bipyridinium molecule sketched in Fig. 1.16 in contact with gold electrodes, the derivative dI/dV of the current I with respect to the applied voltage V has a sharp resonance at $V \simeq -1.8$ V at which the ON differential resistivity ($\simeq 30$ nA/V) exceeds the OFF value by one order of magnitude outside the interval $(-1.9, -1.5)$ V [33].

localized state through the Heisenberg relation:

$$\Delta t \simeq \hbar/2\Delta E. \tag{1.17}$$

The second process is, instead, a Poisson process described by an exponential decay of the probability of finding the electron at the redox site,

$$P(t) = \exp(-t/\tau), \tag{1.18}$$

with a time constant τ depending exponentially on the barrier energy ΔE^*:

$$\tau = \tau_0 \exp(-\Delta E^*/k_B T). \tag{1.19}$$

The interplay between the intrinsic dynamic of the system and the coupling with its environment gives rise to a complicate time dependence, which in most cases can be thought of as responsible for the decoherence of the quantum state to a classical state. Thus, together with the characteristic times Δt and τ above, one has also to consider the decoherence time Γ^{-1}, in turn depending on T. In general, Γ^{-1} can be assumed to be shorter than τ, because thermal activation is a consequence of the loss of coherence; on the contrary, there is no *a priori* relationship between Δt and Γ^{-1}. The following extreme situations are of interest:

- If $\Gamma^{-1} \ll \Delta t$, the system in its environment appears as classic (*mutatis mutandis*, it behaves as a particle in Brownian motion provided that vibron scattering substitutes for molecular collisions) and its stability is controlled by temperature as described by Eq. (1.19).
- If $\Delta t \ll \Gamma^{-1}$, the system has a quantum behavior and its stability is controlled by Eq. (1.17).

Assume now that the redox center is forced in the negatively charged state. In this case, the hole conduction through the wire is increased because the electrostatic potential generated by the negative charge reduces the height of the potential well separating the electrodes and thus increasing (exponentially) the transparency of the barrier. For all applied voltages, the characteristic $I^{(-)}(V)$ is thus expected to be higher (possibly much higher) than $I^{(0)}(V)$:

$$\forall V : I^{(0)}(V) < I^{(-)}(V).$$

In principle, this fact allows the use of the nanoscopic arrangement of Fig. 1.30 as a memory element.

The overall functioning principle would thus be very similar to that of the nonvolatile memories of the MOS technology were it not for the fact that the measurement does certainly perturb the redox center. To evaluate the effects of this perturbation, first we have to understand how a "static current" is felt by an electron localized in the vicinity of the wire where the current is flowing. A current of, say, 1 pA means the passage through the wire of one hole every 1.6×10^{-7} s; assuming a transit time of, say, 10^{-12} s, the flowing current is thus seen by the localized electron as a (presumably) Poisson process of impulsive stabilizing potentials.

Consider now the case, described by Eqs. (1.18) and (1.19), of the time evolution of the electronic state dominated by its interaction with the environment. As discussed in Ref. 99, the continuous observation of a system with time evolution described by Eq. (1.18) confirms the unobserved evolution. The continuous observation, therefore, does not interfere with the proper evolution of the system (in its thermal reservoir).

Consider instead the case of electron evolution dominated by quantum dynamics. In this case, the erratic hole passages through the wire (confirmed macroscopically by the measurement of a static current) behave as repeated measurements with frequency I/e. If the corresponding period is much higher than the characteristic time governing the quantum dynamics, whose lower estimate is given by Eq. (1.17), $e/I \ll \Delta t$, these events are thus responsible for a continuous monitoring of the system, in this way producing a kind of quantum Zeno effect for the position [99]. In the limit of high current, the localized state should become indefinitely stable, thus allowing the permanence of the quantum state even on a macroscopic time scale.

1.6. DESCRIBING SYSTEMS ON THE NANOMETRE LENGTH SCALE: BOTTOM-UP, TOP-DOWN, OR ANYTHING ELSE?

To a large extent [with the remarkable exception of the Bohr's genuine ("orthodox") interpretation of quantum mechanics], physical theories are reductionistic in nature: In ultimate analysis ("in principle"), the properties at a larger length scale *may be reduced to* (i.e., *are completely accounted for by*) the properties at a smaller length scale.

1.6.1. The Bottom-up Description of Nature

Physics may be considered as the study of interacting particles and fields: The motion of a particle in an assigned field is the subject of *mechanics*; the goal of *field theory* is the description of the time evolution of a field in relation to its sources; the complete description of the motion of a particle in the fields it generates is the subject of *field dynamics*. Each theory (mechanics, field theory, and field dynamics) has been developed at both classical and quantum levels; microscopic bodies are adequately described at the quantum level only.

The basic equations usually hold regardless of the quantities that specify the nature of the characters of the theory; in other words, such quantities are external parameters (e.g., quantum mechanics applies to all particles irrespective of their masses; the quantum mechanical description of a given particle contains its mass m as an external parameter). However, nature appears formed by only few elementary constituents. In fact, it is commonly accepted that matter is ultimately

systems
↑
phases
↑
molecules
↑
atoms
↑
nuclei + electrons

FIGURE 1.32. The naive hierarchical organization of ordinary matter.

formed by 12 elementary particles (classified in 2 families: leptons and quarks) interacting via 4, to some extent unified, fields (strong, electromagnetic, weak, and gravitational).

The description of the properties of matter at the laboratory scale, however, does not require consideration of the ultimate constituents of matter. Of the bricks of the Universe, indeed, only few form ordinary matter: protons and neutrons (organized in nuclei) and electrons. Also, the unique fields that are required for the description of the macroscopic properties of practically any system are the electromagnetic and gravitational ones.

Starting from nuclei and electrons, ordinary matter can be imagined hierarchically organized in a bottom-up fashion as shown in Fig. 1.32, where solids are regarded as very large molecules, and liquids are viewed as phases formed by strongly interacting molecules.

Both physics and chemistry aim at reductive descriptions, in which the properties of higher-level objects (in the naive classification of Fig. 1.32) are explained in terms of properties of lower-level objects, assumed as given; the properties of the lowest level are postulated.

The logic of a hierarchical description would be of explaining the properties at each level in terms of properties at the underlying level supposedly given.[10]

1.6.2. The Bottom-up Construction of Physical Theories

It may reasonably be stated that practically all molecular properties are in the reach of a more or less *ab initio* quantum mechanical description. This statement extends (thanks to the properties of the Fourier transform) to single-crystalline, indefinitely

---

[10] However, although it is possible to account for the properties of systems in terms of properties of the constituting phases using *thermodynamics* and for the properties of phases in terms of properties of the constituting molecules using *statistical mechanics*, it is impossible to describe the properties of molecules in terms of those of their constituting atoms. Rather, the properties of both atoms and molecules are accounted for (using *quantum mechanics*) in terms of nuclei and electrons in electromagnetic interaction. Thus, although matter appears hierarchically organized as sketched in the scheme of Fig. 1.32, its theoretical description places atoms and molecules at the same level.

extended solids and surfaces. Although the conceptual structure of the physical description of Nature is thus bottom-up, physical theories have been constructed in a top-down fashion, starting from the description of systems in the laboratory length scale.

Historically, the first physical theory (Aristotelian physics, with its now awkward concepts of motion, proper motion, etc.) was the result of the observation of the world at the human length scale; classical (Newtonian) physics retained the kinematic concepts of Aristotelian physics and resulted from the need that the same laws that hold on the human ("laboratory," since Galilei) length scale hold also on a larger (astronomic) length scale. The application of classical physics to small systems (atoms) failed, but the theory necessary for the description of the atomic world was, nonetheless, expressed in terms of quantities (like energy, momentum) of classical physics as far as possible.

Reiterating and abstracting this procedure, one may conclude that a lower level (LL) theory is constructed from an upper level (UL) theory assuming that UL quantities admit LL analogues and that the latter obey the same laws as the former. In general, this procedure fails and one is forced both to introduce new LL quantities, which do not have UL analogues, and to modify the physical laws (possibly via a change of the mathematical structure of the theory). This procedure is well exemplified by the construction of quantum mechanics from classical mechanics: Quantum mechanics can indeed be seen as a procedure for the transcription of the classical physical laws in terms of new mathematical quantities, as outlined in Table 1.6

This procedure is considered correct only if the LL theory admits a contracted description that is isomorphic to the UL theory:

1. A *LL dynamics* is supposed to be known, and the microscopic state of the system is specified by the N-tuple $\underline{x} = (x_1, \ldots, x_N)$.
2. A *contracted description* of the system is given by the M-tuple $\underline{y} = (y_1, y_2, \ldots, y_M)$, with $y_i f_i(x_1, \ldots, x_N)$, $i = 1, \ldots, M$, and $M \ll N$. Since the correspondence $\underline{x} \to \underline{y}$ is not 1:1 and, in general, several different $\underline{x}$ are associated with the same $\underline{y}$, the contracted description $\underline{y}$ is less detailed than the microscopic description $\underline{x}$.
3. The $\underline{y}$ variables of the contracted description are *secular variables*, which means that $\underline{y}(t)$ varies slowly compared to $\underline{x}(t)$.

TABLE 1.6. Correspondences between classical and quantum mechanical representations of physical concepts.

	Mathematical quantity	
Concept	Classical	Quantum
Particle	Phase space $\mathcal{S}$	Hilbert space $\mathcal{H}$
State	Point of $\mathcal{S}$	Direction in $\mathcal{H}$
Observable	Function on $\mathcal{S}$	Linear operator in $\mathcal{H}$

4. A *UL description* is a closed and causal contracted description of LL dynamics; "closed" means that the contracted description speaks only in terms of the $\underline{y}$ variables, whereas "causal" means that the knowledge at time $t = 0$ specifies $\underline{y}(t)$ completely.

The incomplete specification of the state of the system given by the $\underline{y}$ description is such that, in general, $\langle y_i^2\rangle \neq \langle y_i\rangle^2$, due to fluctuation phenomena.

The process of constructing a macroscopic theory as a contracted description of a microscopic theory is methodologically correct but chronologically false. Indeed, theories are usually built at an upper (= more macroscopic) level of description (thermodynamics preceded statistical mechanics by about half a century) and the existence of a lower (= more microscopic) level of description is inferred from the existence of stochastic elements in the upper description or from experiments taking place in lapses of time shorter than the characteristic time during which $\underline{y}(t)$ varies.

As an example of a quantity that is fundamental at LL but disappears at UL, we mention spin (a quantum observable that does not admit a classical analogue), isospin (in the description of nuclear properties, but mute in atomic physics), color (in quark description of barions, but mute in nuclear physics), and so forth.

1.6.3. Anything Else?

The above analysis seems to indicate that the game of physics is essentially formed by two plays: (i) formulating LL theories for the description of properties at progressively smaller length scale and (ii) constructing a contracted description of LL theories for the accurate description of properties at a larger length scale.

This view is, however, disturbed by the difficulties posed by the measurement problem in quantum mechanics. This theory, indeed, states that the statevector $|\,\Psi\rangle$ evolves in a deterministic, continuous way during an unobserved motion, or has an erratic, discontinuous evolution during a measurement. To avoid an animistic view of Nature (for which the particle does know when it is observed or not) and to make up for the seeming inconsistency, $|\,\Psi\rangle$ cannot be interpreted as a physical quantity.

The "orthodox" interpretation of quantum mechanics specifies the meaning of $|\,\Psi\rangle$ as follows: Microscopic objects, which can be described by commonsense words, obey classical physics (up to relativity). Macroscopic events are changes of state of macroscopic objects. Some rare, carefully prepared events, referred to as measurements (e.g., the "tick" of a Geiger counter, track in a Wilson chamber) are conveniently ascribed to the effects of microscopic particles. These particles are necessarily described by commonsense words (motion, proximity, etc.) and by quantities of classical physics (mass, charge, velocity, etc.). This description is, however, experimentally found to be inadequate—classical physics does not describe the behavior of microscopic particles.

Orthodox interpretation *Quantum mechanics is an algorithm, a machinery, which is superimposed on classical mechanics of the considered microscopic particle and is capable* *of retaining the part of an earlier measurement useful for predicting the results of a later measurement.*

The previous ("antirealistic") statement of the orthodox interpretation is mainly due to Heisenberg [107], who was led to it by his uncertainty principle. The most advanced formulation of ordinary quantum mechanics is perhaps that given by Ludwig in his *Foundations of Quantum Mechanics.* According to him [108, p. vii],

> This book represents a systematic mathematical, as well as conceptual, formulation of the original viewpoint of N. Bohr in which it is assumed that it is necessary to use the classical mode of description in order to describe the measurement process in quantum mechanics.

However, discussing the major results of his formulation, Ludwig argued [108, p. 10]:

> Have we [...] eliminated the measurement problem? Have we eliminated the problem in such a way that the interaction between the microsystem and the measurement apparatus can be analyzed without the need of quantum theory? Of course not. What we have done is to place where it belongs—namely, with the developing.

and concludes [108, p. 11]:

> Quantum mechanics cannot present a closed theory [...] of the preparation and registration processes. This is perhaps disappointing. In fact, it merely demonstrates that quantum mechanics is not a theory which can describe everything from a microsystem to a macrosystem.

Although the orthodox interpretation of quantum mechanics succeeds in removing the inconsistency of the double behavior of Nature at the microscopic level, it reduces $| \Psi \rangle$ to a mere mathematical tool without any physical reality. This fact is particularly disappointing because of the existence of a "classical limit" (Ehrenfest theorem) of quantum mechanics (i.e., of a contracted description of quantum mechanics that in the classical limit is isomorphic with classical mechanics).

Bohr had a somewhat different view: The analysis of a typical measurement experiment shows that it is usually irreversible and involves an amplification process; according to Bohr, this process perturbs the state of the particle and erratically projects it onto an eigenstate [109]. Bohr, however, was unable to formulate his ideas in mathematical terms.

The first idea to circumvent the difficulties of the orthodox interpretation is probably von Neumann's description of the measurement process [110]. According to von Neumann, the measuring apparatus itself is described by quantum mechanics, and the theoretical description of the measurement process proceeds as follows: The measurement of an observable A in the subsystem I is possible through the interaction with another system, II (the "apparatus"), when the following conditions are satisfied:

1. An interaction Hamiltonian H_{int} acts impulsively during the duration of the measure process.

2. If the system I is in one of eigenstates $| a_n\rangle_{\mathrm{I}}$ of A, then immediately after the measurement, system II is in a linear manifold $\mathcal{V}_n$ such that

$$\langle a_n \mid a_{n'}\rangle \Rightarrow \mathcal{V}_n \perp \mathcal{V}_{n'}.$$

(This one-to-one correspondence between eigenstates of the system and linear manifolds of the apparatus is necessary for the apparatus to be able to determine the actual state of the system; the interaction Hamiltonian is necessary to allow the state of the apparatus to be projected into linear manifolds.) These two conditions, however, are not mutually compatible because of the existence of the interference terms. The Von Neumann idea was that the interference terms disappear if the state of the apparatus is perfectly known. To determine the exact state of the apparatus, von Neumann imagined a chain of systems III, IV, . . . , each of which behaves as a measuring apparatus for the previous one. The problem of the knowledge of the exact state of the apparatus shifts, therefore, from II to III, hence to IV and henceforth to *the observer who, because of the self-consciousness of his own state, is able to cancel the interference terms*. Von Neumann's theory, for which the *animate* observer has a crucial role, was surely influenced by Szilard's solution of Maxwell demon paradox. In this solution, Szilard showed that the animate observer (the "demon") participates in the global entropy–information balance, so allowing the Second Law of Thermodynamics not to be violated [111].

Von Neumann's theory, for which the measurement act is described in terms of the Schrödinger equation, and Bohr's ideas, for which the measurement process involves an amplification that destroys the quantum state, have been coalesced in a unique theory by Daneri, Loinger, and Prosperi (DLP) [112]. The DLP theory is first concerned with the foundation of a quantum theory of macrosystems; in principle, this theory would replace classical mechanics as the fundamental theory at the macroscopic level. In this context, the measuring apparatus is regarded as a macrosystem for which macrostates can be defined in a suitable way within the conceptual framework of quantum mechanics. The development ensues by observing that the detection of the state of a microsystem by a macroscopic apparatus always involves an amplification process that can be seen as the irreversible transition of the amplifying apparatus to a condition of stable equilibrium. This irreversible transition permits the linear superposition of Hilbert-space vectors, representing the quantum state of the "microsystem + macrosystem" after the measurement, to be replaced by a statistical distribution of quantum microstates. The aim of the DLP theory is to show that this statistical distribution coincides with the distribution of eigenstates predicted by the algorithm of quantum mechanics. This approach avoids the characterization of the macrosystem by a state vector as made by von Neumann in his regressive analysis of the measure process, and it confers a sound physical ground to the Bohr interpretation. In fact, the Bohr discrimination between classical and quantum objects, described by classical and quantum physics, respectively, is replaced in the quantum theory of ergodicity and measurement by the distinction between microscopic and macroscopic objects, both ruled by quantum mechanics, the macroscopicity of the measuring apparatus being accounted for by a description in terms of quantum statistical mechanics.

To capture the physical concepts on which the DLP theory is based, it is useful to consider a typical experiment for the visualization of the path of charged particles. A Wilson chamber may operate as a measurement apparatus when all paths therein may be visualized. The actual path is detected through the following:

- The *interaction* of the particle with molecules of the apparatus resulting in the formation of an ionized column around the particle path.
- The *irreversible formation* of an invisible (mesoscopic) track around the primary ionization events.
- The *magnification* of the mesoscopic track up to the formation of a visible track.

The primary events of the measurement are the formation of cation–anion pairs (the anion being possibly an electron, in which case one speaks of ionization) produced by a low-impact-parameter collision of the particle with a molecule in the Wilson chamber; the mesoscopic track results from the formation of clusters around the ions before they recombine; and the magnification is produced by the condensation around the clusters of droplets that persist even after getting electroneutrality. These events are hierarchically controlled by the corresponding driving forces (involving energies of the order of 30 eV for ionization or heterolytic dissociation, 10 eV for cluster formation, and 1 eV for droplet condensation).

Since prefix "micro" to "scopic" has a different (often contradictory) meaning from "micro" to "meter," it is useful to refer to events on the sub-NLS as "quantum events." With this change of terms, the appearance of a quantum phenomenon at the macroscopic world involves several length scales:

- It is initiated at the sub-NLS (the *quantum scale*).
- It proceeds irreversibly at the NLS (the *mesoscopic scale*).
- It is magnified and stabilized at a length scale much larger than the NLS (the *macroscopic scale*).

The correspondence among events, phenomena, and length scales is given in Table 1.7.

TABLE 1.7. The events concurring to a measurement and their length scales.

Event	Phenomenon	Length scale
Quantum	Interaction	$\lesssim 10^{-1}$ nm
Mesoscopic	Irreversible evolution	≈ 1 nm
Macroscopic	Amplification	$\gg 1$ nm

What is worth noting is that the irreversible formation of a mesoscopic track (*latent measurement*) occurs via the irreversible dispersion of a great energy over many degrees of freedom, whereas the magnification of the mesoscopic track to make it macroscopically observable (*actual measurement*) is performed by the amplification provided by the apparatus prepared in a metastable state. As argued by Bohr and formalized by DLP, interaction, irreversibility, and metastability seem to play a fundamental role in the measurement. This scheme, however, does not exhaust all possible measurements: The technological progress in the cryogeny has indeed shown that the metastability of the measurement apparatus is not a fundamental prerequisite; and molecular electronics seems to make the quantum world accessible to the measurement without the condition of irreversibility. The first point is discussed in Ref. 113; the second was the matter of the discussion of Section 1.5.3.

1.7. CONCLUSIONS: PRELIMINARY, TENTATIVE, PROVISIONAL

It is quite embarrassing trying a conclusion from a study on molecular electronics, a field just at the beginning. We conclude this chapter by limiting our synthesis effort to trying to give an answer to the following question: *Does indeed hybrid molecular electronics overcome the limits of microelectronics?* The birth of molecular electronics has been hypothesized because it is expectedly able to overcome the barriers of physical, technological, and economic nature, which sooner or later are presumed to limit the exponential development of microelectronics.

To what extent hybrid molecular electronics is able to solve the physical problem, providing molecules by themselves suitable for computation, was discussed in Section 1.5.

That silicon–organic structures capable of computation may be prepared with geometries on the NLS, without the use of extreme lithography (or involves it in a few strata only) has been demonstrated in Section 1.4.

About the last point, first we observe that because the process architecture hypothesized in Section 1.4.4 does not involve extreme lithography, it does not impact seriously the transistor production cost. The inspection of Fig. 1.20 shows that the S^nPT allows the integration of approximately 50 transistors in an area of a contact. This suggests that a mature hybrid molecular electronics may multiply by a factor of at least 10^2 the current bit density of ICs, the additional production cost being controlled by the cost of functional molecules.

We try to estimate this additional cost with the following argument. Let n_{bit} be the number of guest molecules in a single crosspoint; for a linewidth of 30 nm, n_{bit} is of the order of 10^3. The molecular weight M of the functional molecule depends largely on the considered molecule, but may be assumed to be lower than (although close to) 10^3 g mol^{-1}. The mass of functional molecules forming each memory element is thus $(n_{\mathrm{bit}}/N_{\mathrm{A}})M \approx 10^{-18}$ g. The cost of chemicals is largely variable too, however with a spectrum ranging from 10^{-3} \$ g^{-1} for commodities like hydrocarbons to 10^3 \$ g^{-1} for specialties like antibiotics. Assuming that the functional molecules of interest for molecular electronics are just in the upper part of this spectrum, the cost of chemicals forming the memory element is of the order of 10^{-15} \$.

In the absence of yield information, assessing anything about production cost is certainly risky. The above value, however, is so low (lower than the estimated transistor cost at the end of Roadmap, 10^{-9} \$, by six order of magnitude—see Fig. 1.3) that even the economic factor seems to be favorably impacted by hybrid molecular electronics.

REFERENCES

1. G.E. Moore, Obituary: Robert N. Noyce, *Phys. Today* **44**(1), 83 (1991).
2. G.E. Moore, Cramming more components onto integrated circuits, *Electronics* **38**, 114–117 (1965).
3. G.E. Moore, Progress in digital integrated electronics, *IEDM Tech. Dig*. 11–13 (1975).
4. J. Birnbaum and R.S. Williams, Physics and the information revolution, *Phys. Today* **53**(1), 38–42 (2000).
5. E. Ross, The dramatic shift to semiconductor foundries: From zero to 50% in 30 years, *Future Fab Int.* Issue 17 (2001), available at http://www.future-fab.com/
6. P. Gargini, Sailing with the ITRS into nanotechnology, *Semicon West* (July 2004), available at http://www.intel.com/research/silicon/nanotechnology.htm.
7. G.F. Cerofolini, The evolution of IC technology and the need of new substrates, in *Chemistry for Innovative Materials*, edited by G.F. Cerofolini, R.M. Mininni, and P. Schwarz, EniChem, Milano, 1991, pp. 64–83.
8. G.F. Cerofolini, L. Meda, and M. Sparpaglione, Substrates: The material bases of microelectronics and nanoelectronics, *Prog. Quant. Electr*. **17**, 273–298 (1993).
9. A.D. Wilson, Lithography requirements in complex VLSI device fabrication, in *The Physics of VLSI*, edited by J.C. Knights, American Institute of Physics, New York, 1984, pp. 69–91.
10. J.D. Plummer and P.B. Griffin, Material and process limits in silicon VLSI technology, *Proc. IEEE* **89**, 240–258 (2001).
11. A.I. Kingon, J.-P. Maria, and S.K. Streiffer, Alternative dielectrics to silicon dioxide for memory and logic devices, *Nature* **406**, 1032–1038 (2000).
12. T. Homma, Low dielectric constant materials and methods for interlayer dielectric films in ultralarge-scale integrated circuit multilevel interconnections, *Mater. Sci. Eng. R* **23**, 243–285 (1998).
13. G.F. Cerofolini, Strategies for ultralow-κ dielectrics for integrated-circuit interconnects, *Mater. Sci. Semicond. Process*. **5**, 265–270 (2003).
14. S.P. Murarka, Multilevel interconnections for ULSI and GSI era, *Mater. Sci. Eng. R* **19**, 87–151 (1997).
15. J.A. Davis, R. Venkatesan, A. Kaloyeros, M. Beylansky, S.J. Souri, K. Banerjee, K.C. Saraswat, A. Rahman, R. Reif, and J.D. Meindl, Interconnect limits on gigascale integration (GSI) in the 21st century, *Proc. IEEE* **89**, 305–324 (2001).
16. Semiconductor Industry Association (SIA), International Technology Roadmap for Semiconductors. 2005 Edition; available at http://public.itrs.net.
17. J.A. Hutchby, G.I. Bourianoff, V.V. Zhirnov, and J.E. Brewer, Extending the road beyond CMOS, *IEEE Circuits Dev. Mag*. **18**(2), 28–41 (2002).
18. Y.-K. Choi, T.-J. King, and C. Hu, A spacer patterning technology for nanoscale CMOS, *IEEE Trans. Electron Devices* **49**, 436–441 (2002).
19. C.M. Garner, Technology challenges & chemicals (2004), available at http://www.intel.com/research/silicon/nanotechnology.htm.

20. R.W. Keyes, Physical limits in semiconductor electronics, *Science* **195**, 1230–1235 (1977).
21. L. Baldi, and G.F. Cerofolini, La legge di Moore e lo sviluppo dei circuiti integrati, *Mondo Digitale* **1**(3), 3–15 (2002) (in Italian).
22. P.A. Packan, Pushing the limits, *Science* **285**, 2079–281 (1999).
23. P.S. Peercy, The drive to miniaturization, *Nature* **406**, 1023–1026 (2000).
24. R.W. Keyes, Fundamental limits of silicon technology, *Proc. IEEE* **89**, 227–239 (2001).
25. D.J. Frank, R.H. Dennard, E. Nowak, P.M. Solomon, Y. Taur, and H.-S.P. Wong, Device scaling limits of Si MOSFETs and their application dependencies, *Proc. IEEE* **89**, 259–288 (2001).
26. T. Ito and S. Okazaki, Pushing the limits of lithography, *Nature* **406**, 1027–1031 (2000).
27. L.R. Harriott, Limits of lithography, *Proc. IEEE* **89**, 366–374 (2001).
28. S. Lloyd, Ultimate physical limits to computations, *Nature* **406**, 1047–1054 (2000).
29. Y. Aharonov and D. Bohm, Time in quantum theory and the uncertainty relation for the time and energy domain, *Phys. Rev.* **122**, 1649–1658 (1961).
30. V.V. Zhirnov, R.K. Cavin, J.A. Hutchby, and G.I. Bourianoff, Limits to binary logic switch scaling—a *gedanken* model, *Proc. IEEE* **91**, 1934–1939 (2003).
31. J.D. Meindl and J.A. Davis, *The* fundamental limit on binary switching energy for terascale integration (TSI), *IEEE J. Solid-State Circuits* **35**, 1515–1516 (2000).
32. R.W. Landauer, Irreversibility and heat generation in the computing process, *IBM J. Res. Dev.* **5**, 183–191 (1961).
33. D.I. Gittins, D. Bethell, D.J. Schiffrin, and R.J. Nichols, A nanometre-scale electronic switch consisting of a metal cluster and redox-addressable groups, *Nature* **408**, 67–69 (2000).
34. S. Becker and K. Müllen, Nanochemistry—Architecture at the mesoscale, in *Stimulating Concepts in Chemistry*, edited by F. Vögtle, J.F. Stoddart, and M. Shibasaki, Wiley–VCH, Weinheim, 2000, pp. 317–337.
35. B.A. Mantooth and P.S. Weiss, Fabrication, assembly, and characterization of molecular electronic components, *Proc. IEEE* **91**, 1785–1802 (2003).
36. M.A. Reed, Prospects for molecular-scale electronics, *MRS Bull.* **26**(2), 113–120 (2001).
37. A. Aviram and M. Ratner, Molecular rectfiers, *Chem. Phys. Lett.* **29**, 277–283 (1974).
38. N. van Gulick, Theoretical aspects of the linked ring problem, *New J. Chem.* **17**, 619–625 (1993); originally presented at the Reaction Mechanisms Conference held in Princeton, NJ, in 1960.
39. H.L. Frisch and E. Wasserman, Chemical topology, *J. Am. Chem. Soc.* **83**, 3789–3795 (1961).
40. D.M. Walba, Topological chemistry, *Tetrahedron* **41**, 3161–3212 (1985).
41. J.-P. Sauvage, Interlacing molecular threads on transition metals: Catenands, catenates, and knots, *Acc. Chem. Res.* **23**, 319–327 (1990).
42. A. Credi, V. Balzani, S.J. Langford, and J.F. Stoddard, Molecular logic. An XOR gate based on a mechanical molecular machine, *J. Am. Chem. Soc.* **119**, 2679–2681 (1997).
43. V. Balzani, M. Gomez-Lopez, and J.F. Stoddart, Molecular machines, *Acc. Chem. Res.* **31**, 405–414 (1998).
44. A.S. Martin, J.R. Sambles, and G.J. Ashwell, Molecular rectifier, *Phys. Rev. Lett.* **70**, 218–221 (1993).
45. C. Joachim, J.K. Gimzewski, and A. Aviram, Electronics using hybrid-molecular and mono-molecular devices, *Nature* **408**, 541–548 (2000).
46. C.P. Collier, E.W. Wong, M. Belohradsky, F.M. Raymo, J.F. Stoddart, P.J. Kuekes, R.S. Williams, and J. R. Heath, Electronically configurable molecular-based logic gates, *Science* **285**, 391–394 (1999).
47. C.P. Collier, G. Mattersteing, E.W. Wong, Y. Luo, K. Beverly, J. Sampaio, F.M. Raymo, J.F. Stoddart, and J.R. Heath, A [2]Catenane-based solid state electronically reconfigurable switch, *Science* **289**, 1172–1175 (2000).

48. M.R. Stan, P.D. Franzon, S.C. Goldstein, J.C. Lach, and M.M. Ziegler, Molecular electronics: From devices and interconnect to circuits and architecture, *Proc. IEEE* **91**, 1940–1957 (2003).
49. Y. Luo, C.P. Collier, J.O. Jeppesen, K.A. Nielsen, E. Delonno, G. Ho, J. Perkins, H.-R. Tseng, T. Yamamoto, J.F. Stoddart, and J.R. Heath, Two-dimensional molecular electronics circuits, *Chem. Phys. Chem.* **3**, 519–525 (2002).
50. L.J. Guo, Recent progress in nanoimprint technology and its applications, *J. Phys. D: Appl. Phys.* **37**, R123–R141 (2004).
51. Y. Chen, D.A.A. Ohlberg, X. Li, D.R. Stewart, J.O. Jeppesen, K.A. Nielsen, J.F. Stoddart, D.L. Olynick, and E. Anderson, Nanoscale molecular-switch devices fabricated by imprint lithography, *Appl. Phys. Lett.* **82**, 1610–1612 (2003).
52. Y. Chen, G.Y. Jung, D.A.A. Ohlberg1, X. Li, D.R. Stewart, J.O. Jeppesen, K.A. Nielsen, J.F. Stoddart and R.S. Williams, Nanoscale molecular-switch crossbar circuits, *Nanotechnology* **14**, 462–468 (2003).
53. D.R. Stewart, D.A.A. Ohlberg, P. Beck, Y. Chen, R.S. Williams, J.O. Jeppesen, K.A. Nielsen, and J.F. Stoddart, Molecule-independent electrical switching in Pt/organic monolayer/Ti devices, *Nano Lett.* **4**, 133–136 (2004).
54. C.N. Lau, D.R. Stewart, R.S. Williams, and D. Bockrath, Direct observation of nanoscale switching centers in metal/molecule/metal structures, *Nano Lett.* **4**, 569–572 (2004).
55. R.F. Service, Next-generation technology hits an early midlife crisis, *Science* **302**, 556–559 (2003).
56. G. F. Cerofolini and G. Ferla, Toward a hybrid micro-nanoelectronics, *J. Nanoparticle Res.* **4**, 185–191 (2002).
57. J. M. Tour, Molecular wires and devices, in *Stimulating Concepts in Chemistry*, edited by F. Vögtle, J.F. Stoddart, and M. Shibasaki, Wiley–VCH, Weinheim, 2000, pp. 237–253.
58. Anon., Patent office takes small step to get the measure of nanotech, *Nature* **432**, 8 (2004).
59. Y.-K. Choi, J. Zhu, J. Grunes, J. Bokor, and G.A. Somorjai, Fabrication of sub-10-nm silicon nanowire arrays by size reduction lithography, *J. Phys. Chem. B* **107**, 3340–3343 (2003).
60. G.F. Cerofolini, G. Arena, M. Camalleri, C. Galati, S. Reina, L. Renna, D. Mascolo, and V. Nosik, Strategies for nanoelectronics, *Microelectr. Eng.* **81**, 405–419 (2005).
61. G.F. Cerofolini, G. Arena, M. Camalleri, C. Galati, S. Reina, L. Renna, and D. Mascolo, A hybrid approach to nanoelectronics, *Nanotechnology* **16**, 1040–1047 (2005).
62. W.R. Hunter, T.C. Holloway, P.K. Chatterjee, and A.F. Tasch, Jr., New edge-defined approach for submicrometer MOSFET fabrication, *IEEE Electron Device Lett.* **2**, 4–6 (1981).
63. D.C. Flanders and N.N. Efremow, Generation of $<$ 50 nm period gratings using edge defined technique, *J. Vac. Sci. Technol. B* **1**, 1105–1108 (1983).
64. U. Hilleringmann, T. Vieregge, and J.T. Horstmann, Masking and etching of siilicon and related materials for geometries down to 25 nm, *IECON99 Conference Proceedings* Vol. 1, IEEE Industrial Electronics Society, Piscataway, NJ, 1999, pp. S. 23–28.
65. J.T. Horstmann, U. Hilleringmann, and K. Goser, Matching analysis of NMOS-transistors with a channel lenght down to 30 nm, *IECON99 Conference Proceedings* Vol. 1, IEEE Industrial Electronics Society, Piscataway, NJ, 1999, pp. S. 56–61.
66. G.F. Cerofolini, Nano-mesoscopic interface: Hybrid devices, in *Encyclopedia of Nanoscience and Nanotechnology*, edited by J.A. Schwarz, C. Contescu, and K. Putyera, Dekker, New York, 2004, pp. 2211–2219.
67. S.F. Bent, Organic functionalization of group IV semiconductor surfaces: Principles, examples, applications, and prospects, *Surf. Sci.* **500**, 879–903 (2002).
68. J.M. Buriak, Organometallic chemistry on silicon and germanium surfaces, *Chem. Rev.* **102**, 1271–1308 (2002).
69. G. Cleland, B.R. Horrocks, and A. Houlton, Direct functionalization of silicon *via* the self-assembly of alcohols, *J. Chem. Soc. Faraday Trans.* **91**, 4001–4003 (1995).
70. A. Ulman, Self-assembled monolayers of alkyltrichlorosilanes: Building block for future organic materials, *Adv. Mater.* **2**, 573–582 (1990).

71. C.A. Roth, Silylation of organic chemicals, *Ind. Eng. Chem. Prod. Res. Dev.* **11**, 134–139 (1972).
72. M.P. Stewart, F. Maya, D.V. Kosynkin, S.M. Dirk, J.J. Stapleton, C.L. McGuiness, D.L. Allara, and J.M. Tour, Direct covalent grafting of conjugated molecules onto Si, GaAs and Pd surfaces from aryldiazonium salts, *J. Am. Chem. Soc.* **126**, 370–378 (2004).
73. A.B. Sieval, R. Linke, H. Zuilhof, and E.J.R. Sudhölter, High-quality alkyl monolayers on silicon surfaces, *Adv. Mater.* **12**, 1457–1460 (2000).
74. H. Ubara, T. Imura, and A. Hiraki, Formation of Si–H bonds on the surface of microcrystalline silicon covered with SiO_x by HF treatment, *Solid State Commun.* **50**, 673–675 (1984).
75. J.S. Hovis and R.J. Hamers, Structure and bonding of ordered organic monolayers of 1,5-cyclooctadiene on the silicon(100) surface, *J. Phys. Chem. B* **101**, 9581–9585 (1997).
76. H. Liu and R. J. Hamers, An X-ray photoelectron spectroscopy study of the bonding of unsaturated organic molecules to the Si(100) surface, *Surf. Sci.* **416**, 354–362 (1998).
77. J. Terry, M.R. Linford, C. Wigren, R. Cao, P. Pianetta, and C.E.D. Chidsey, Electronic structure of alkyl monolayers on Si(111), *J. Appl. Phys.* **85**, 213–221 (1999).
78. A. Lehner, G. Steinhoff, M.S. Brandt, M. Eickhoff, and M. Stutzmann, Hydrosilylation of crystalline silicon (111) and hydrogenated amorphous silicon surfaces: A comparative X-ray photoelectron spectroscopy study, *J. Appl. Phys.* **94**, 2289–2294 (2003).
79. A.B. Sieval, A.L. Demirel, J.W.M. Nissink, M.R. Linford, J.H. van der Maas, W.H. de Jeu, H. Zuilhof, and E.J.R. Sudholter, Highly stable Si–C linked functionalized monolayers on the silicon (100) surface, *Langmuir* **14**, 1759–1768 (1998).
80. A.B. Sieval, V. Vleeming, H. Zuilhof, and E.J.R. Sudholter, An improved method for the preparation of organic monolayers of 1-alkenes on hydrogen-terminated silicon surfaces, *Langmuir* **15**, 8288 (1999).
81. M. Kosuri, H. Gerung, Q. Li, S.M. Han, B.C. Bunker, and T.M. Mayer, Vapor-phase adsorption kinetics of 1-decene on H-terminated Si(100), *Langmuir* **19**, 9315–9320 (2003).
82. F.A. Cotton and G. Wilkinson, *Advanced Inorganic Chemistry*, 5th ed., Wiley, New York, 1988, p. 1255.
83. G.F. Cerofolini, C. Galati, S. Reina, and L. Renna, Functionalization of the (100) surface of hydrogen-terminated silicon via hydrosilation of 1-alkyne, *Mater. Sci. Eng. C* **23**, 253–257 (2003).
84. G.F. Cerofolini, C. Galati, S. Reina, and L. Renna, The addition of organic functional groups to silicon via hydrosilation of 1-alkynes to hydrogen-terminated, 1 × 1 reconstructed, (100) silicon surfaces, *Semicond. Sci. Technol.* **18**, 423–429 (2003).
85. G.F. Cerofolini, C. Galati, S. Reina, L. Renna, O. Viscuso, G.G. Condorelli, and I.L. Fragalà, X-ray photoemission spectroscopy study at different takeoff angles of hydrosilation of 1-alkynes at hydrogen-terminated 1 × 1-reconstructed (100)-oriented silicon, *Mater. Sci. Eng. C* **23**, 989–994 (2003).
86. G.F. Cerofolini, C. Galati, S. Reina, L. Renna, F. Giannazzo, and V. Raineri, Hydrosilation of 1-alkyne at nearly flat, terraced, homogeneously hydrogen terminated, (100) silicon surfaces, *Surf. Interf. Anal.* **36**, 71–76 (2004).
87. G.F. Cerofolini, C. Galati, S. Reina, L. Renna, G.G. Condorelli, I.L. Fragalà, G. Giorgi, A. Sgamellotti, and N. Re, Functionalization of atomically flat, dihydrogen terminated, 1 × 1 (100) silicon with 1-alkyne, *Appl. Surf. Sci.* **246**, 52–67 (2005).
88. Y. Huang, X. Duan, Y. Cui, L.J. Lauhon, K.-H. Kim, and C.M. Lieber, Logic gates and computation from assembled nanowire building blocks, *Science* **294**, 1313–1317 (2001).
89. Z. Zhong, D. Wang, Y. Cui, M.W. Bockrath, and C.M. Lieber, Nanowire crossbar arrays as address decoders for integrated nanosystems, *Science* **302**, 1377–1379 (2003).
90. M.M. Ziegler and M.R. Stan, CMOS/nano co-design for crossbar-based molecular electronic systems, *IEEE Trans. Nanotechnol.* **2**, 217–230 (2003).
91. A. DeHon, Array-based architecture for FET-based, nanoscale electronics, *IEEE Trans. Nanotechnol.* **2**, 23–32 (2003).

92. M. Forshaw, R. Sadler, D. Crawley, and K. Nikolić, A short review of nanoelectronic architectures, *Nanotechnology* **15**, S220–S223 (2004).
93. G.F. Cerofolini and L. Meda, *Physical Chemistry of, in and on Silicon*, Springer-Verlag, Berlin, 1989, Chap. 10.
94. A.M. Stoneham, The challenges of nanostructures for theory, *Mater. Sci. Eng. C* **23**, 235–241 (2003).
95. J.K. Gimzewski and C. Joachim, Nanoscale science of single molecules using local probes, *Science* **283**, 1683–1688 (1999).
96. R. Brown, J. Gallop, and M. Milton, Review of techniques for single molecule detection in biological application, NPL Report COAM 2, National Physical Laboratory, Teddington, UK, 2001.
97. M. Orrit, Single-molecule spectroscopy: The road ahead, *J. Chem. Phys.* **11**, 10,938–19,946 (2002).
98. S. Datta, Electrical resistance: An atomistic view, *Nanotechnology* **15**, S433–S451 (2004).
99. E. Joos, in *Decoherence and the Appearance of a Classical World in Quantum Theory*, edited by D. Giulini, E. Joos, C. Kiefer, J. Kupsch, I.-O. Stamatescu, and H. D. Zeh, Springer–Verlag, Berlin, 1996, Chap. 3.
100. P. Pavan, R. Bez, P. Olivo, and E. Zanoni, Flash memory cells—An overview, *Proc. IEEE* **85**, 1248–1271 (1997).
101. R. Camerlenghi, A. Modelli, A. Visconti, and R. Bez, Introduction to flash memory, *Proc. IEEE* **91**, 489–502 (2003).
102. D. Segal, A. Nitzan, M. Ratner, and W.B. Davis, Activated conduction in microscopic molecular junctions, *J. Phys. Chem. B* **104**, 2790–2793 (2000).
103. D. Segal and A. Nitzan, Steady state quantum mechanics of thermally relaxing systems, *Chem. Phys.* **268**, 315–335 (2001).
104. D. Segal and A. Nitzan, Heating in current carrying molecular junctions, *J. Chem. Phys.* **117**, 3915–3927 (2002).
105. A.S. Alexandrov and AM. Bratkovsky, Memory effect in a molecular quantum dot with strong electron-vibron interaction, *Phys. Rev. B* **67**, 235312-1–8 (2003).
106. C. Joachim and M. A. Ratner, Molecular wires: Guiding the super-exchange interactions between two electrodes, *Nanotechnology* **15**, 1065–1075 (2004).
107. W. Heisenberg, Über den anschaulichen Inhalt der quantentheoretischen Kinematik und Mechanik, *Z. Physik* **43**, 172–198 (1927). English translation: The physical content of quantum kinematics and mechanics, in *Quantum Theory and Measurement*, edited by J.A. Wheeler and W.H. Zurek, Princeton University Press, Princeton, NJ, 1983, pp. 62–84.
108. G. Ludwig, *Foundations of Quantum Mechanics. I*, Springer–Verlag, New York, 1983.
109. N. Bohr, The quantum postulate and the recent development of atomic theory, *Nature* **121**, 580–590 (1928); reprinted in N. Bohr, *Atomic Theory and the Description of Nature*, Cambridge University Press, Cambridge, 1934, pp. 52–91.
110. J. von Neumann, *Mathematischen Grundlagen der Quantenmechanik*, Springer–Verlag, Berlin, 1932, pp. 184–237. English translation: *Mathematical Foundations of Quantum Mechanics*, Princeton University Press, Princeton, NJ, 1955, pp. 347–445.
111. L. Szilard, Über die Entropieverminderung in einem thermodynamischen System bei Eingriffen intelligenter Wesen, *Z. Physik* **53**, 840–856 (1929). English translation: On the decrease of entropy in a thermodynamic system by the intervention of intelligent beings, *Behav. Sci.* **9**, 301–310 (1964).
112. A. Daneri, A. Loinger, and G.M. Prosperi, Quantum theory of measurement and ergodicity conditions, *Nucl. Phys.* **33**, 297–319 (1962).
113. G.F. Cerofolini, The calorimetric detection of rare events in the cryogenic regime, *Appl. Phys. A* **51**, 467–475 (1990).

2

From SOI Basics to Nano-Size MOSFETs

Sorin Cristoloveanu*

2.1. INTRODUCTION

Silicon-on-insulator (SOI) technology was initiated half a century ago for the fabrication of radiation-hard circuits. Several SOI materials and structures have been conceived for dielectrically separating the thin, active device volume from the silicon substrate (Fig. 2.1) [1, 2]. The background idea is that in a bulk silicon metal–oxide–semiconductor (MOS) transistor, only the superficial layer (0.1–0.2 μm thick) is actually useful for electron transport, whereas the substrate is causing undesirable effects to occur.

However, the overwhelming success of bulk-Si complementary MOS (CMOS) confined SOI technology to niche applications. More recently, several factors have attracted renewed interest for SOI: invention of new SOI materials (Unibond, Eltran, ITOX), need for lower-power and higher-speed circuits, and visible limitations of bulk CMOS scaling. SOI transistors are now recognized as unchallenged devices to push forward the frontiers of ultimate CMOS scaling.

This chapter is designed to offer a comprehensive tutorial on state-of-the-art SOI technologies. The next two sections describe the key advantages of SOI circuits and the synthesis of the main SOI materials. The physical mechanisms involved in the operation of fully depleted and partially depleted SOI MOSFETs are discussed in Section 2.4. Characterization techniques, recommended for SOI materials and devices, are briefly presented in Section 2.5.

Sections 2.6 and 2.7 are dedicated to more advanced concepts related to the MOS field-effect transistor (MOSFET) miniaturization. We show that SOI is the *necessary* technology for CMOS scaling [3,4]. The SOI MOSFET is the smallest

* Institute of Microelectronics, Electromagnetism and Photonics (UMR CNRS-INPG-UJF)
ENSERG, BP 257, 38016 Grenoble Cedex 1, France
sorin@enserg.fr

device conceivable in the CMOS world. We focus on the unusual dimensional effects that may take place in a small volume. The short-channel, narrow-channel, and thin-body effects are addressed by including the impact of the gate and buried insulators. We finally show that three-dimensional (3-D) coupling effects govern the operation and optimization of advanced multiple-gate MOSFETs.

2.2. PRINCIPLES OF SOI TECHNOLOGY

An SOI circuit is composed of single-device islands, dielectrically isolated from each other and from the underlying substrate (Fig. 2.1). The *lateral* isolation offers a more compact design and simplified technology than in bulk silicon, whereas the *vertical* isolation achieves thin films (Fig. 2.2) and eliminates most of the detrimental substrate effects (latch-up, punch-through, etc). Since the source and drain regions extend down to the buried oxide (BOX), the junction surface, leakage currents, and junction capacitances are naturally reduced. The implications are improved speed, lower-power dissipation, and higher temperature of operation [1,2].

The very thin drain and source regions make SOI devices less affected by short-channel effects, originated from "charge sharing" and from drain-induced barrier lowering. SOI MOSFETs are also rather tolerant to transient radiation effects.

The SOI circuits present a definite advantage in the highly competitive domain of low-power and low-voltage circuits. A small gate voltage gap is suited to

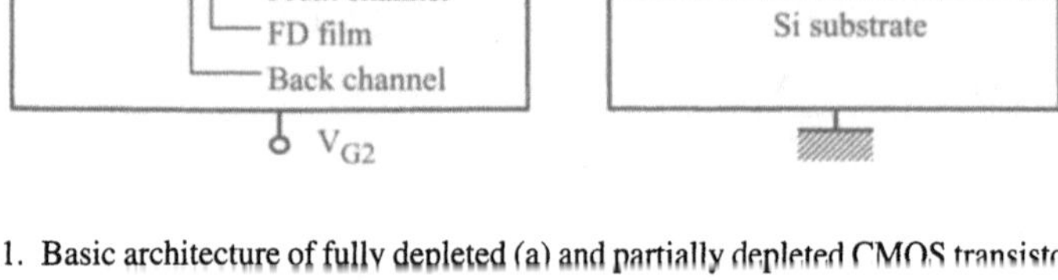

FIGURE 2.1. Basic architecture of fully depleted (a) and partially depleted CMOS transistors on SOI

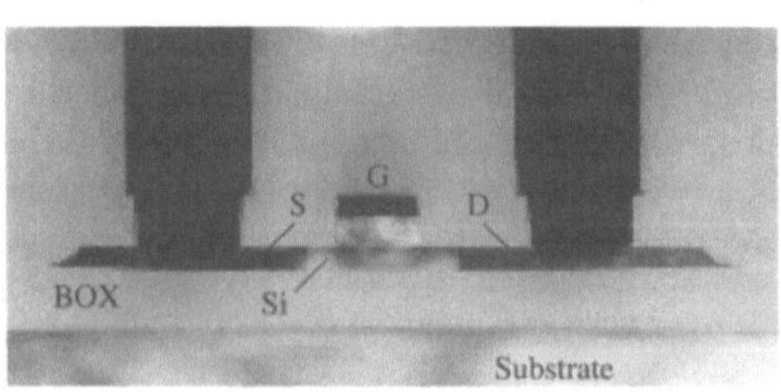

FIGURE 2.2. Cross section of a thin-film SOI MOS transistor.

switch a transistor from the off to on state. Fully depleted SOI offers a quasi-ideal subthreshold slope (60 mV/decade at room temperature).

Operation at equal *voltage* consistently shows about 20–30% increase in performance for SOI as compared to bulk silicon circuits. Conversely, operation at equal *low-power* dissipation may yield more than 100% performance gain. It is admitted that SOI circuits perform like bulk-Si circuits from the *next* generation.

The SOI circuits were handicapped by the cost of SOI wafers and by the unavailability of dedicated libraries for design. In the past, SOI technology was utilized only for circuits with high added value, because there was no need for an alternative technology. However, in the late 1990s, SOI has been included in the *International Technology Roadmap of Semiconductors* as the best suited option for most families of integrated devices.

CMOS circuits High-end microprocessors are being fabricated on SOI (IBM, AMD, Motorola, etc.). Radio-frequency (RF) SOI devices also show unchallenged capability. Extremely low-power circuits for mobile communication, portable processors operated with one-battery supply (0.5–1.2 V), and even battery less watches are currently fabricated with SOI technology. The SOI versatility has been taken advantage of for conceiving capacitor less dynamic random access memories (DRAMs). Globally, SOI is an ideal substrate for systems-on-chip. Fully depleted CMOS/SOI circuits operate successfully at temperatures beyond 300°C (for aeronautics and automobiles): The leakage currents are much smaller and the threshold voltage is less temperature sensitive than in bulk Si [5]. SOI circuits can also be radiation-hardened, able to sustain doses above 10 Mrad for the space industry.

High-voltage devices The outstanding advantage is the dielectric isolation. Power transistors can have lateral (LD–MOSFETs essentially for lightening) or vertical configuration. A vertical power device (IGBT, LDMOS, VMOS, etc.) can be accommodated in a thick SOI film *or* in the non-SOI section of mixed SOI-bulk wafers (see Fig. 2.3c_5). They can be rendered "smart" by being located next to a low-power SOI CMOS [6].

Innovative devices Novel device architectures and functions are based on the unique SOI features such as the adjustment of the thickness of the Si overlay and BOX, and the implementation of additional layers underneath the BOX. SOI is an ideal material for microsensors and micro (opto)-electromechanical systems (MEMS, MOEMS), where very thin membranes are required; the interface between the BOX and substrate is a perfect etch-stop mark. Many transducers for detection of pressure, acceleration (airbags), gas flow, temperature, radiation, magnetic field, and so forth, have successfully been integrated on SOI [1, 7].

Three-dimensional circuits containing consecutive thin silicon and BOX layers have been demonstrated with Epitaxial Lateral Overgrowth (ELO) and Zone Melting Recrystallization (ZMR) methods (Fig.2.3). For example, a 3-D image-signal processor contains arithmetic units and memories in bulk-Si bottom level, fast A/D converters in an intermediate SOI

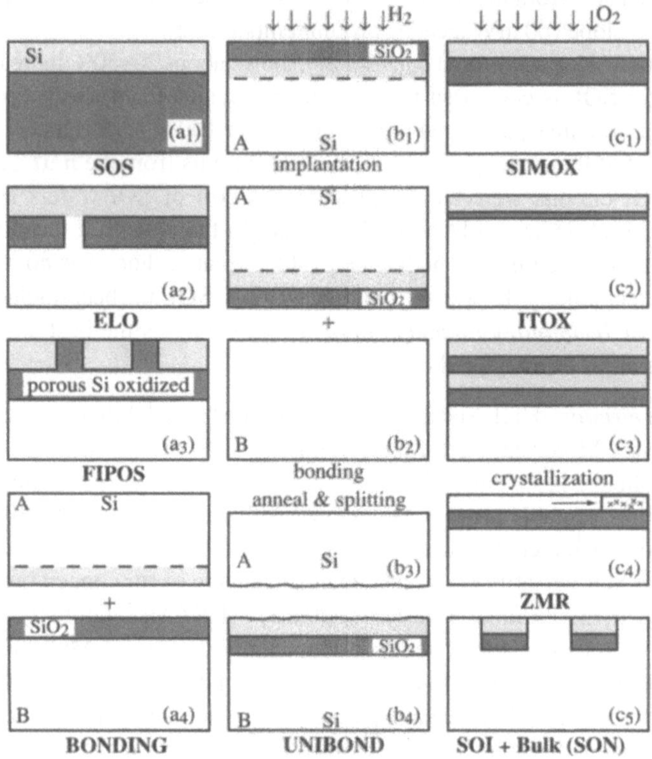

FIGURE 2.3. Main SOI materials: (a) SOS, ELO, FIPOS, and wafer bonding, (b) Unibond processing sequence, (c) SIMOX and ITOX variants, multiple-layer SOI for 3-D circuits, ZMR, and SON. The acronyms are defined in the text.

layer, and photodiode arrays in an upper SOI layer [8]. The SOI family also includes optical devices: switches, waveguides, and modulators.

Finally, most of the Si nanoelectronic devices (single-electron transisters, tunneling transistors, quantum dots, and wires) have used SOI, either for processing ease or for ultrathin-film capability [9, 10].

2.3. SOI WAFER TECHNOLOGIES

Various SOI technologies are illustrated in Fig. 2.3. Details about the fabrication methods for SOI wafers are discussed in Ref. 4. The basics are summarized in the following subsections.

2.3.1. Wafer Bonding

The synthesis of an SOI structure is a two-step process [11]. The first step consists in mating two Si wafers, one of which is oxidized (Fig. 2.3a$_4$). Bonding occurs via hydrophilic or hydrophobic bonds and Van der Waals forces. Impurities and air bubbles must be carefully eliminated prior to wafer bonding.

In the second step, one side of the bonded structure is thinned down (by mechanical grinding and chemical etching) in order to reach the target thickness of the silicon film (dotted line in Fig. 2.3a$_4$). Several types of etch-stop layer are being used: P^+/P^- or P/N junctions formed by implantation or epitaxy, Ge-rich sacrificial layers, porous silicon, and so forth.

Wafer bonding is attractive for the fabrication of sensors, power components, and other thick-film devices. This process does not produce ultra thin films (<100 nm) with enough uniformity and quality for advanced CMOSs.

2.3.2. Unibond

Unibond belongs to the family of bonded structures, except that the thinning phase is substituted by the revolutionary Smart-Cut process (commercialized by SOITEC) [12]. Hydrogen atoms are deeply implanted into the oxidized Si wafer A (Fig. 2.3b$_1$) in order to generate microcavities. Then wafer A is bonded to a support wafer B and annealed to enhance the bonding strength (Fig. 2.3b$_2$). This causes the hydrogen-induced microcavities to coalesce laterally. The two wafers naturally separate from each other—not at the bonded interface but at a depth defined by the location of microcavities (Fig. 2.3b$_3$). Finally, the rough SOI structure (Fig. 2.3b$_4$) is polished and thinned down.

Smart-Cut features outstanding advantages [4]:

1. The etch-back step is avoided.
2. The premium wafer A defines the quality of the SOI film (wafer B serving as a mechanical support) and can be repeatedly recycled (Fig. 2.3b$_3$). Unibond is virtually a single-wafer, cost-effective process.
3. The thickness of the silicon film and/or buried oxide can be adjusted (by tuning the implant energy and oxidation time) in a wide range: $t_{si} \simeq$ 20 nm to 1.5 μm and $t_{box} \simeq 2$ nm to 5 μm. Unibond is adaptable to most architectures from ultra thin CMOS to thick-film devices.
4. The crystal quality, defect density, and electrical properties of the Si film and thermal buried oxide are remarkable [13]. Note that the more defective bonded interface is situated underneath the BOX.

The Smart-Cut process is universal and flexible. It is currently applied to fabricate new SOI structures, where the Si film is replaced by strained Si, Si/SiGe stacks, Ge, or compound semiconductors (for optoelectronic and high-temperature circuits). In parallel, the BOX can be quartz, glass, Al_2O_3, or polymer (for flexible electronics). Smart-Cut allows bonding and de-bonding, enabling the transfer of already fabricated CMOS circuits from one substrate to another.

2.3.3. Eltran

Eltran means *epitaxial layer transfer* from a seed wafer to a handle wafer using porous silicon [14]. An electrochemical reaction is used to form porous silicon, which is mechanically weak but still preserving a crystalline quality. The thin Si film to be transferred is epitaxially grown on the porous layer. The key to epitaxy success is to seal the pores at the surface by high-temperature annealing in hydrogen ambient.

The surface of the epitaxial Si film is then oxidized before the wafer is bonded to a handle wafer. Wafer splitting is achieved using a fine and powerful water jet. The cleavage can be optimized by forming two porous layers with different porosities: fine pores at the surface to improve epitaxy and coarser pores in the volume for easy splitting. Finally, the residual porous silicon is etched away from the SOI wafer and the surface is smoothed by hydrogen annealing. The donor wafer can be recycled. Eltran features the same basic advantages as those listed for Smart-Cut. The main difference is the splitting mechanism.

2.3.4. SIMOX

The SIMOX (Separation by IMplanted OXygen) buried oxide is synthesized by internal oxidation during the deep implantation of high doses of oxygen into a Si wafer (Fig. 2.3c_1) [15]. Long annealing at high temperature ($\simeq$1320°C) is mandatory to recover a suitable crystalline quality of the film. High-current implanters (100 mA) were especially conceived. The 8–12- in. SIMOX wafers feature good thickness uniformity, low defect density (except threading dislocations: 10^4–10^6 cm^{-2}), robust BOX, sharp Si–SiO_2 interface, and high carrier mobility and lifetime [16].

Typical SIMOX variants are presented in Fig. 2.3c:

- Low-dose SIMOX: the implantation [(3–4) $\times$ 10^{17} O^+/cm^2] can be completed with an oxygen-rich anneal for enhanced BOX integrity (ITOX process). The BOX is 50–150 nm thick, whereas the Si film is 0.1 μm thick or less (Fig. 2.3c_2). Higher implant dose (1.8 $\times$ 10^{18} O^+/cm^2) results in a thicker BOX (0.4 μm). The film thickness is controlled by adjusting the implant energy or, for very thick films, by subsequent epitaxy.
- Double SIMOX (Fig. 2.3c_3) requires two implantation steps at different energies. The Si layer sandwiched between the two oxides can serve for interconnects, wave guiding, additional gates, or electric shielding.
- Laterally isolated single-transistor islands, formed by implantation through a patterned oxide.
- Interrupted oxides (Fig. 2.3c_5), which are synthesized by localized implantation and can be viewed as SOI regions integrated into a bulk Si wafer.
- Implantation through a strained Si/SiGe stack enables the formation of strained SOI.

2.3.5. Other SOI Materials

Silicon-On-Nothing (SON) process consists in growing by selective epitaxy a sacrificial SiGe layer in STI (shallow trench isolation) predefined regions of a bulk-Si wafer. A silicon film with suitable thickness (20 nm or less) is then epitaxially grown. Selective etching of the SiGe layer leaves an empty space (air gap) underneath the film. The suspended Si membrane can be used to fabricate gate-all-around transistors. Alternatively, the air gap can be filled with a dielectric in order to form a localized SOI structure, integrated in the bulk-Si wafer (Fig. 2.3c_5) [17, 18].

Silicon-On-Sapphire (SOS) is formed by heteroepitaxial growth (Fig. 2.3a_1). The interface region contains crystallographic defects due to the Si–Al_2O_3 mismatch. The electrical properties are affected by the lateral stress, in-depth inhomogeneity of SOS films, and interface traps [19]. However, the quasi-infinite thickness of the insulator makes SOS attractive for RF applications [20]. SOS has been improved by SPER (*solid-phase epitaxial regrowth*), which consists in implanting Si ions such as to amorphize the film. Subsequent annealing allows the epitaxial regrowth of the film, starting from the "seeding" surface toward the Si–Al_2O_3 interface. Another solution is to remove, by sacrificial oxidation, most of the SOS film and to complete the thickness by a second epitaxial growth [21].

ZMR and ELO. ZMR material is fabricated by depositing a poly-Si layer on an oxidized wafer. *Zone Melting Recrystallization* requires high-energy sources (lamps, lasers, beams, or strip heaters) scanned across the wafer. The ZMR process can be seeded or unseeded (Fig. 2.3c_4), but it is basically limited by the small lateral extension of single-crystal regions, free from grain subboundaries and associated defects. A variant is the *metal-induced lateral crystallization* (MILC) of a deposited amorphous Si layer. In the *epitaxial lateral overgrowth* (ELO) method, the single-crystal Si film is grown on a seeded oxide (Fig. 2.3a_2). The epitaxial growth proceeds in both vertical and lateral directions (with a ratio of 1 to 10–100), hence ELO process is finished by thinning the Si film. ELO and ZMR serve for the integration of 3-D stacked circuits: Once the first circuit layer is completed, the epitaxial growth can be repeated.

FIPOS The FIPOS method (*full isolation by porous oxidized silicon*) makes use of the very fast oxidation of porous silicon (Fig. 2.3a_3), which has a large surface-to-volume ratio ($10^3\ cm^2\ cm^{-3}$). Selected p-type regions of the Si wafer are converted into porous silicon, their thickness and porosity being adjusted by anodic reaction. FIPOS has been proposed for combining, in a single chip, photoluminescent and electroluminescent porous Si devices with fast SOI–CMOS circuits.

The SOI family contains more materials, some fading (silicon-on-zirconia) and others, emerging (strained Si or Ge). Novel SOI wafers are needed as the basic ingredient for the fabrication of advanced structures (quantum dots, wires, single-electron circuits, tunneling devices, etc.). New materials, such as stacked layers for 3-D circuits, light transmission, and optical and power devices, are being explored. The combination of strained layers (Si, SiGe) or compound semiconductors (III–V, II–VI) on various substrates (SiC, diamond, glass, air, flexible polymers, etc.)

opens exciting prospects in nanoelectronics and microelectronics. All of these emerging materials, still labeled SOI *for Semiconductor On Insulator* (instead of Silicon On Insulator), can be synthesized by adapting the Smart-Cut process.

2.4. SOI MOSFETS

Partially depleted (PD) SOI MOSFETs are currently used for fabricating high-end microprocessors, whereas fully depleted (FD) MOSFETs are unavoidable for advanced scaling. In SOI MOSFETs (Fig. 2.1), two inversion channels can be activated, one at the front Si–SiO_2 interface and the other at the back Si–BOX interface (via substrate, back-gate bias V_{G_2}). *Full depletion* means that the depletion region covers the whole thickness of the transistor body.

2.4.1. *Fully Depleted SOI MOSFETs*

In FD SOI MOSFETs (Fig. 2.1a), the depletion charge does not vary with the gate voltage, which enables enhanced control of the inversion charge and current. The front- and back-surface potentials are coupled, meaning that the electrical characteristics of one channel vary with the bias applied to the opposite gate. As a consequence, the front-gate characteristics depend on the back-gate bias and quality of the BOX and interface.

Specific $I_D(V_G)$ relations apply to FD SOI MOSFETs. The characteristics of the front-channel transistor are schematically illustrated in Fig. 2.4 for three distinct bias conditions of the back interface (inversion, depletion, and accumulation).

2.4.1.1. Threshold Voltage

The lateral shift of $I_D(V_{G_1})$ curves (Fig. 2.4a) is explained by the variation of the front-channel threshold voltage, $V_{T_1}^{dep}$, with back-gate bias. The potential coupling causes $V_{T_1}^{dep}$ to decrease linearly with increasing V_{G_2} (Fig. 2.4a), between two

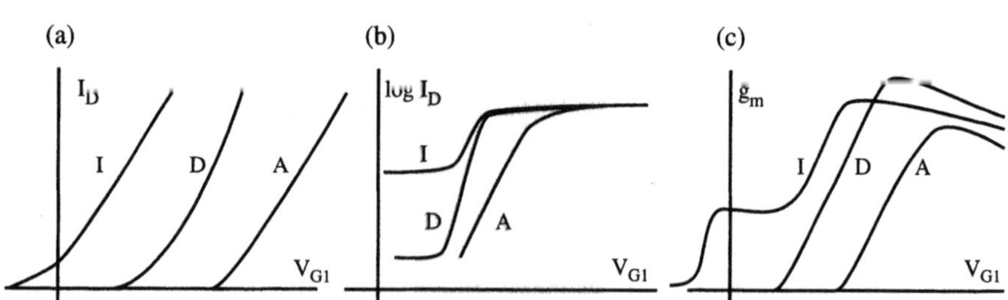

FIGURE 2.4. Generic front-channel characteristics of a FD n-channel SOI MOSFET for various back-gate biases, leading to accumulation (A), depletion (D), and inversion (I) at the back interface. The characteristics are (a) $I_D(V_{G_1})$ in strong inversion, (b) $\log I_D(V_{G_1})$ in weak inversion, and (c) transconductance $g_m(V_{G_1})$.

plateaus corresponding respectively to accumulation $V_{T_1}^{acc}$ and inversion at the back interface [1,22]:

$$V_{T_1}^{dep} = V_{T_1}^{acc} - \frac{C_{si} C_{ox_2} \left(V_{G_2} - V_{G_2}^{acc} \right)}{C_{ox_1} (C_{ox_2} + C_{si} + C_{it_2})}, \tag{2.1}$$

where C_{si}, C_{ox}, and C_{it} are the capacitances of the FD film, oxide, and interface traps, respectively. The subscripts 1 and 2 hold for the front- or the back-channel parameters. They can be interchanged in order to account for the variation of the back-channel threshold voltage V_{T_2} with V_{G_1}.

The threshold voltage decreases in thinner films because the depletion charge is reduced. A V_T rebound is observed in ultrathin films ($t_{si} \leq 10\,\text{nm}$, Fig. 2.5a), where quantum effects arise and lead to the formation of a 2-D subband system [23].

2.4.1.2. Subthreshold Slope

The subthreshold swing, $S = dV_G/dI_D$, is minimum (next to the theoretical limit of 60 mV/decade at room temperature):

$$S_1^{dep} = 2.3 \frac{kT}{q} \left(1 + \frac{C_{it_1}}{C_{ox_1}} + \alpha_1 \frac{C_{si}}{C_{ox_1}} \right) \tag{2.2}$$

for depletion at the back interface (Fig. 2.4b) [24]. The interface coupling coefficient α_1,

$$\alpha_1 = \frac{C_{ox_2} + C_{it_2}}{C_{si} + C_{ox_2} + C_{it_2}} < 1, \tag{2.3}$$

accounts for the influence of back interface traps and BOX thickness C_{ox_2}.

By accumulating the back channel, the front inversion channel becomes decoupled from back interface defects. However, coefficient α_1 tends to unity (as in bulk-Si or PD MOSFETs), causing an overall degradation of the swing (Fig. 2.4b).

The above simplified equations are valid only when the BOX is thick enough so that substrate effects occurring underneath the BOX can be ignored. Since the capacitances of the BOX and Si substrate are connected in series, the swing for thin BOXs also depends on the density of traps and surface charge at the *third* interface: BOX–Si substrate [25]. The subthreshold slope normally improves for thinner silicon films (Fig. 2.5b) and thicker BOXs [25]. Film thinning leads to a lower subthreshold swing only in the case of few states at the silicon layer/BOX interface (Fig. 2.5b).

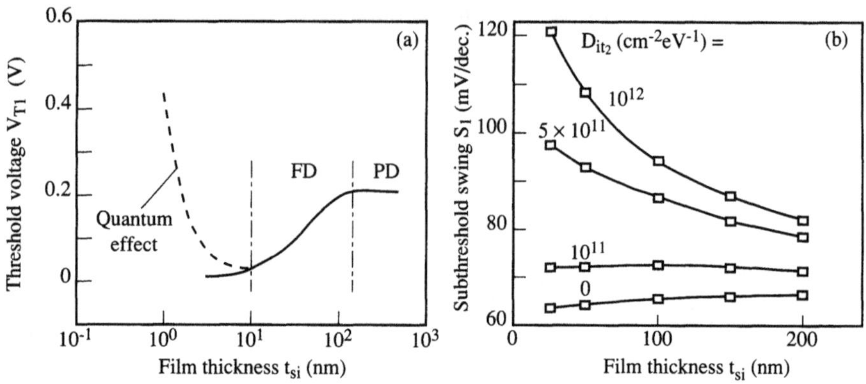

FIGURE 2.5. (a) Threshold voltage versus film thickness (film doping $\simeq 10^{17}$ cm^{-3}). For films thinner than 10 nm, energy quantization leads to an increase in V_T. (b) Subthreshold swing variation with film thickness for various densities of back interface traps ($t_{ox_1} = 27$ nm, $t_{ox_2} = 350$ nm, film and substrate doping $= 10^{15}$ cm^{-3}). Adapted from Ref. 25.

2.4.1.3. Transconductance

For MOSFET operation in a strong inversion and ohmic region, the front-channel drain current and transconductance are given by

$$I_D = \frac{C_{ox_1} W V_D}{L} \frac{\mu_1}{1 + \theta_1 (V_{G_1} - V_{T_1}) + \theta_2 (V_{G_1} - V_{T_1})^2} \left(V_{G_1} - V_{T_1}(V_{G_2})\right), \tag{2.4}$$

$$g_{m_1} = \frac{C_{ox_1} W V_D}{L} \frac{\mu_1 \left[1 - \theta_2 (V_{G_1} - V_{T_1})^2\right]}{\left[1 + \theta_1 (V_{G_1} - V_{T_1}) + \theta_2 (V_{G_1} - V_{T_1})^2\right]^2}, \tag{2.5}$$

where μ_1 is the mobility of front-channel carriers and $\theta_{1,2}$ are the mobility attenuation coefficients. Coefficient θ_2 reflects the surface roughness scattering and is relevant for ultra thin gate oxides.

The transconductance of FD MOSFETs (Fig. 2.4c) is complicated by the influence of the back-gate bias on front-channel threshold voltage, $V_{T_1}(V_{G_2})$, and carrier mobility. The effective mobility and transconductance are maximum for depletion at the back interface, due to the combined effects of reduced vertical field and series resistances [26]. The unusual distortion of the transconductance (curve I, Fig. 2.4c) originates from the premature activation of the back channel. While the front interface is still depleted, increasing V_{G_1} reduces the back threshold voltage and opens the back channel before the front channel. The transconductance exhibits a characteristic plateau [26].

2.4.1.4. Volume Inversion

The simultaneous activation of front and back channels induces the onset of *volume inversion* by continuity [27]: The inversion charge covers the whole volume of thin

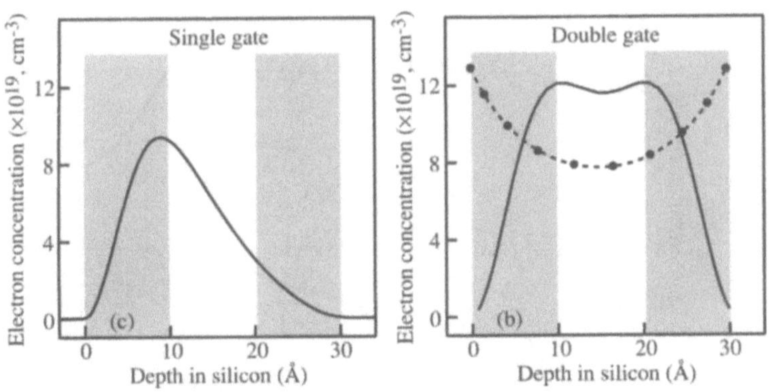

FIGURE 2.6. Minority carrier distributions in 3-nm-thick SOI MOSFETs operated in (a) single-gate mode and (b) double-gate mode with volume inversion. The dotted line results from nonquantum simulations (Poisson equation), which are inappropriate in ultrathin films. In the gray regions, the carrier mobility is drastically reduced by surface roughness scattering and by the imperfect crystallinity of the near-interface monolayers.

and low-doped films. The maximum density of the inversion charge is located away from the interface (Fig. 2.6a). For double-gate operation (Fig. 2.6b), the electric field cancels in the middle of the film, enabling the carrier mobility to increase. Note that some degree of volume inversion subsists in single-gate MOSFETs if the film is ultrathin (Fig. 2.6a).

The key advantages of volume inversion are: increased current drive and transconductance, attenuated influence of interface defects (traps, fixed charges, roughness), and reduced $1/f$ noise. We will show in Section 2.7 that volume inversion is a basic mechanism in multiple-gate MOSFETs.

2.4.1.5. Defect Coupling

We have seen that in FD MOSFETs, carriers flowing at one interface are influenced by the presence of defects at the opposite interface. This mechanism of *defect coupling* is responsible for an apparent degradation of the front-channel properties, which is actually induced by the BOX damage. This situation is frequently observed after back interface degradation via radiation or hot-carrier injection [28].

2.4.1.6. Metastable Dip

This recent effect is a combination of coupling and transient mechanisms [29]. The front-channel transconductance is measured from accumulation to strong inversion (direct scan) with the back channel biased next to inversion. A surprising metastable dip (MSD) is observed for $-2.5 < V_{G1} < -1$ V in Figure. 2.7.

In "accumulation," the generation of majority carriers is not instantaneous, so that interface coupling occurs. As V_{G1} increases, the back channel opens first, the current increases linearly, and the transconductance shows a typical plateau (as in

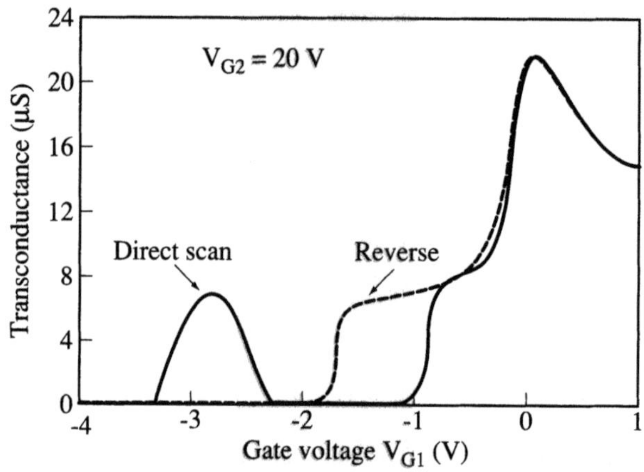

FIGURE 2.7. Transconductance versus front-gate voltage from accumulation to inversion (direct scan: MSD effect) and vice versa (no dip). The dip subsists in small-geometry FD MOSFETs; its lateral extension depends on the front-and back-gate voltages, delay and hold times, and so forth. Adapted from Ref. 29.

Fig. 2.4c, curve I). As soon as the accumulation layer is completed ($V_{G1} \simeq -2.5$ V), the coupling stops: The back-channel current is blocked and the transconductance falls to zero. For $V_{G1} > -1$ V, the accumulation layer is no longer sustained by the gate and the FD coupling-related plateau resumes. Finally, the transconductance recovers a classical shape dominated by the front-channel conduction.

Remark that for the reverse direction of V_{G1} scan, equilibrium is reached faster and there is no dip. This hysteresis can be the basis for novel capacitorless DRAMs [29].

2.4.2. *Partially Depleted SOI MOSFETs*

In PD SOI MOSFETs, a neutral region subsists that cancels the interface coupling effects (Fig. 2.1b). When the body is grounded (via independent body contacts or body-source ties), PD SOI transistors become similar to bulk-Si MOSFETs and the standard $I_D(V_G, V_D)$ equations apply. However, the body contacts are far from being ideal (intrinsic resistance, excess noise, increased die size). If body contacts are not supplied, *floating-body* effects arise, leading to rather detrimental consequences. Minor floating-body effects are also observed in FD transistors [30].

2.4.2.1. Kink Effect

The classical floating-body mechanisms are triggered by the charging of the film body with majority carriers generated by impact ionization [31]. This leads to a reduction in the threshold voltage. The kink effect denotes a sudden jump in drain current (Fig. 2.8a) and in low-frequency noise [32].

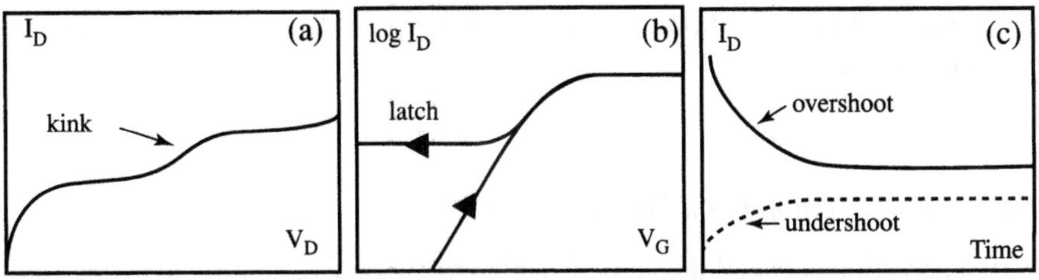

FIGURE 2.8. (a) Typical floating-body effects in PD SOI MOSFETs: (a) kink in output characteristics $I_D(V_D)$, (b) hysteresis and latch in subthreshold curves $I_D(V_G)$, and (c) drain current transients.

2.4.2.2. Hysteresis and Latch

Body charging is particularly effective in weak inversion, where the current depends exponentially on potential. As the current increases, more electrons are involved in impact ionization, leading to abnormally low values of the swing (below 60 mV/decade) and hysteresis. For very high drain bias, the body charge is considerable and sustains the inversion channel even when the gate is turned off (Fig. 2.8b). This extreme effect is called transistor latch [33].

2.4.2.3. Parasitic Bipolar Transistor

The lateral bipolar transistor becomes effective in short-channel PD or FD MOSFETs. It is triggered by the impact ionization current, which leads to the forward biasing of the base/emitter (body/source) junction. A large collector (drain) current is generated, enhancing the original contribution of the MOS channel. The parasitic bipolar induces a premature breakdown, in both PD and FD transistors [34], as compared to bulk silicon MOSFETs.

The solutions for attenuating the parasitic bipolar action are based on lifetime killing and source engineering: low-doped drain (LDD) to reduce the impact ionization and bipolar gain, source silicidation, and so forth [35].

2.4.2.4. Transient and History Effects

Drain current transients are observed when the floating-body potential is forced out of the equilibrium condition. We will see in Section 2.5.6 that the current can temporarily increase (overshoot) or decrease (undershoot) (Fig. 2.8c). The current transients depend on channel length and drain bias, because the impact ionization represents an additional source of majority carriers. As a general rule, the current variation is proportional to the difference between the final and initial body charges, and the time constant depends on the generation–recombination process.

During high-frequency switching of integrated circuits, the transistor body is not able to reach equilibrium rapidly enough. The charging and discharging of the body is an iterative process that may cause "history" effects and dynamic instabilities. In a ring oscillator, the switching delay of an inverter is governed by the amount of available current, which can be higher (overshoot) or lower (undershoot) than at equilibrium. The switching speed depends on the number of previous switches.

2.4.3. *Transition from Partial to Full Depletion*

Partial depletion occurs if the vertical depletion region w_D, controlled by the gate, does not cover the whole body ($w_D < t_{si}$). This classical definition does not apply to very short devices, in which the lateral depletion regions of the source and drain junctions come into play. The coupling between the lateral and vertical depletion regions can indeed enhance the overall depletion [36]. The junctions cause a lowering of the *effective* doping seen by the gate, letting the vertical depletion region extend deeper. As a final effect, the transition from PD to FD operation is not exclusively controlled by the doping/thickness ratio; the channel length contributes via a *length-to-doping* transformation.

The dark gray region in Fig. 2.9 represents the classical full-depletion domain, whereas the line between the light gray and white regions indicates the revisited FD–PD boundary. The consequence is dramatic for the survival of the PD technology. In order to maintain PD operation in a 50-nm-long, 100-nm-thick MOSFET, a doping level (2×10^{18} cm^{-3}) more than one order of magnitude higher than in a 0.25-μm device (10^{17} cm^{-3}) is needed [36]. This conclusion is independent of other doping criteria related to threshold voltage adjustment.

The above discussion is valid for SOI MOSFETs without pockets. In the case of higher doping levels localized near the source and drain (pockets), the transition from PD to FD can show an opposite trend: Shorter transistors exhibit a higher effective doping, making them more partially depleted [37].

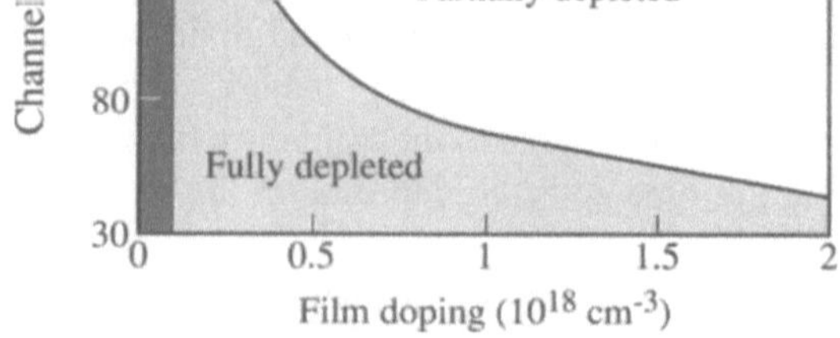

FIGURE 2.9. Transition between partial depletion and full depletion in 100-nm-thick SOI MOSFETs as a function of channel length and doping. The vertical dark gray region corresponds to full depletion in the classical limit of long channels. Adapted from Ref. 36.

2.5. ELECTRICAL CHARACTERIZATION TECHNIQUES FOR SOI

The characterization of SOI structures is hampered by several problems: thinness of the film, presence of the BOX, stacked interfaces, and typical defects (strain, in-depth inhomogeneities, dislocations, precipitates, etc.). Most physical-chemical (RBS, SIMS, Auger, ellipsometry, X–rays) and microscopic (TEM, AFM, etc.) techniques can be adapted to SOI. The measurement of surface/interface roughness or the identification and count of defects with different nature requires special expertise.

The electrical properties are of uppermost interests, as they directly impact on the performance of integrated circuits. A number of conventional methods are no longer applicable in thin films, but, fortunately, novel techniques can be implemented.

2.5.1. *Wafer Characterization: Ψ-MOSFET*

The pseudo-MOS transistor (Ψ-MOSFET) takes advantage of the the upside-down MOS structure inherent in SOI materials (Fig. 2.10). The Si substrate acts as a gate terminal and can be biased to induce a conduction channel (inversion or accumulation) at the interface. The BOX plays the role of a gate oxide and the Si film represents the transistor body. For *in situ* Ψ-MOSFET operation, low-pressure probes are placed on the film to form source and drain point contacts [38]. The Ψ-MOSFET is very much like the point-contact transistor that Shockley's group attempted to fabricate half a century ago. The experimental setup is composed of a two-probe or four-probe system, with adjustable pressure, and a picoameter. The measurement can be conducted on the whole wafer, but it is more accurate if small Si islands ($5 \times 5\,cm^2$) are separated by etching.

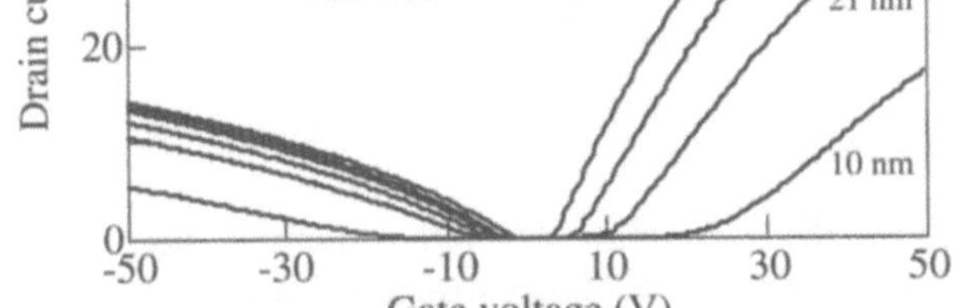

FIGURE 2.10. Configuration of the Ψ-MOS transistor and typical $I_D(V_G)$ characteristics for electron and hole channels in ultrathin film, state-of-the-art Unibond materials.

Very pure MOSFET-like characteristics are produced (Fig. 2.10). In strong inversion/accumulation and ohmic region, the drain current is given by

$$I_{DS} = f_g C_{ox} V_D \frac{\mu}{1+\theta_1(V_{G_1} - V_{T_1})}(V_G - V_{T,FB}), \tag{2.6}$$

where $f_g \simeq 0.75$ is a form factor that accounts for the nonparallel distribution of current lines.

The slope of $I_D/\sqrt{g_m}$ versus V_G curves yields the effective mobility μ of electrons *and* holes, whereas the intercept with the V_G axis gives the threshold (V_T) or the flat band (V_{FB}) voltages. The density of traps at the film–BOX interface is calculated from the subthreshold slope in weak inversion, the fixed charge density from V_{FB}, and the film doping from the difference $V_T - V_{FB}$. The carrier lifetime can be evaluated by recording the transient drain current after the gate is pulsed in strong inversion [39].

Recent measurements in state-of-the-art SOI show that the Ψ-MOSFET remains effective for the characterization of ultrathin and strained films (Fig. 2.10). Equation (2.6) needs, however, to be revisited for films thinner than 30 nm [see, e.g., Eq. (2.5)] [40].

Ψ-MOSFET measurements may be correlated with other methods [1]:

- *Hall effect* for independent determination of the carrier mobility and concentration
- *Four-probe* measurements of the film resistivity
- *Spreading resistance* for evaluation of the resistivity profile
- Nondestructive detection of pinholes by using the decomposition of $CuSO_2$ solution when a leakage current flows through the BOX
- *Surface photovoltage* measurement to determine the diffusion length of minority carriers
- *Photoconductivity* for extraction of the carrier recombination lifetime
- *Photo-induced current transient spectroscopy* (PICTS) for investigation of deep-level traps

We have to keep in mind that thin semiconductor films are in general fully depleted. For transport measurements, it is essential to bias the substrate, as in Ψ-MOSFETs, in order to activate a conduction channel at the film–BOX interface.

2.5.2. MOSFET Characteristics

Most frequently, the properties of SOI structures are inferred from the static/dynamic characteristics of MOS transistors. Parameters like mobility, threshold voltage, swing, and lifetime are determined for the front and back channels, separately or in the coupled mode. The parameter extraction is similar in SOI and bulk-Si MOSFETs and will not be evoked explicitly [1].

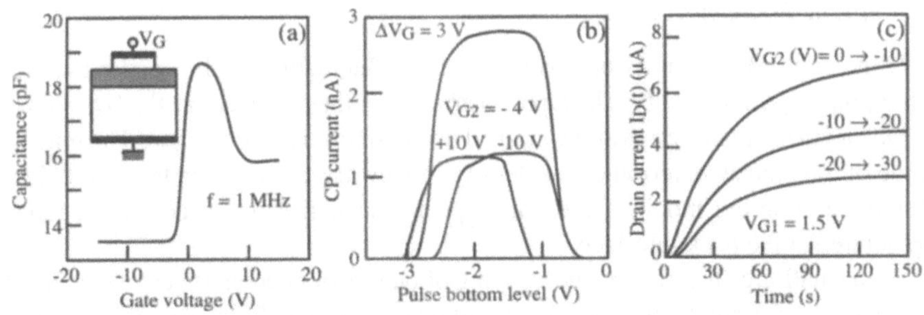

FIGURE 2.11. Usual characterization techniques in SOI devices: (a) SIS structure (silicon–insulator–silicon) and typical $C(V_G)$ curve at high frequency; (b) front-gate charge pumping current versus base level of the gate pulse for various back-gate voltages; and (c) front-channel drain current transients after pulsing the back gate into accumulation.

2.5.3. SIS and MOS Capacitance

The capacitance and conductance techniques can still be applied to SOI structures. In a silicon-insulator-silicon (SIS) capacitor, the film is biased through a top metal contact (gate) while the substrate is grounded. The conventional theory for bulk MOS capacitors is modified to account for the special C(V) curve (Fig. 2.11a), due to the formation of depletion regions on each side of the buried oxide. The limitation of SIS capacitance and conductance measurements comes from the large number of parameters to be determined from data at low and high frequency.

Even more delicate is the case of plain MOS–SOI capacitors where two oxides and three interfaces coexist. An additional contact to the film allows probing independently the front and back interfaces. However, the large series resistance of the film makes the interpretation complex.

2.5.4. Charge Pumping Technique

Charge pumping (CP) is a very sensitive method for the characterization of interface traps. The adaptation of CP to SOI transistors requires a body contact or gate-controlled p-i-n diodes [1]. The gate is repeatedly switched from inversion (where the minority carriers are trapped on the interface states) to accumulation (where the trapped carriers recombine with majority carriers). This recombination gives rise to an *average* CP current I_{CP} in the body terminal, which is proportional to the trap density and frequency.

A typical curve is obtained by varying the bottom level of the pulse while keeping the pulse magnitude ΔV_G and frequency fixed. Figure 2.11b illustrates the *rectanglelike* CP curve: The left- and right-hand edges correspond to $(V_T - \Delta V_G)$ and V_{FB}, respectively, whereas the plateau level gives the trap concentration.

This CP signature is altered by interface coupling, when the back interface goes from inversion to depletion and accumulation (Fig. 2.11b). In depletion, the CP current increases because the front and back interface traps are pumped simultaneously.

2.5.5. *Low-Frequency Noise*

$1/f$ noise in MOS transistors originates from fluctuation in the carrier number (trapping) and/or mobility. The typical variation of the noise factor, S_{I_D}/I_D^2, shows a plateau in weak inversion and a marked decrease in strong inversion. The magnitude of the plateau determines the density of slow traps located in the gate oxide. Interface coupling modifies the noise spectra primarily by an additional noise generated at the back interface.

In small-area transistors, the trapping of one carrier becomes detectable in the time domain: A small pulse is superposed on the average drain current value. The duration of such a *random telegraph signal* (RTS) reflects the time constant of the trap. A comprehensive view of the trapping mechanism is obtained by performing both back-channel and front-channel measurements.

2.5.6. *Drain Current Transients*

This technique, especially developed for SOI transistors, consists in inducing a temporary excess or deficit of majority carriers. In a long-channel SOI MOSFET, equilibrium is reached by recombination–generation mechanisms.

When the transistor is switched from depletion to strong inversion, the depletion region extends and the majority carriers accumulate at the bottom of the film. The body potential increases; hence, the threshold voltage is lowered and induces an excess current (overshoot) (see Fig. 2.8c). Subsequent carrier recombination causes the current to decrease toward the equilibrium value. The duration of the transient is processed with Zerbst-like techniques and provides the recombination carrier lifetime [1]. The opposite effect (deficit of majority carriers and current undershoot) is governed by carrier generation and occurs when the gate is switched from strong to weak inversion.

An experimental variant consists in using the two gates. The front gate is biased in inversion, whereas the back gate is pulsed into accumulation. The film potential drops, instantly lowering the inversion charge and current. The current relaxes back to equilibrium (undershoot) through carrier generation within the film and at the interfaces (Figs. 2.11c and 2.14a). Compact models are available for extracting the generation–recombination rates [41]. The roles of the two gates can be interchanged for probing different regions of the film. In short-channel SOI MOSFETs, the junctions contribute substantially to the extraction/supply of majority carriers, thus reducing the transient time.

2.6. DIMENSIONAL EFFECTS IN SOI MOSFETs

In the following subsections, more advanced concepts, related to various size effects, will be addressed.

2.6.1. Short Channels

In bulk-Si MOSFETs, length scaling is enabled by increasing the body doping and forming shallow junctions. Ultralarge doping degrades the carrier mobility and gives rise to band-to-band tunneling current. The advantage of SOI is that film thickness can entirely govern the device scaling. The scaling rules are reviewed in detail in Refs. 2–4 and will not be developed here.

Drain-induced barrier lowering (DIBL) and charge sharing between the gate and source/drain contacts are well known short-channel effects, which occur in both bulk-Si and SOI MOSFETs. In SOI, they are better controlled by reducing the film thickness [3, 4, 42].

The major short-channel effect in SOI is due to the penetration of the electric field from the drain into the BOX and substrate (inset of Fig. 2.12). The fringing field tends to increase the surface potential at the film–BOX interface: *drain-induced virtual substrate biasing* (DIVSB) [42, 43]. Since the front and back interfaces are naturally coupled in FD films, the front-channel properties become degraded. In particular, the threshold voltage V_T is lowered with increasing drain bias, very much as in DIBL, although DIVSB is totally distinct.

Figure 2.12 illustrates the scaling strategy in SOI. It compares V_T lowering (by DIVSB and DIBL) in highly doped and undoped MOSFETs. A thick undoped film is clearly not suitable. However, film thinning gradually erases the advantage of heavy doping. For 15-nm-thick film, ΔV_T becomes reasonable and the doping effect disappears. It is concluded that an *undoped and ultrathin* MOSFET is exceptionally robust to short-channel effects. Other benefits are the high carrier mobility and excellent subthreshold slope.

Solutions for further improvement of ΔV_T aim at reducing the fringing field penetration in the BOX and substrate: thinner BOX with lower permittivity, double-

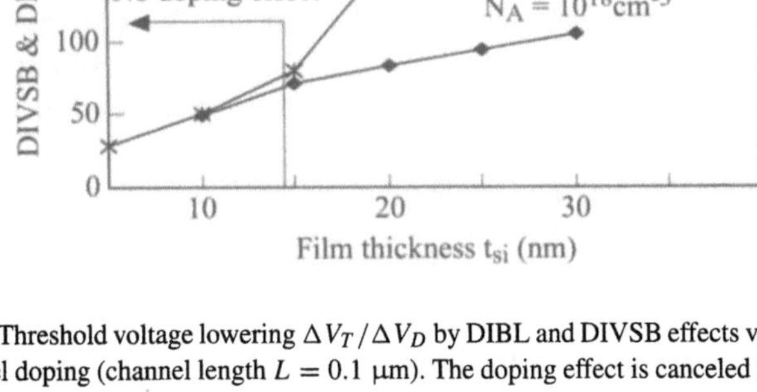

FIGURE 2.12. Threshold voltage lowering $\Delta V_T/\Delta V_D$ by DIBL and DIVSB effects versus film thickness and channel doping (channel length $L = 0.1$ μm). The doping effect is canceled for films thinner than 15 nm.

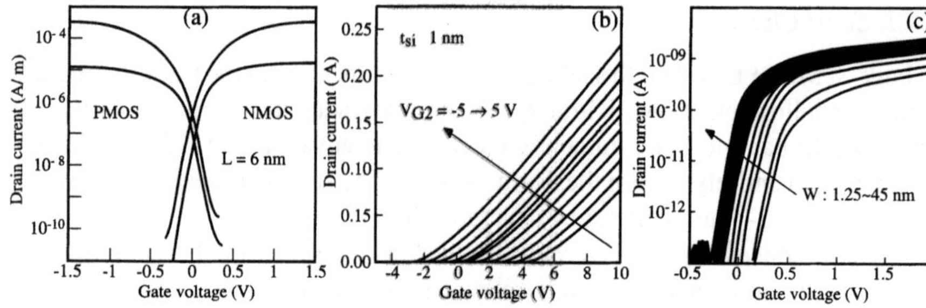

FIGURE 2.13. Drain current versus gate voltage in ultra-miniaturized SOI MOSFETs: (a) sub-10-nm-long transistor (after Ref. 48), (b) 1-nm-thick transistor (after Ref. 42), and (c) supernarrow transistors including the 1-nm-wide MOSFET (after Ref. 51).

gate structure, or ground plane (highly doped region or metal layer underneath the BOX) [43,44].

Electrostatic considerations demonstrate that the minimum channel length is proportional to the film thickness: $L_{min} \simeq 4t_{si}$ [45]. This guiding rule makes it clear that SOI MOSFETs can break the 10-nm-length barrier... if films thinner than 3 nm can be routinely manufactured [46]. As a confirmation, 2.5-nm-long transistors with acceptable characteristics have been simulated for 1-nm-thick SOI films [47]. On the practical side, a 6-nm-long SOI MOSFET (Fig. 2.13a) has been fabricated [48].

Emerging technologies like strained-Si-, SiGe-, and Ge-based MOSFETs do not compete with SOI. They must be SOI-like: Regardless of the semiconductor, the electrostatic problems are more or less the same. The semiconductor film should be ultrathin and placed on an insulator. Otherwise, the short-channel effects will become an issue again in extremely small MOSFETs.

2.6.2. *Narrow Channels*

Standard narrow-channel effects are well documented [49,50]. A parasitic channel forms on the edges of the transistor because the gate control differs over the main channel (central region) and edges. When the threshold voltage is lower on the edges, reverse narrow-channel effects occur: V_T decreases with width. Note that the parasitic conduction is normally masked by overdoping the edges such as to increase the local V_T value. This strategy becomes difficult in ultrathin films.

According to the lateral isolation process (STI, LOCOS, mesa), the side walls may feature a different crystal orientation, additional defects, and variable thickness. In general, by reducing the width, the carrier mobility and subthreshold swing are degraded, and V_T may increase (normal narrow-channel effects). The coupling between width and thickness appears to be more important than the width–length coupling: Short-channel effects hardly depend on width, whereas the narrow-channel effects are clearly reduced in thinner films [49].

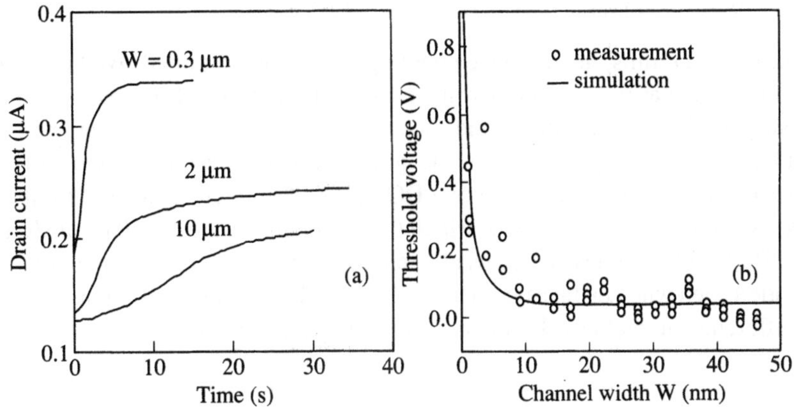

FIGURE 2.14. (a) Drain current undershoot in devices with variable width, measured after switching the back gate from depletion to accumulation: A shorter transient reflects a lower carrier lifetime (after Ref. 49). (b) Threshold voltage increase in the ultranarrow SOI MOSFETs shown in Fig. 2.13 (after Ref. 51).

Detailed measurements revealed the attenuation of the floating-body effects (FBE) in narrower MOSFETs [49]: The breakdown voltage, snap-back voltage and subthreshold swing in saturation are all increased, whereas the gain of the bipolar transistor decreases. This implies that the accumulation of the majority carriers in the body (origin of FBE, see Section 2.4.2) is tempered when the width is scaled down. Several reasons have been identified [49]:

- Carrier lifetime degradation near the side walls, which helps the recombination of the majority carriers in the body (Fig. 2.14a).
- Impurity outdiffusion into the isolation oxide (along the channel side walls), which lowers the source/body potential barrier, making the removal of majority carriers from the body easier.
- Local thinning of the Si film edges that leads to a FD-like behavior with attenuated FBE.

Extremely narrow SOI MOSFETs are subject to special effects. *Lateral* quantum confinement and subband splitting lead to a remarkable increase of the threshold voltage for widths below 10 nm (Figs. 2.13c and Fig. 2.14b) [51]. This is similar to the effect of *vertical* confinement shown in Fig. 2.5a. If the transistor is also ultrathin, the confinement exhibits a 2-D pattern.

2.6.3. Channel Thickness

The film thickness defines the capability of SOI MOSFETs to withstand short-channel effects. For ultimate scaling, the body must be very thin, which, in turn, induces interesting thickness effects. Figure 2.13b demonstrates that film thinning down to a few monolayers can be achieved [42].

2.6.3.1. Supercoupling

In ultrathin FD films, the interface coupling effects are amplified. We know from Section 2.4.1 that the threshold voltage of the front channel decreases with increasing the back-gate voltage and vice versa [22]. Drawing the characteristics $V_{T1}(V_{G2})$ and $V_{G1}(V_{T2})$ on the same graph normally results in two different curves (Fig. 2.15). The intercept point DG defines the unique couple of front- and back-gate voltages (V_{T1}, V_{T2}) for which the two channels are *simultaneously* activated [52]. These values can be used to emulate a perfectly balanced double-gate operation even when using asymmetrical MOSFETs, because they compensate for the difference in thickness between the front and back oxides:

$$V_{G2} - V_{T2} = \frac{t_{ox2}}{t_{ox1}}(V_{G1} - V_{T1}). \tag{2.7}$$

An intriguing feature is that, in sub-10-nm-thick films, the two curves tend to coincide [52,53]. This implies that an arbitrary back-gate bias V_{G2} is promoted as threshold voltage V_{T2} as soon as the front gate is biased at threshold ($V_{G1} = V_{T1}$): When one channel reaches strong inversion, the opposite channel is also dragged into inversion. This effect, called *supercoupling*, reflects the fact that the body potential is quasi-flat between the two gates. The film behaves as a rigid quasi-rectangular well: When the potential at one interface is modified by the gate, the potential of the entire film follows.

The notion of front and back channels becomes obsolete and needs to be replaced by the concept of *volume inversion* [27]. The minority (or majority) carriers are no longer confined at the interface and spread across the film thickness. Volume inversion is a key mechanism for double-gate MOSFETs to succeed (Section 2.7.1).

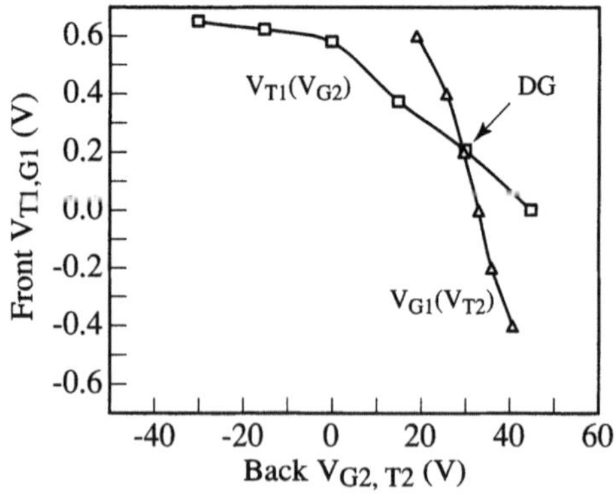

FIGURE 2.15. Front- (back-) channel threshold voltage versus back- (front-) gate bias in a 47-nm-thick SOI MOSFET ($L = 10\,\text{m}$, $W = 10\,\text{m}$) (after Ref. 52). Point DG indicates the appropriate bias for balanced channels in pseudo-double-gate operation.

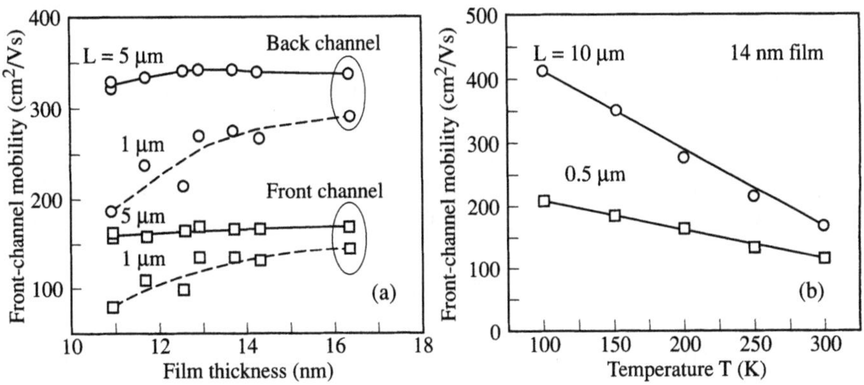

FIGURE 2.16. (a) Front- and back-channel electron mobility versus film thickness in short and long SOI MOSFETs. (b) Front-channel electron mobility versus temperature (after Ref. 57).

The front and back channels cannot be separated. This means that yet another conventional formulation like "front-channel mobility" should be translated into "mobility seen from the front gate."

2.6.3.2. Mobility Issues

In sub-10-nm-thick films, vertical quantum confinement and sub-band splitting become relevant [54,55]. Not only does the threshold voltage increase for thinner films (see Fig. 2.5a) but also the carrier mobility is affected by competing effects: lower effective mass and enhanced phonon scattering. The centroid of the inversion charge is moved toward the volume of the film; hence, the surface roughness scattering is also reduced [54]. Monte Carlo simulations suggest that the mobility is maximum in 3–5-nm-thick films [56]. However, there is no experimental support to date. Early measurements actually showed the opposite trend: mobility degradation in thinner films.

The mobility-thickness correlation can be investigated by probing different wafers (with variable thinning) or the same wafer, where thickness variations were locally caused by fluctuations in the material topology. The latter approach has the advantage of being free from the impact of the more or less aggressive thinning process [57]. Figure 2.16a shows that in short-channel, 10–20-nm-thick transistors, the mobility does decrease for thinner films, whereas in long MOSFETs, the thickness effect is irrelevant. This difference implies that the mobility degradation in thinner films might be an artifact, due to the series resistance. The series resistance effect is marked in short-channel transistors and does increase in thinner films ($R_S \sim 1/t_{si}$).

The accurate extraction of the carrier mobility in ultrathin SOI transistors may also be affected by the following [57]:

- Film thinning by sacrificial oxidation generates defects (stacking faults, dislocations, etc.), which subsequently may degrade the quality of the gate oxide interface. As a matter of fact, the mobility can decrease in thinner films (more aggressive oxidation) and the back-channel mobility can exceed the front-channel mobility.

- Polysilicon depletion and quantum confinement lower the effective gate capacitance. In MOSFETs with ultrathin gate oxide, the front-channel mobility can be underestimated by as much as 35%.
- The substrate acts as a back gate with infinite overlap (see Section 2.7.1), lowering the series resistance of the back channel (as compared to the front channel).
- A fair comparison requires that the back surface potential (not the back-gate bias !) remains constant. This is not trivial because the back-channel threshold voltage changes with thickness. Keeping the back interface accumulated is not suitable because gate-induced floating-body effects (see next subsection and Fig. 2.17b) may jeopardize the comparison.

The mobility behavior at low temperature may look different according to the channel length (Fig. 2.16b) and substrate bias. Care is needed to understand if a particular trend has physics grounds or is merely an artifact. For example, the attenuated mobility-temperature dependence in short devices is not due to a stronger Coulomb scattering than in the longer transistor; it is actually a series resistance effect.

The *geometrical magnetoresistance* is an attractive opportunity to determine the carrier mobility and clarify the debate [58]. A high magnetic field B is applied perpendicular to the chip. If the MOSFET is short and wide, the Hall field is short-circuited and the magnetoresistance becomes significant: $R_B = R_0(1 + \mu^2 B^2)$. The carrier mobility is extracted from the linear $R_B(B^2)$ plot, without needing to know the effective channel length. The other advantage of the geometrical magnetoresistance method is that the mobility can be measured even when the split C-V (capacitance-voltage) method fails, this is in weak inversion, in very short transistors, and in the back channel.

2.6.4. *Ultrathin Gate Dielectric*

The standard FBEs are due to impact ionization, which charges the body with majority carriers, thus increasing the body potential (see Section 2.4.2). A totally different FBE takes place in MOSFETs with ultrathin (<2 nm) gate oxide. The body is charged by the gate tunneling current, giving rise to *gate-induced floating-body effect* (GIFBE). GIFBE occurs even at low drain voltage and is not related to impact ionization. The typical feature is a second peak in transconductance, which shows a second peak (Fig. 2.17) [52,59].

The body potential is defined by the balance between the incoming gate tunneling current (body charging) and the outgoing current (body discharging via junction leakage and/or carrier recombination). In PD MOSFETs, the increase in body potential directly lowers the threshold voltage [59], giving rise to a "kink" in the drain current [60] and a second g_m peak (Fig. 2.17a). The model differs for FD transistors (Fig. 2.17b), where GIFBE increases the potential at the back interface

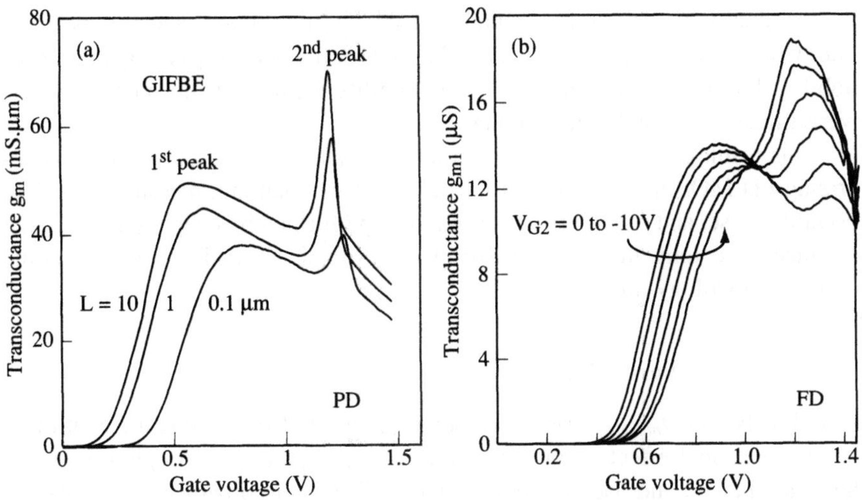

FIGURE 2.17. Transconductance modification by GIFBE in SOI MOSFETs: (a) second peak in normalized transconductance ($g_m \times L$) curves for short and long PD transistors and (b) second peak transformation as a function of back-gate bias in FB SOI MOSFETs ($L = W = 10$ μm, $V_D = 0.1$ V). After Ref. 61.

(film–BOX); the front-channel threshold voltage is *indirectly* lowered by interface coupling effect (Section 2.4.1) [61].

Gate-induced FBE is a dimensional effect that decreases in shorter MOSFETs because the tunneling current ($\sim LW$) is reduced while the junction leakage ($\sim t_{si} W$) is rather constant. GIFBE also decreases in narrower MOSFETs but for a different reason: degraded carrier lifetime and lower source–drain barrier near the side walls (see Section 2.6.2). In addition, GIFBE depends on the scanning mode (speed and direction) of the $I_D(V_G)$ characteristics. For slower measurements, the second peak of the transconductance appears at a lower gate voltage. The asymmetry between gradual body charging (for increasing V_G) and body discharging (for decreasing V_G) is summarized by a hysteresis in $I_D(V_G)$ curves [52, 59].

The transient effects (drain current overshoot and undershoot, see Section 2.5.6) and history effects are drastically modified by GIFBE. First, the tunneling current enables a faster recovery of the equilibrium body charge. Second, the charge stored in the body can prevent the body potential to fall when the gate is turned off. Although expecting an undershoot, one may observe an overshoot [52]. In general, GIFBE allows reaching the steady-state more rapidly. This means faster "history effects" in digital circuits [52].

Another consequence is the modification of the low-frequency noise spectrum. An excess GIFBE-related Lorentz noise ($1/f^2$) is superimposed on the conventional $1/f$ noise, increasing the total noise by one to two orders of magnitude [62].

In FD MOSFETs, the second peak in transconductance is markedly amplified when the back interface is driven toward accumulation (by substrate biasing or radiation effects). An interesting consequence is shown in Fig. 2.17b: The GIFBE peak gradually distorts and eventually offsets the mobility-related first peak. In other words, the transconductance maximum is no longer governed by the carrier mobility, but by the body potential. The mobility extracted from such a curve is overestimated and totally meaningless.

This result carries the signal for a radical change in the strategy for characterizing FD MOSFETs. In order to avoid interface coupling, measurements like mobility, noise, and reliability tests used to be performed by keeping the opposite interface in accumulation [1]. GIFBE increases in accumulation and renders the above methodology improper.

2.6.5. Innovative Buried Insulators

Very thin BOXs ($t_{box} < 20\,\text{nm}$) are being prospected for a superior control of the short-channel effects. Counterarguments are; (i) stronger coupling between the front-channel and back interface defects, (ii) increased parasitic capacitance, (iii) degraded subthreshold swing, (iv) higher impact of the depleted region underneath the BOX and of the more defective BOX–substrate interface, and so forth. Although BOX thinning is unavoidable, excessive thinning may be detrimental.

The BOX needs also optimization to address the crucial problem of self-heating in SOI transistors. In thin SOI, the heat path through the source/drain regions is squeezed: The thermal resistance increases and the body temperature increases dramatically (even by more than 100°C) [63]. The mobility and threshold voltage are lowered, leading to a drop in performance [1, 2]. Self-heating is primarily due to the poor thermal conductivity of the BOX, which blocks the heat dissipation in the silicon substrate. A reasonable solution is to modify the generic SOI structure by replacing the standard SiO_2 BOX with buried alumina [64, 65] or other dielectric [66] with improved thermal conductivity. These new structures are still SOIs except that the letter I in is no longer restricted to SiO_2 and recovers the general meaning of the buried insulator.

Various dielectrics (Al_2O_3, SiC, diamond, quartz, etc.) were evaluated by comparing the total thermal conductance and the self-heating of corresponding MOSFETs [66]. The thermal conductance was deduced from the equivalent thermal circuit of the transistor, whereas the temperature increase was calculated by 2-D simulations [65]. It was found that the thermal conductance of the BOX governs the thermal behavior of SOI MOSFETs. By comparison, the heat flow through the front-gate stack or source/drain terminals plays a marginal role.

The superiority of the novel dielectrics over SiO_2, illustrated in Fig. 2.18a, increases in shorter-channel MOSFETs. There is no BOX thickness effect if the BOX is either extremely conductive (diamond, SiC) or isolating (air). BOX thinning from 400 nm down to 50 nm or less makes sense only for Al_2O_3, quartz, and SiO_2 (Fig. 2.18a).

Reducing the self-heating from 100°C to 50°C represents an immediate gain of more than 25% in mobility ($\mu \sim T^{-1.5}$). This improvement applies simultaneously to electrons and holes and corresponds to the gain in speed expected from the next generation of CMOS scaling. It follows that the mobility engineering can

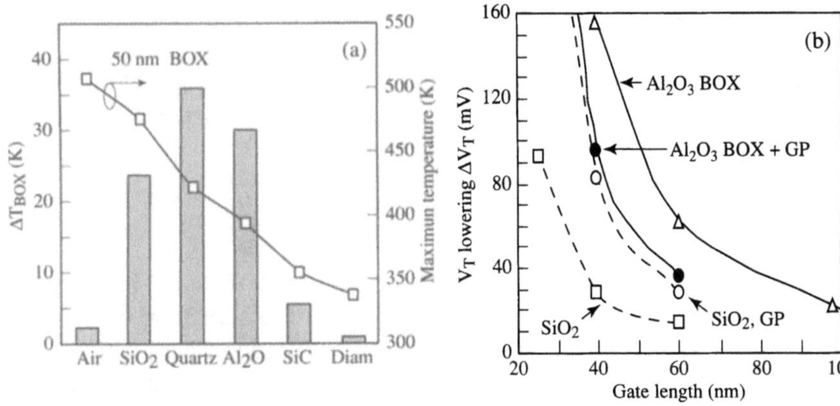

FIGURE 2.18. (a) Temperature difference ΔT_{BOX} (columns) between thick SiO_2 BOX (400 nm) and thin BOX (50 nm) and maximum body temperature (line) for several BOX dielectrics (50 nm long MOSFETs, after Ref. 66). (b) Threshold voltage roll-off, induced by DIBL and DIVSB, versus channel length in several FD MOSFETs (SiO_2 or Al_2O_3 BOX, ground plane or no ground plane, after Ref. 65.

be achieved not only by using strained silicon but also by preventing excessive self-heating. It is clear that these novel dielectrics are even more attractive for temperature management in high-power SOI devices.

The BOX nature also impacts on the electrostatic behavior and performance of the transistor. The change in the dielectric constant impedes the 2-D distributions of the electric potential in the transistor. In particular, the coupling and short-channel effects are modified. For example, the coupling between the front and back channels in MOSFETs with buried Al_2O_3 is three times higher than for SiO_2; it is still weak enough so long as the capacitances of the film and gate oxide exceed the BOX capacitance.

Systematic simulations [65, 66] show that the classical short-channel effects (charge sharing and DIBL) are marginally degraded for a high-K BOX. A 25-nm-long, alumina–BOX MOSFET exhibits only a 25% larger threshold voltage roll-off, which can be further attenuated by thinning the BOX [65]. The control of the fringing fields (DIVSB, see Section 2.6.1) is variable: excellent ($\Delta V_T / V_D \simeq$ 100 mV/V) for diamond, quartz, SiO_2, and air, or modest (250 mV/V) for SiC and Al_2O_3. Quartz and diamond are best suited dielectrics for *thick* BOX. For Al_2O_3, the device architecture can be revisited by including a ground plane (GP).

Figure 2.18b shows combined solutions: ultrathin Si film (5–10 nm), thin BOX (50 nm), and GP. Without GP, the DIVSB effect increases exponentially in MOSFETs shorter than 50 nm. The acceptable limits for threshold voltage lowering (100 mV/V) and subthreshold swing ($\leq$ 100 mV/decade) are no longer fulfilled, more so for buried alumina MOSFETs. The performance improvement due to a GP is more effective for shorter channels and alumina BOX. The conclusion is twofold: The slight electrostatic disadvantage of alumina BOX is practically erased in GP MOSFETs, and it is minor compared to the huge thermal advantage.

2.7. MULTIPLE-GATE SOI MOSFETs

Innovative transistors with two or more gates are currently being explored for enhanced performance and functionality. Our discussion is restricted to dimensional effects and coupling between the various gates or channels.

2.7.1. Double-Gate MOSFETs

Double-gate (DG) MOSFETs are ideal devices for electrostatic integrity and ultimate scaling below the 10-nm channel length. The formation of front and back inversion channels induces *volume inversion* (Section 2.4.1), which brings enhanced drain current and transconductance. Gauss' law teaches us that the total inversion charge in the DG mode is roughly twice the inversion charge in the single-gate (SG) mode. Moreover, the subthreshold swing is ideal: 60 mV/decade at room temperature. The essential aspect is that the minority carriers flow in the middle of the film and experience less surface scattering; hence, the mobility [67, 68], radiation hardness [2, 69], and $1/f$ noise are improved. Ernst et al. [42] reported an outstanding transconductance increase, by more than 250%, for a 3-nm-thick DG-mode transistor.

The two gates collaborate to provide an excellent control of the potential and inversion charge, so that short-channel effects (DIBL, DIVSB, punch-through) are being reduced. The minimum channel length is definitely smaller in DG than in SG transistors: 15–30-nm-long DG–MOSFETs have already been fabricated [70, 71]. Numerical simulations including quantum effects, band-to-band tunneling, and direct source-to-drain tunneling anticipate surprisingly good characteristics for DG MOSFETs as short as 2–8 nm [47]. The recommended body thickness-to-length ratio is roughly $1/2$, a condition less stringent than in SG MOSFETs.

The main difficulty resides in devising a realistic and pragmatical technology. Several solutions are shown in Fig. 2.19. The planar process is suitable in many respects but cannot guarantee the self-alignment of the two gates. The DG technology can be greatly simplified if a reasonable degree of gate misalignment is tolerable. A possibility is to design a longer bottom gate, whereas the channel length is still defined by the source/drain implantation through the shorter top gate (Fig. 2.19c). Numerical simulations show that such nonideal DG MOSFETs should not be disregarded: The subthreshold swing is minimum and, surprisingly, the transconductance and drive current may be higher than in an "ideal" DG transistor with symmetrical self-aligned gates, [72]. The reason is the dual action of the longer gate, which contributes to volume inversion in the body and, simultaneously, to accumulation in the source/drain regions. This *field-effect junction* mechanism contributes to the dynamic lowering of the series resistance. The gain in transconductance compensates for the parasitic overlapping capacitance.

Such asymmetrical DG transistors have recently been fabricated starting from an SOI wafer and using a wafer-bonding technology. The bottom gate was made on top of the SOI film. This wafer is turned upside-down and bonded to a support

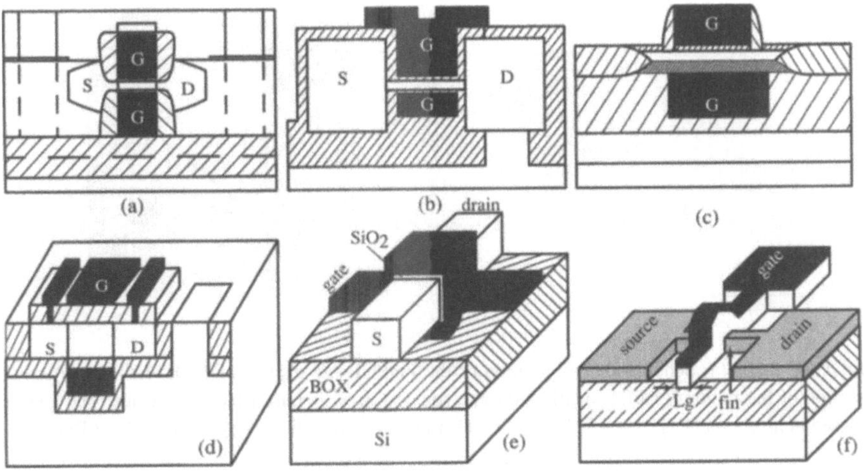

FIGURE 2.19. (a) Technological solutions tested for multiple-gate SOI MOSFETs: (a) self-aligned sacrificial SiGe/Si-body/SiGe stack, where the SiGe layers are subsequently replaced by the gates and the source/drain terminals are formed by selective epitaxy; (b) tunnel epitaxy through the empty space (sacrificial layer) between the gates; (c) wafer bonding; (d) epitaxial lateral overgrowth after formation of the bottom gate; (e) gate-all-around; (f) FinFET.

Si wafer. After etching the substrate and BOX of the handling SOI wafer, the front gate was formed on the denuded side of the film, roughly aligned to the bottom gate (Fig. 2.19c). Preliminary measurements confirm the anticipated trends [73].

A totally different approach is to adopt a nonplanar technology. In fully vertical DG MOSFETs, the source–body–drain stack as well as the current flow are perpendicular to the wafer surface. These devices are attractive because the channel length (i.e., body thickness) can be controlled by body epitaxy, instead of electron-beam lithography. They suffer, however, from the asymmetry of the source and drain terminals and from the difficulty in achieving tiny pillars with ultrasmall intergate distances.

The FinFETs is a more pragmatic nonplanar DG transistor with a relatively easy-to-implement process. In DG FinFETs (Figs. 2.10f and Fig. 2.20a), the gate covers three sides of the body (fin), but the top channel is deactivated by using a thicker dielectric. The FinFET is a semi-vertical device because the current is controlled by the two vertical gates and flows horizontally along the body side walls.

A more advanced alternative is to etch off the top gate and provide independent contacts to the lateral gates. The advantage of this device, called MIGFET or MIG–FinFET, is that two gates can play different functions. It is expected that MIGFETs can considerably reduce the complexity of digital circuits [74].

Although the FinFET performance is very promising, two critical scaling issues need to receive attention: (i) the control of the crystal quality and orientation on the side walls by wet etching and (ii) the trimming of the transistor body (intergate distance) in order to control the short-channel effects [75,76].

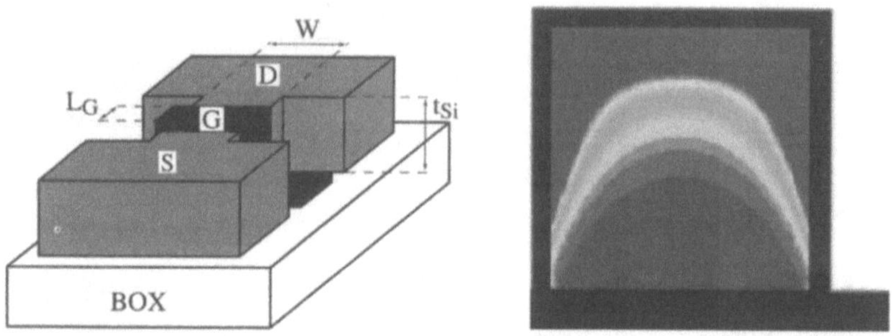

(a) (b)

FIGURE 2.20. (a) Schematic configuration of the FinFET and triple-gate MOSFET and (b) cross section of the minority carrier distribution in a square (20 × 20 nm) triple-gate device. The gate is biased in inversion ($V_{G1} = +0.5$ V) and the substrate in accumulation ($V_{G2} - 10$ V). After Ref. 78.

2.7.2. Triple-Gate MOSFETs

A FineFET with an active top gate is abusively named triple-gate MOSFET (TG MOSFET). Actually, one single gate controls three different sections of the channel: two vertical and one horizontal (Fig. 2.20a). The gate dielectric should be equally thin on the three sides of the body in order to avoid multiple threshold voltages. The performance is encouraging. The magnitude of the current can be adjusted via the fin width, which governs the contribution of the top surface channel. This advantage is debatable because volume inversion and scaling capability are lowered in wider fins.

In TG FinFETs, the discrimination of the different channels is possible by using the back-gate action and the variable fin width. It was found that the carrier mobility is significantly degraded on the fin side walls compared to the top and bottom channels [77]. Process refinements [75,76] and crystal orientation are under investigation to improve FinFET performance.

At the device level, it is important to understand the size effects as well as the interaction between the different channels [78]. The coupling effects depend on the fin height, t_{si}, and width, W. A square TG FinFET ($t_{si} = W = 20$ nm, Fig. 2.20b) features an inhomogeneous vertical variation of the electron concentration and surface potential on the lateral sides. The 'measured' threshold voltage represents the lowest value of the position-dependent V_T. In addition, if a larger negative substrate bias is applied, the accumulation layer can block the inversion of all three channels.

A narrow and tall fin exhibits two distinct regions [78]. At the bottom of the device, the carrier distribution is inhomogeneous (2-D), similar to that of the square fin (Fig. 2.20b). In the upper region, the carrier profile becomes vertically homogeneous and quasi-1-D in the lateral direction; the potential variation along the height of the fin is negligible, except next to the top gate, where corner effects may appear. The front channel and the upper regions of the lateral channels are in strong inversion, fully controlled by the gate. The substrate-to-body coupling

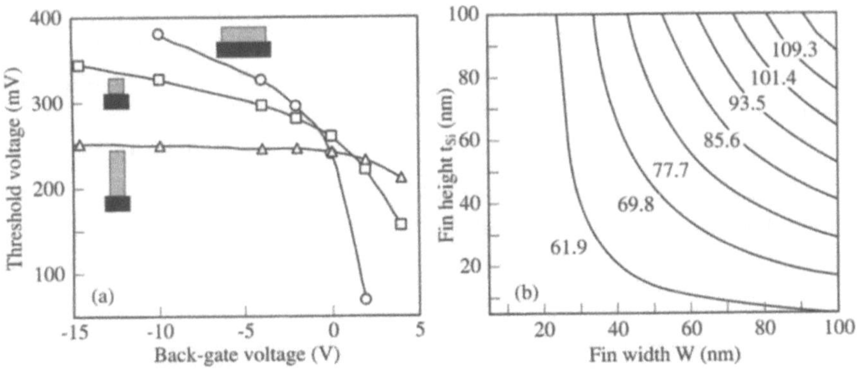

FIGURE 2.21. (a) Threshold voltage as a function of substrate bias in TG FinFETs with wide, square, and tall configurations (aspect ratios: $t_{si}/W = 20/80,\ 20/20,\ 80/20$, respectively). (b) Contours of fixed subthreshold swing versus the aspect ratio of TG FinFETs. After Ref. 78.

effects are weak and restricted to the bottom corners. The lateral coupling prevails, reducing the back-gate action.

For thin and wide fins ($t_{si} \ll W$, as in FD MOSFETs), the lateral gates do not control the body well enough. Instead, the back-gate coupling is strong and modulates the front-channel conduction.

The variation of the threshold voltage with back-gate bias depends on the aspect ratio (t_{si}/W) of the fin (Fig. 2.21a). For wide fins, the coupling effect $V_T(V_{G2})$ is strong due to the classical 1-D vertical coupling [22] between the front channel and the back gate. In tall fins, the electrostatics is controlled by the lateral gates, which inherently tends to reduce or suppress the coupling to the bottom gate. The coupling coefficient saturates for an aspect ratio of $t_{si}/W \simeq 4$ [78].

The square fin shows intermediate coupling intensity. The geometry optimization aims to reach more current per fin (wider transistors) while avoiding too much sensitivity to substrate effects. Note that even for a grounded back gate, a virtual substrate biasing can be induced by radiation, hot-carrier injection, or DIVSB effects.

Further simulations (Fig. 2.21b) show the equivalent role played by W and t_{si} in the suppression of the short-channel effects. In order to achieve a low subthreshold swing (below 75 mV/decade, Fig. 2.21b), both dimensions W and t_{si} should be reasonable small (TG case) or one dimension must be *very* small.

If a very narrow body is manufacturable, the fringing fields are controlled by the lateral gates and FinFETs are suitable. The lateral gates define the back-surface potential, blocking the penetration of the fringing field from the drain (DIVSB, Section 2.6.1). This control can be enhanced by letting the lateral gates extend vertically into the BOX (π gate) and laterally underneath the film (Ω gate) [2]. The π-gate and Ω-gate architectures do relax the constraint of ultra narrow fins, but, in turn, require a thick enough BOX. Can these configurations survive as the SOI materials evolve toward a very thin BOX ?

The opposite situation is when neither the body width nor the gate architecture enable good enough control of the back-surface potential. Here, very thin films are needed to prevent short-channel effects. In this case, the top and bottom gates dominate, so that FD planar devices are more attractive [78].

At this point, it is essential to underline the 3-D nature of the coupling effects:

- *Lateral* coupling between the side gates
- *Vertical* coupling between the top gate and the bottom gate
- *Longitudinal* coupling between the drain and the body via the fringing fields (DIVSB).

An ultimate and spectacular size effect is related to the transistor *volume*. FinFET technology is capable of producing devices with all dimensions (thickness, width, length) in the 10-nm range (see Fig. 2.13). A 10^{-18} cm^3 body volume raises interesting fundamental questions. What doping level is induced by one single impurity? Does the impurity position matter? Should atomistic simulations include the silicon atoms one by one?

2.7.3. Gate-All-Around MOSFETs

The gate-all-around (GAA) MOSFETs have been invented by Colinge [2]. The GAA technology is complex: (i) formation of a small-size Si membrane, (ii) thermal oxidation, and (iii) wrapping a homogeneous gate. The membrane can be processed on SOI by etching part of the BOX underneath the silicon film [2] or in bulk Si by SON technology [17]: epitaxy of a sacrificial layer of SiGe, epitaxy of the thin Si film, and, finally, removal of the SiGe layer (see Figure 2.3c$_5$, Section 2.3.5). GAA MOSFETs can also have a vertical pillar configuration. The formation of a pillar with small enough diameter is a very challenging operation.

The structure of GAA MOSFETs is conceptually simple and suitable for investigating the coupling, corner, and quantum effects. The corners are intrinsic to FinFET and GAA architectures. In each corner, the electrostatic coupling between the adjacent gates creates favorable conditions for the accumulation of minority carriers [79]. The corner regions have a lower threshold voltage and turn on earlier than the main channel. This causes an increase of the leakage current in the off state and poor subthreshold characteristics.

The corner effect increases for square bodies with high doping. Corner rounding equalizes the minority carrier distribution and suppresses the activation of the parasitic channels. The control of the corner radius is a delicate process. A simpler solution for attenuating the corner effect is to leave the body *undoped*; the threshold voltage can be adjusted with a midgap metal gate.

Quantum simulations for GAA MOSFETs with very small cross sections, show that quantization leads to electron repulsion from the interface and corners into the body (right diagram in Fig. 2.22) [78]. This repulsion opposes the electrostatic effect of corner attraction, resulting in relatively low surface and corner concentrations. Quantum repulsion and volume inversion lead to the formation of

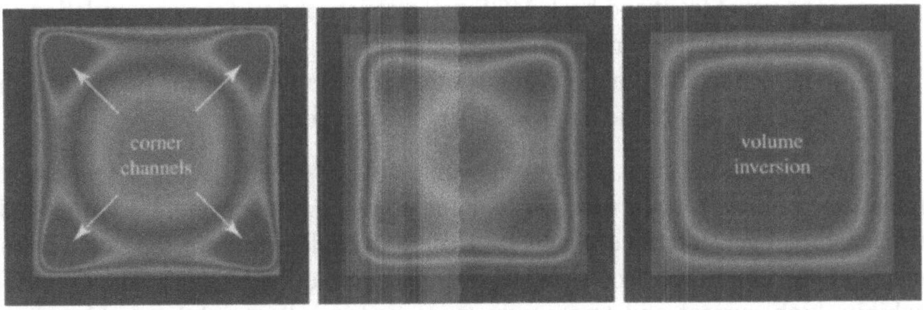

FIGURE 2.22. Minority carrier distribution in a 10-nm square GAA transistor. From left to right: strong inversion with pronounced corner effects, moderate inversion where the corners start forming, and weak inversion with volume inversion and no trace of corners. After Ref. 78.

a channel in the center of the device. The subthreshold corner effect is suppressed; hence, the off-state current and swing are excellent.

For higher gate bias (left diagram in Fig. 2.22), the electrostatic effects are gradually taking over the quantum effects. Four electron filaments are formed along the corners, in strong inversion, but do not degrade the device performance. Nano-size and undoped fins show clear advantage for downscaling and for avoiding parasitic effects.

2.7.4. Four-Gate FET

Unlike GAA and TG MOSFETs, the four-gate FET (G^4-FET) is a genuine four-gate transistor, operated in accumulation/depletion modes [80]. Figure 2.23 shows an inversion-mode, p-channel SOI MOSFET with two N^+ body contacts. The same device becomes a G^4-FET when the current is driven by electrons in the perpendicular direction. The majority carriers flow between the body contacts, which play the role of source and drain for the G^4-FET (Fig. 2.23a). There are four *independent* gates:

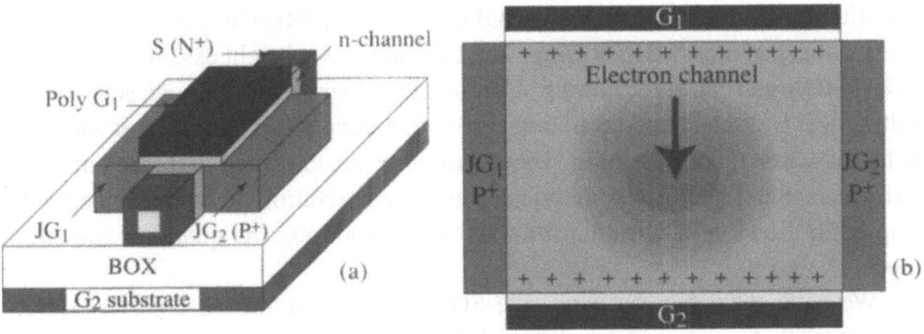

FIGURE 2.23. Basic configuration of the four-gate transistor and carrier distribution in the cross section for operation in volume mode (depletion-all-around) with inverted interfaces.

- The usual front and back MOS gates govern the surface accumulation or vertical depletion regions.
- The two lateral junctions control the effective width of the body through the extension of the horizontal depletion regions.

The conduction path is modulated by mixed MOS–JFET (junction field-effect transistor) effects: from a wire-like volume conduction to strongly accumulated front and/or back interface channels. Different models explain the conduction mechanisms in surface accumulation or pure volume modes [81]. The G^4-FET exhibits high current and transconductance and excellent subthreshold swing. Each gate has the capability of switching the transistor on and off. The independent action of the four gates opens promising perspectives for novel applications: mixed-signal circuits, nanoelectronic devices (quantum wires), four-level schemes enabling logic functions with a reduced number of transistors, and so forth.

Note that the G^4-FET accommodates naturally to scaling. As the gate length of CMOS circuits decreases, the width of the G^4-FET is reduced, increasing the junction gate action. However, the G^4-FET will not compete for minimum size records; it will be more suitable for innovative circuit designs.

A most exciting aspect is the depletion-all-around (DAA) mode of operation. The majority carrier channel is surrounded by depletion regions. A quantum wire can be formed (Fig. 2.23b), the dimensions of which are vertically and laterally biased controlled, not lithographically controlled. The volume–conduction channel benefits from a double shielding effect; it is separated from the interface, first by the depletion regions and second by the inversion layers. In the DAA mode, the device features maximum mobility, minimum noise, and unchallenged radiation-hardness capability [82]. Note that the G^4-FET structure makes possible the independent cross-conduction of majority carriers (in the volume) and minority carriers (at the interfaces), which looks like a source for revolutionary devices.

2.8. CONCLUSIONS

In this chapter, we have presented the principles of SOI technologies and their potential for ultimate scaling. Only SOI devices can pretend expanding the imminent frontiers of the CMOS scaling. The SOI horizon is bright essentially because bulk CMOS can hardly continue. Continuous progress in material science and technology is already incorporating novel semiconductors (strained layers) and dielectrics in SOI configuration. These structures are expected to infuse enhanced performance and new functionalities. The characterization techniques need to be adapted to the case of ultrathin films in order to address the quantum and strain effects.

We have seen that the nano-size SOI MOSFET stands as a perfect device for a smooth transition from microelectronics to nanoelectronics. Recent results for state-of-the-art SOI MOS transistors were reviewed and demonstrated the mixed flavors of device scaling. Thin tunneling oxides turn on remarkable GIFBEs. In nanometer-thick SOI films, the coupling effects are amplified, leading to supercoupling and interesting quantum effects. The self-heating issue can be alleviated by thinning the BOX and replacing it with a different dielectric that offers improved thermal conductivity. A ground plane or additional gates avoid degrading the electrostatic behavior of the nano-MOSFET.

The family of size effects in SOI is very rich because each dimension of the transistor plays a specific role. More importantly, a given size effect (length, width, thickness) is modulated by the other dimensions. The control of these 3-D coupling effects is vital for the MOSFET scaling beyond the 10-nm channel-length barrier. What is certain is that all dimensions will be reduced concomitantly. The semiconductor body will presumably be the thinnest layer in the device. In parallel, the transistor architecture will evolve to multiple gates, opening a wide space for new circuit topologies.

Acknowledgments. Most of this work has been performed at the Center for Projects in Advanced Microelectronics (CPMA), operated by CNRS, LETI, and universities. Special thanks to our colleagues from Grenoble and elsewhere (in quasi-alphabetical order): R. Ritzenthaler, M. Bawedin, O. Faynot, J. Pretet, F. Allibert, M. Cassé, T. Poiroux, A. Ohata, K. Oshima, N. Bresson, H. Iwai, S. Deleonibus, B. Dufrene, B. Blalock, K. Akarvardar, F. Dauge, C. Gallon, A. Vandooren, J-H. Lee, C. Mazuré, and P. Gentil.

REFERENCES

1. S. Cristoloveanu and S.S. Li, *Characterization of Silicon-On-Insulator Materials and Devices*, Kluwer, Boston (1995).
2. J-P. Colinge, *Silicon-On-Insulator Technology: Materials to VLSI*, 3rd ed., Kluwer, Boston (2004).
3. D.J. Frank, R.H. Dennard, E. Nowak, P.M. Solomon, Y. Taur, and H.S.P. Wong, *IEEE Proc.* **89**(3), 259–288 (2001).
4. G.K. Celler and S. Cristoloveanu, *J. Appl. Phys.* **93** 4955–4978 (2003).
5. S. Cristoloveanu and G. Reichert, 1998 High Temperature Electronic Materials, *Devices and Sensors Conference Proceedings* (1998).
6. T. Ohno, S. Matsumoto, and K. Izumi, *IEEE Trans. Electron Devices* **40**, 2074 (1993).
7. H. Vogt, in *SOI Technology and Devices VI*, Electrochemical Society, Pennington, NJ, 1994, p. 430.
8. T. Nishimura, Y. Inoue, K. Sugahara, S. Kusonoki, T. Kumamoto, S. Nakagawa, M. Nakaya, Y. Horiba, and Y. Akasaka, *Technical Digest IEDM*, p. 111 (1987).
9. A. Zaslavsky, C. Aydin, S. Luryi, et al., *Appl. Phys. Lett.* **83**(8), 1653–1655 (2003).
10. Y. Ono, Y. Takahashi, K. Yamazaki, et al., *Technical Digest IEDM*, Piscataway, NJ, 1999, pp. 367–370.
11. J.B. Lasky, *Appl. Phys. Lett.* **48**, P. 78 (1986).
12. M. Bruel, *Electron. Lett.* **31**, p. 1201 (1995).
13. D. Munteanu, et al., *Microelectro. Eng.* **36**, 1–4, 395–398 (1997).

14. T. Yonehara, in *Silicon Wafer Bonding Technology for VLSI and MEMS Applications*, edited by S.S. Iyer and A.J. Auberton-Hervé, INSPEC, London, 2002, Chap. 4, p. 53.
15. K. Izumi, M. Doken, and H. Ariyoshi, *Electron. Lett.* **14**, p. 593 (1978).
16. S. Cristoloveanu, *J. Electrochem. Soc.* **138**, p. 3131 (1991).
17. T. Skotnicki, in *Silicon-On-Insulator Technology and Devices X*, Electrochemical Society, Pennington, NJ, 2001, Vol. 2001–3, pp. 391–402.
18. J. Pretet, S. Monfray, S. Cristoloveanu, and T. Skotnicki, *IEEE Trans. Electron Devices* **51** (2), 240–245 (2004) .
19. S. Cristoloveanu, *Rep. Prog. Phys.* **3** p. 327 (1987).
20. I. Lagnado and P.R. de la Houssaye, in *Silicon-On-Insulator Technology and Devices X*, Electrochemical Society, Pennington, NJ, 2001, Vol. 2001–3, pp. 265–270.
21. Y. Moriyasu, T. Morishita, M. Matsui, and A. Yasujima, in *Silicon-On-Insulator Technology and Devices IX*, Electrochemical Society, Pennington, NJ, Vol. 99–3, 1999, pp. 137–142.
22. H.K. Lim and J.G. Fossum, *IEEE Trans. Electron Devices*, **30**, 1244–1251 (1983).
23. Y. Omura, T. Ishiyama, M. Shoji, and K. Izumi, in *SOI Technology and Devices VII*, Electrochem. Society, Pennington, NJ, 1996 p. 199.
24. B. Mazhari, S. Cristoloveanu, D.E. Ioannou, and A.L. Caviglia, *IEEE Trans. Electron Devices* **38**, 1289 (1991).
25. F. Balestra, M. Benachir, J. Brini, and G. Ghibaudo, *IEEE Trans. Electron Devices* (USA) **37**, 2303 (1990).
26. T. Ouisse, S. Cristoloveanu, and G. Borel, *Solid-State Electron.* **35**, 141 (1992).
27. F. Balestra, S. Cristoloveanu, M. Bénachir, J. Brini, and T. Elewa, *IEEE Electron Device Lett.* **8**, 410 (1987).
28. S. Cristoloveanu, *Microelectron. Reliab.* **37**, 1003 (1997).
29. M. Bawedin, S. Cristoloveanu, and D. Flandre, IEEE International SOI Conference., Charleston, SC, 2004.
30. J.Y. Choi and J.G. Fossum, *IEEE Trans. Electron Devices* **38**, 1384 (1991).
31. I.M. Hafez, G. Ghibaudo, and F. Balestra, *IEEE Trans. Electron Devices* **37**, 818 (1990).
32. J. Jomaah, F. Balestra, and G. Ghibaudo, *Physica Status Solidi (a)*, **142**, 533 (1994).
33. F. Balestra, J. Jomaah, G. Ghibaudo. et al., *IEEE Trans. Electron Devices* **41**, 109 (1994).
34. M. Yoshimi, M. Takahashi, T. Wada, K. Kato, S. Kambayashi, M. Kemmoshi, and K. Natori, *IEEE Trans. Electron Devices* **37**, 2015 (1990).
35. E.P. Ver Ploeg, T. Watanabe, N.A. Kistler, J.C.S. Woo, and J.D. Plummer, *Technical Digest IEDM*, 337 (1992).
36. F. Allibert, J. Pretet, G. Pananakakis, and S. Cristoloveanu, *Appl. Phys. Lett.* **84**, 1192–1194 (2004).
37. S. Zaouia, S. Goktepeli, A.H. Perera, and S. Cristoloveanu, in *Silicon-On-Insulator Technology and Devices XII*, Electrochemical Society, Pennington, NJ, 2005.
38. S. Cristoloveanu and S. Williams, *IEEE Electron Device Lett.* **13**(2), 102–104 (1992).
39. S. Cristoloveanu, D. Munteanu, and M. Liu, *IEEE Trans. Electron Devices* **47**(5), 1018–1027 (2000).
40. N. Bresson and S. Cristoloveanu, *Microelectronic Eng.* **72**(1–4), 357–361 (2004).
41. D. Munteanu, D.A. Weiser, S. Cristoloveanu, O. Faynot, J–L. Pelloie, and J.G. Fossum. *IEEE Trans. Electron Devices* **45**(8), 1678–1683 (1998).
42. S. Cristoloveanu, T. Ernst, D. Munteanu, and T. Ouisse, *Int. J. High Speed Electron. Syst.* **10**(1), 217–230 (2000).
43. T. Ernst, C. Tinella, and S. Cristoloveanu, *Solid-State Electron.* **46**(3), 373–378 (2002).
44. H–S. Wong, D.J. Franck, and P.M. Solomon, *Technical Digest IEDM*, 407 (1998).
45. R–H. Yan, A. Ourmazd, and K.F. Lee, *IEEE Trans. Electron Devices* **39**(7), 1704–1710 (1992).
46. D. Franck, S. Laux, and M. Fischetti, *Technical Digest IEDM*, 553 (1992).
47. K.K. Likharev, *Nano and Giga Challenges in Microelectronics*, Elsevier, Amsterdam, 2003 pp. 27–68.

48. B. Doris, et al., *Technical Digest IEDM*, IEEE, Piscataway, NJ, 2003, pp. 27.3.1–4.
49. J. Pretet, D. Ioannou, N. Subba, S. Cristoloveanu, W. Maszara, and C. Raynaud, *Solid-State Electron.* **46**(11), 1699–1707 (2002).
50. T. Elewa, B. Kleveland, S. Cristoloveanu, B. Boukriss, and A. Chovet, *IEEE Trans. Electron Devices* **39**(4), 874–882 (1992).
51. H. Majima, H. Ishikuro, and T. Hiramoto, *Technical Digest IEDM*'99, 1999, pp. 379–382.
52. J. Pretet, A. Ohata, F. Dieudonné, et al., in *Silicon Nitride and Silicon Dioxide Thin Insulating Films VII*, Electrochemical Society, Pennington, NJ, 2003, Vol. PV-2003-02, pp. 476–487.
53. A. Ohata, J. Pretet, S. Cristoloveanu, and A. Zaslavsky, *IEEE Trans. Electron Devices* **52**(1), 124–125 (2005).
54. T. Ernst, S. Cristoloveanu, G. Ghibaudo, T. Ouisse, S. Horiguchi, Y. Ono, Y. Takahashi, and K. Murase, *IEEE Trans. Electron Devices* **50**, 830–838 (2003).
55. C. Fiegna and A. Abramo, International Conference on Simulation of Semiconductor Processes and Devices, SISPAD'97, 1997, pp. 93–96.
56. F. Gamiz, J.B. Roldan, J.A. Lopez-Villanueva, et al., in *Silicon-On-Insulator Technology and Devices X*, Electrochemical Society, Pennington, NJ, Vol. PV-2001-3, 2003, pp. 157–168.
57. A. Ohata, M. Cassé, S. Cristoloveanu, and T. Poiroux, Proc. ESSDERC 2004, IEEE, 2004 pp. 109–112.
58. C. Gallon, et al., IEEE International SOI Conference, Charleston, SC, (2004), pp. 153–155.
59. J. Pretet, T. Matsumoto, T. Poiroux, et al., Proc. ESSDERC'02, University of Bologna, 2002, pp. 515–518.
60. A. Mercha, J.M. Rafi, E. Simoen, E. Augendre, and C. Claeys, *IEEE Trans. Electron Devices* **50**, (7), 1675–1682 (2003).
61. M. Cassé, J. Pretet, S. Cristoloveanu, et al., *Solid-State Electron.* **48**(7), 1243–1247 (2004).
62. F. Dieudonné, S. Haendler, J. Jomaah, and F. Balestra, *Solid-State Electron.* **48**(6), 985–997 (2004).
63. L.T. Su, K.E. Goodson, D.A. Antoniadis, M.I. Flik, and J.E. Chung, *Technical Digest IEDM*, 357 (1992).
64. S. Bengtsson, M. Bergh, M. Choumas, et al., *Jpn. J. Appl. Phys.* **35**, 4175–4181 (1996).
65. K. Oshima, S. Cristoloveanu, B. Guillaumot, H. Iwai, and S. Deleonibus, *Solid-State Electron.* **48**, 907–917 (2004).
66. N. Bresson, S. Cristoloveanu, K. Oshima, et al., IEEE International SOI Conference, Charleston, SC 2004.
67. F. Gamiz, J.B. Roldan, J.A. Lopez-Villanueva, et al., *Silicon-On-Insulator Technology and Devices X*, Electrochemical Society, Pennington, NJ, 2001, pp. 157–168.
68. D. Esseni, M. Mastrapasqua, G.K. Celler, et al., *IEEE Trans. Electron Devices* **50**(3), 802–808 (2003).
69. C.R. Cirba, S. Cristoloveanu, R.D. Schrimpf, et al., in *Silicon–on–Insulator Technology and Devices XI*, Electrochemical Society, Pennington, NJ, 2003, Vol. 2003–05, pp. 493–498.
70. D. Hisamoto, W-C. Lee, J. Kedzierski, et al., *Technical Digest IEDM*,'98 1998, pp. 1032–1034.
71. D. Hisamoto, W-C. Lee, J. Kedzierski, et al., *IEEE Trans. Electron Devices* **47**(12), 2320–2325 (2000).
72. F. Allibert, A. Zaslavsky, J. Pretet, and S. Cristoloveanu, Proc. ESSDERC'2001, Frontier Group 2001 pp. 267–270.
73. J. Widiez, F. Daugé, M. Vinet, et al., IEEE International SOI Conference, Charleston, SC, 2004, pp. 185–186.
74. L. Chang, M. Ieong, and M. Yang, *IEEE Trans. Electron Devices* **51**(10), 1621–1627 (2004).
75. Y.X. Liu, et al., *IEEE Electron Device Lett.* **25**, 510–512 (2004).
76. W. Xiong, G. Gebara, J. Zaman, et al., *IEEE Electron Device Lett.* **25**(8), 541–543 (2004).
77. F. Daugé, J. Pretet, S. Cristoloveanu, et al., *Solid-State Electron.* **48**, 535–542 (2004).

78. S. Cristoloveanu, R. Ritzenthaler, A. Ohata, and O. Faynot, *Int. J. High Speed Electronics Syst.* **16**(1), 9–30 (2006).
79. J.G. Fossum, J.W. Yang, and V.P. Trivedi, *IEEE Electron Device Lett.* **24**(12), 745–747 (2003).
80. B.J. Blalock, S. Cristoloveanu, B. Dufrene, et al., *Frontiers in Electronics—Future Chips*, World Scientific, Singapore, 2002, Vol. 26, pp. 305–314.
81. K. Akarvardar, B. Dufrene, S. Cristoloveanu, et al., Proc. ESSDERC'03, Lisbon 2003, pp. 127–130.
82. K. Akarvardar, in *Silicon–On–Insulator Technology and Devices XII*, Electrochemical Society Pennington, NJ, 2005, pp. 99–106.

3

Strategies of Nanoscale Semiconductor Lasers

Samuel S. Mao

3.1. INTRODUCTION

Semiconductor lasers are in many ways second only to transistors as to their impact on today's high-tech industries. The unique characteristics, such as narrow emission wavelength, high-frequency modulation, and device integratibility, make semiconductor lasers ideal photon sources for applications as diverse as telecommunication, signal processing, material characterization, and medical diagnostics. Advances in material growth technologies, particularly molecular-beam epitaxy, metal–organic chemical vapor deposition, and a suite of innovative chemical and physical synthesis techniques, make the fabrication of high-quality nanoscale semiconductor structures possible. Thanks to the quantum size effects that drastically modify the energy spectra of confined electrons in reduced dimensions, the population inversion necessary for lasing action occurs more efficiently as the active semiconductor gain medium is scaled down from the bulk to the nanometer scale. Consequently, semiconductor lasers built with nanoscale active media are expected to exhibit extraordinary features such as great color range, high optical gain, and low lasing threshold. Indeed, miniaturized lasers using nanoscale semiconductor gain media—two-dimensional quantum wells, one-dimensional quantum wires, and zero-dimensional quantum dots—have shown significant improvements in device performance. This chapter provides an overview of the physics and technologies behind the rapid progress in the miniaturization of semiconductor lasers.

Lawrence Berkeley National Laboratory and Department of Mechanical Engineering, University of California, Berkeley, CA 94720 ssmao@lbl.gov

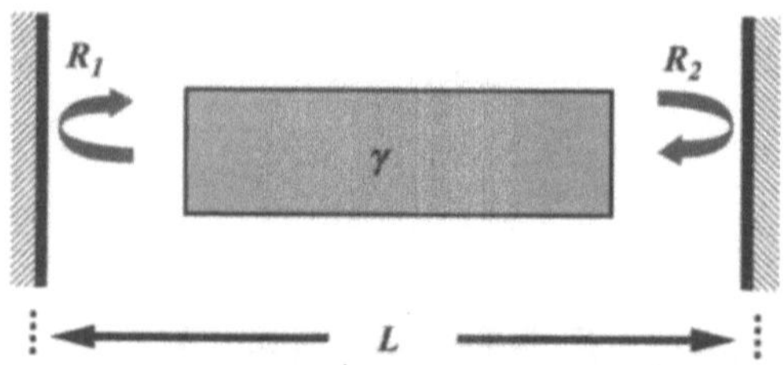

FIGURE 3.1. Basic elements of a laser with a Fabry–Perot-type cavity: gain medium and optical cavity.

3.1.1. Semiconductor Laser Fundamentals

The concept of a semiconductor-based lasing device may be traced back to 1950s when von Neumann [1] considered the possibility of light amplification in semiconductors. In 1962, four groups independently succeeded in demonstrating the first semiconductor laser devices [2–5], which, although operated under low temperatures and high-current pulses, opened up an entirely new field of laser electronics. Semiconductor lasers in modern electronics incorporate heterostructures, an innovative design that was first suggested in 1963 by Kroemer [6] and Alferov [7]. In 1967, heterostructures of GaAs and AlGaAs were successfully grown [8] by the liquid phase epitaxy technique, and by 1970, Alferov's group [9] and Hayashi and Panish [10] reported double heterostructure lasers that could continuously operate at room temperature.

Two basic elements necessary for realizing a semiconductor laser (Fig. 3.1) include an *active gain medium* that provides optical gain by stimulated emission and an *optical resonant cavity* that confines the photons to create positive optical feedback. Pumped by either electrical or optical energy, electrons and holes within the semiconductor gain material can be excited to nonequilibrium energy levels so light radiation can be amplified (positive gain). If the resulting gain is sufficient to overcome the losses of the optical cavity that provides the necessary feedback of the radiation, lasing oscillation can be established at a well-established threshold.

3.1.1.1. Electrons in Semiconductor

For semiconductor materials [11], in the simplest case, there are two bands of allowed energy states (Fig. 3.2): the valence band and the conduction band, which are separated by a defined band gap E_g. In an intrinsic (undoped) semiconductor with no external excitation and at a temperature of $T = 0$ K, the conduction band is completely empty and the valence band is completely filled with electrons. For $T > 0$ K, under the thermodynamic equilibrium condition, electrons and holes (the vacancy of electrons) are distributed over a range of energies according to Fermi–Dirac statistics. The probability of whether a state with the energy level E

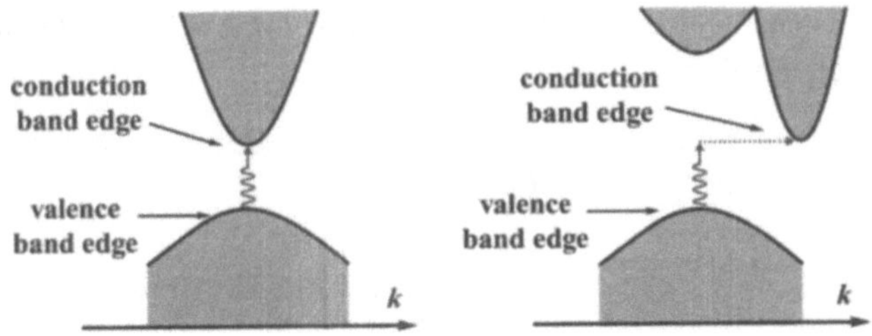

FIGURE 3.2. Energy band structures of a direct semiconductor (left) and an indirect semiconductor.

is occupied by an electron is expressed as

$$f(E) = \frac{1}{e^{(E-E_F)/k_B T} + 1}, \tag{3.1}$$

where k_B (8.617347×10^{-5} eV/K) is the Boltzmann constant and E_F is the Fermi level. In undoped semiconductors, the Fermi level is located in the middle, between the conduction and valence band edges. At $T = 0$ K, $f(E)$ is a step function that has a value of 1 (all states are filled) below the Fermi level.

A free electron has a kinetic energy of $E = p^2/2m_0$, where m_0 (9.109534×10^{-31} kg) is the free-electron rest mass and p is the momentum. When treated as a quantum mechanical particle, the momentum $p = \hbar k$ is proportional to the wave number k, where $\hbar = h/2\pi$ (6.582173×10^{-16} eV s) is the reduced Planck constant. Thus, for a free electron, the dependence of energy versus wave number has the form $E(k) = \hbar^2 k^2/2m_0$. Similarly, the electron energies in the conduction band $E_C(k)$ and the valence band $E_V(k)$ of a semiconductor have the form

$$E_C(k) = E_g + \frac{\hbar^2 k^2}{2m_e}, \tag{3.2a}$$

$$E_V(k) = -\frac{\hbar^2 k^2}{2m_h}, \tag{3.2b}$$

where m_e and m_h are electron and hole effective masses defined by accounting for the interactions of the carriers with the semiconductor material lattice. In direct band-gap semiconductors (e.g., GaAs), the valence band maximum and the conduction band minimum have the same k, whereas for indirect semiconductors (e.g., silicon), the minimum and maximum have different k values (Fig. 3.2).

The motion of electrons in a conduction or valence band is governed by the Schrödinger equation. For a bulk semiconductor with liner dimensions L_x,L_y, and L_z, volume $V = L_x L_y L_z$, the wave function of an electron in a given band,

$\Psi(\boldsymbol{r}) \sim e^{i\boldsymbol{k}\cdot\boldsymbol{r}}$, is characterized by wave vector **k** (k_x, k_y, k_z), which can only take quantized discrete values,

$$\begin{aligned} k_x &= \frac{2\pi n_x}{L_x}, \\ k_y &= \frac{2\pi n_y}{L_y}, \\ k_z &= \frac{2\pi n_z}{L_z}, \end{aligned} \tag{3.3}$$

where n_x, n_y, and n_z are integers. There is one allowed wave vector (electron state) in each volume element $(2\pi)^3/V$ of **k** space, so in a sphere of volume $4\pi k^3/3$, the total number of electron states is

$$N(k) = 2\frac{4\pi k^3/3}{(2\pi)^3/V} = \frac{V}{3\pi^2}k^3, \tag{3.4}$$

where the factor 2 comes from the two allowed spin quantum states for each allowed **k**. Using the kinetic energy relation, $E(k) = \hbar^2 k^2/2m$, the number of electron states can be expressed as a function of energy:

$$N(E) = \frac{V}{3\pi^2}\left(\frac{2mE}{\hbar^2}\right)^{3/2}. \tag{3.5}$$

Therefore, the density of state defined as the number of electron states per unit energy per volume is given by

$$\rho(E) = \frac{1}{V}\frac{dN(E)}{dE} = \frac{1}{2\pi^2}\left(\frac{2m}{\hbar^2}\right)^{3/2} E^{1/2}. \tag{3.6}$$

3.1.1.2. Photons in Semiconductor

The valence band has lower energy, in which any electron that absorbs energy greater than the band-gap energy ($\hbar\omega > E_g$) may move upward into the conduction band, leaving behind a hole in the valence band. Equivalently, electrons in the conduction band may release similar amount of energy and move downward to the valence band. Fundamental to semiconductor laser operation is radiative interband transitions in which generation and recombination of electron–hole pairs are achieved with the absorption or emission of photons. In such interband transitions, conservation of energy and momentum must be fulfilled. For light wavelength of interest to semiconductor lasers, the momentum of photons with energy E_{ph}, $\hbar k = \hbar\omega/c = E_{\mathrm{ph}}/c$, can be neglected compared to the momentum of the electronic carriers (electrons and holes) in semiconductors. Radiative transition between an electron in the conduction band with energy E_2 and a hole in the

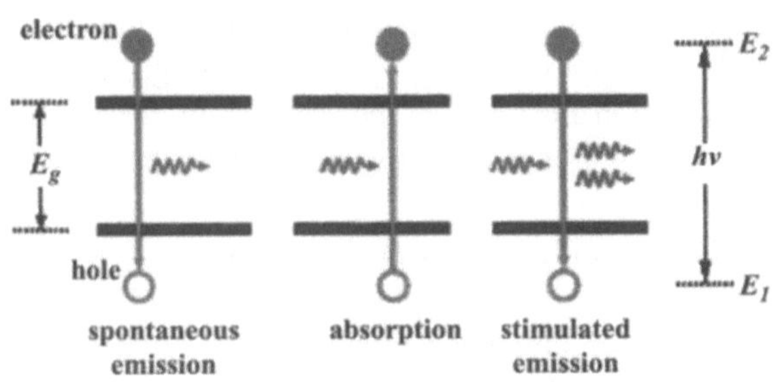

FIGURE 3.3. Schematic illustration of three possible routes of radiative band-to-band transition in a semiconductor material.

valence band with energy E_1 essentially takes place at the same wave vector **k**.

$$E_{\mathrm{ph}} = \hbar\omega = E_2(k) - E_1(k). \tag{3.7}$$

In energy band diagrams such as those in Fig. 3.2, such transition can be depicted by a vertical arrow.

There are three types of radiative band-to-band transitions in semiconductors, as illustrated in Fig. 3.3. The first process is spontaneous emission, where a recombination of an electron–hole pair leads to the emission of a photon, random in direction, phase, and time. Because this process depends on the existence of an electron at E_2 and a hole at E_1 simultaneously, the transition rate for spontaneous emission R_{sp} is proportional to the product of the electron density at E_2 and the hole density at E_1. The electron density at the energy level E_2 is the product of the density of states $\rho(E_2)$ and the probability that they are occupied by electrons given by the Fermi distribution function, $f(E_2)$. Similarly, the hole density at the energy level E_1 is the product of the density of states $\rho(E_1)$ and the probability of not being occupied by electrons, $1 - f(E_1)$. The transition rate per volume for spontaneous emission of photons is therefore

$$R_{\mathrm{sp}} = A\,\rho(E_2) f(E_2)\,\rho(E_1)\,[1 - f(E_1)], \tag{3.8}$$

where A is a constant for spontaneous emission.

The second process is (stimulated) absorption; an electron–hole pair is generated as the result of the absorption of an incoming photon. This is a three-particle process and the transition rate R_{12} is proportional to the product of the density of unoccupied states $\rho(E_2)[1 - f(E_2)]$ in the conduction band, the density of occupied states $\rho(E_1) f(E_1)$ in the valence band, and the density of photons $\rho(\hbar\omega)$ with energy $\hbar\omega = E_2 - E_1$:

$$R_{12} = B_{12}\,\rho(\hbar\omega)\,\rho(E_1)\,f(E_1)\,\rho(E_2)\,[1 - f(E_2)], \tag{3.9}$$

where B_{12} is a constant for absorption.

The third process is stimulated emission; a recombination of an electron–hole pair is stimulated by a photon, with a second photon generated simultaneously that has the same direction and phase as the first photon. Similar to absorption, the transition rate R_{21} for stimulated emission can be described as

$$R_{21} = B_{21}\,\rho(\hbar\omega)\,\rho(E_2)\,f(E_2)\,\rho(E_1)\,[1 - f(E_1)], \tag{3.10}$$

where B_{21} is a constant for stimulated emission.

For absorption and stimulated emission, the spectral energy density of photons in a medium of refractive index n is governed by Planck's blackbody radiation theory:

$$\rho(\hbar\omega)\,d(\hbar\omega) = \frac{n^3(\hbar\omega)^2}{\pi^2\hbar^3c^3}\,\frac{1}{e^{\hbar\omega/k_BT} - 1}\,d(\hbar\omega). \tag{3.11}$$

The equilibrium condition requires that absorption and emission must be balanced: $R_{12} = R_{21} + R_{sp}$. Therefore, through simple derivation,

$$B_{12} = B_{21} \equiv B,$$
$$A = \frac{n^3(\hbar\omega)^2}{\pi^2\hbar^3c^3}B. \tag{3.12}$$

3.1.1.3. Semiconductor *p–n* Junction

Different from intrinsic semiconductor materials, doping can generate energy states within the band gap. Doping with "donor" atoms gives new energy states close to the conduction band, as each donor atom has an electron that can easily be excited into the conduction band. Negatively charged electrons are the majority carrier and the material is said to be *n*-type. In contrast, doping with "acceptor" atoms gives new energy states close to the valence band, as each acceptor atom requires an extra electron to complete its bonds, which is usually obtained from a valence band electron, leaving behind a hole in the valence band. Positively charged holes are the majority carriers and the material is said to be *p*-type.

When two semiconductor materials are brought together (one *p*-type and one *n*-type), a *p–n* junction is formed (Fig. 3.4). Without an applied external voltage,

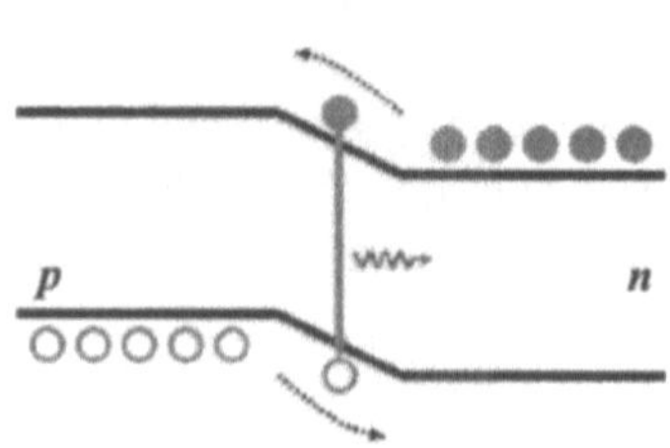

FIGURE 3.4. A semiconductor *p–n* junction under forward bias.

holes in the p-type material diffuse to the n-type region and electrons in the n-type material diffuse to the p-type region. Diffusion of holes from the p-type material leaves behind negatively charged ionized acceptors and diffusion of electrons from the n-type material leaves behind positively charged ionized donors. An electric field is therefore created in the narrow "depletion" region across the p–n junction, where the majority carriers are absent. The electric field causes the bending of the energy bands at the junction and serves to move the minority carriers across the junction and create a drift current, which balances with the diffusion current in equilibrium to give an overall zero current density.

When an external voltage is applied across the p–n junction, with the p-type semiconductor connected to the positive electrode (forward bias) as shown in Fig. 3.4, diffusion of majority carriers is easier due to the reduced potential barrier. Current flows from the p-side to the n-side, with the injected electrons flowing in the opposite direction. As a result, within a depletion region where electrons and holes co-exist, there is a large concentration of electrons in the conduction band and holes in the valence band. Photons could be easily generated by radiative recombination of electron–hole pairs if the material is a direct band-gap semiconductor (Fig. 3.4).

Quasi-Fermi levels can be introduced with a hole Fermi level (E_{FV}) for the p-type region and an electron Fermi level (E_{FC}) for the n-type region. With the injection of electrons and holes induced by the forward bias, E_{FC} and E_{FV} are no longer equal across the junction. In the transitional (depletion) region, the carrier distribution cannot be described by a single equilibrium Fermi distribution function. Instead, separate quasi-Fermi distribution functions are necessary for the electrons in the conduction band $f_C(E)$ and for the holes in the valence band $f_V(E)$:

$$
\begin{aligned}
f_C(E) &= \frac{1}{e^{E-E_{\mathrm{FC}}/k_BT}+1},\\
f_V(E) &= \frac{1}{e^{E-E_{\mathrm{FV}}/k_BT}+1}.
\end{aligned}
\tag{3.13}
$$

The ratio between the rate of radiative band-to-band transition of absorption (R_{12}) and that of stimulated emission (R_{21}) can be applied to determine whether an optical wave with photon energy $\hbar\omega$ is absorbed or amplified by stimulated emission. Following the above discussions,

$$
\frac{R_{12}}{R_{21}} = \frac{f_V(E_1)\,[1-f_C(E_2)]}{f_C(E_2)\,[1-f_V(E_1)]} = e^{[\hbar\omega-(E_{FC}-E_{FV})]/k_BT}. \tag{3.14}
$$

A necessary condition for semiconductor laser operation is that the rate of stimulated emission be larger than the absorption rate; therefore,

$$
E_{\mathrm{FC}} - E_{\mathrm{FV}} > \hbar\omega; \tag{3.15}
$$

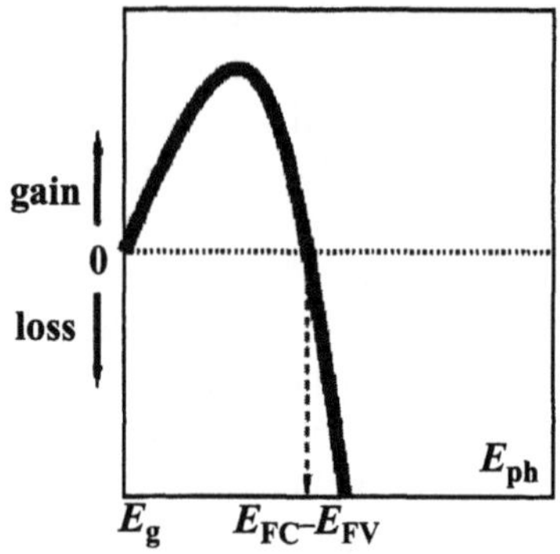

FIGURE 3.5. A schematic illustration of gain as a function of photon energy.

that is, only light-wave radiations with photon energies ($\hbar\omega$) smaller than the separation of quasi-Fermi levels (and larger than the band gap E_g) are amplified. This condition was first derived by Bernard and Duraffourg [12] and Basov et al. [13] in 1961. Laser operation requires pumping to build up and maintain a nonequilibrium carrier distribution (thus, E_{FC} and E_{FV}) in the semiconductor material. Pumping can be achieved by either optical or electronic excitation. A schematic illustration of the gain dependence on photon energy for a fixed pumping level is shown in Fig. 3.5. There is no gain for $\hbar\omega < E_g$, as no electronic transitions exist at these energies. The gain becomes zero and changes the sign at $\hbar\omega = E_{FC} - E_{FV}$, when absorption dominates at higher energies (frequencies).

It is noted that in addition to band-to-band radiative recombination, nonradiative recombination processes [11] in which electron–hole recombination does not lead to photon emission are always present. Auger recombination is one such process that involves electrons and holes recombining and transferring the energy to another electron or hole. Electrons or holes can also move nonradiatively into energy states that exist within the band gap known as traps. Such traps are created either by impurities or defects introduced at the time of fabrication or intentionally by ion implantation. In order for a semiconductor material to achieve high optical gain, defect-free crystalline quality of the material is usually expected.

3.1.2. The Scope

Dramatic progress in the development of nanoscale crystal growth and fabrication technologies has driven the miniaturization of semiconductor lasers, a trend also motivated by the desire to achieve greater color range, higher optical gain, and lower lasing threshold. This chapter provides a general picture of the current status of semiconductor laser technology based on two-, one-, and zero-dimensional nanoscale quantum structures and offers an introduction to some of the essential characteristics of quantum confinement in nanoscale semiconductor materials. A particular focus here is semiconductor lasers with one- and zero-dimensional nanoscale structures (nanoscale quantum wires and quantum dots) as the active gain media. The additional carrier confinement in quantum wires and quantum dots over conventional two-dimensional quantum well structures is expected to result in semiconductor lasers with characteristics superior to quantum well lasers.

3.2. LASING FROM NANOSTRUCTURES: CARRIER CONFINEMENT

The primary motivations behind developing nanoscale semiconductor lasers are substantial performance improvements [14–17] as the size is scaled down to the nanometer regime. This section intends to provide a basic understanding of the emerging physical phenomena associated with carrier confinement in the active semiconductor gain media with reduced dimensions. The physics behind improved lasing characteristics—greater color range, higher material gain, and lower lasing threshold—in semiconductor lasers based on nanoscale gain media will be illustrated. In the discussion of the benefits of a reduced lasing threshold, direct current injection will be assumed as the excitation mechanism.

3.2.1. Greater Color Range

One attractive feature of nanoscale materials stems from the fact that electrons and holes are confined in a small region to produce discrete energy levels. The use of nanoscale materials with size-dependent discrete transition energy levels provides a means of tuning the wavelength of semiconductor light emission. In principle, any color emission, from the far-IR (infrared) to the near-UV (ultraviolet), is possible by changing materials as well as their size. Semiconductor lasers with widely tunable wavelength ranges are essential for applications like high-capacity, wavelength division multiplexed (WDM), fiber communication systems [18].

3.2.1.1. Spherical Nanocrystals: An Example

One classical example of changes in the light emission wavelengths resulting from small variations in the size of the nanostructure is the color of nanoscale CdSe [19], one of the most developed nanocrystals. CdSe exhibits a red color for a "large" (e.g., 3 nm) nanocrystal but turns yellow at approximately half the size. Quantum confinement generally shifts the optical transitions to higher energies (blue shift, shorter wavelength). Figure 3.6 shows the lasing spectra of semiconductor lasers using CdSe nanocrystals of different radii as the active gain media [17].

Light emission characteristics of semiconductor materials are strongly influenced by the transitions between electron and hole energy levels; a simple quantum mechanical model of electrons and holes can be used to illustrate the size-dependent behavior of nanoscale semiconductor materials. The motion of electrons is controlled in nanoscale structures by enforcing a restriction on them in terms of potential barriers, which prevent electrons and holes from moving in

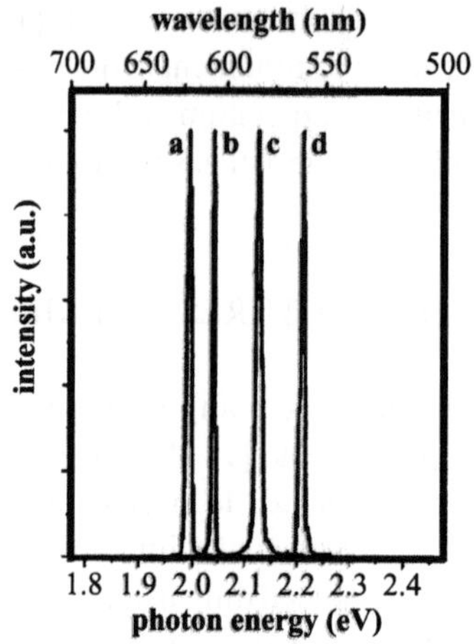

FIGURE 3.6. Lasing spectra (80 K) of nanocrystal lasers with different crystal radii: 2.7 nm (a), 2.4 nm (b), 2.1 nm (c), and 1.7 nm (d). The active lasing media are CdSe nanocrystals except for (d), which are CdSe/ZnS core/shell nanocrystals. After Ref. 17.

certain or all directions. For a spherical nanocrystal surrounded by an infinite potential barrier, by solving the three-dimensional Schrödinger equation, the electron and hole energy levels, characterized by an angular momentum quantum number l, can be written as [20]

$$E_{n,l} = \frac{\hbar^2 \phi_{n,l}^2}{2ma^2}, \tag{3.16}$$

with $n = 1, 2, 3 \ldots$ and $l = 0, 1, 2 \ldots$. In the above expression, m represents the electron or hole effective mass, a is the nanocrystal radius, $\phi_{n,l}$ is the nth root of the spherical Bessel function of order l, $j_l(\phi_{n,l}) = 0$, with $\phi_{1,0} = \pi$. The energy levels of the nanocrystals are discrete, as dictated by the quantum number n and l. The energies of the lowest electron and hole levels scale with the square of the inverse radius of the nanocrystal (a). An increase of nanocrystal band-gap energy with respect to the bulk material band gap can be obtained as

$$E_g^{\text{nanocrystal}} = E_g^{\text{bulk}} + \frac{\hbar^2 \pi^2}{2m_r a^2}, \tag{3.17}$$

where $m_r = m_e m_h/(m_e + m_h)$ is the reduced effective mass of the electron–hole pair. As a consequence, the photon energy of the band-edge optical transition increases, which gives rise to a dramatic color change in nanocrystals. For CdSe nanocrystals, the energy gap can shift from 1.8 eV (bulk value) to approximately 3.0 eV, passing through almost the whole visual part of the optical spectrum.

Strictly speaking, one needs to include Coulomb interactions between the electron and the hole inside a nanoscale material and adopt the concept of exciton: the electrically bonded electron–hole pair. In doing so, the lowest excitation energy needs to be modified [21, 22] as

$$\left\{\frac{\hbar^2\pi^2}{2m_r a^2}\right\}_{\text{free}} \to \left\{\frac{\hbar^2\pi^2}{2m_r a^2} - \frac{1.8e^2}{\varepsilon a}\right\}_{\text{bond}}, \tag{3.18}$$

where e is the electric charge and ε is the dielectric constant of the semiconductor. In the rest of this section, we will primarily apply the free-electron and hole model to illustrate basic physics concepts underlying the improvement of semiconductor laser performance resulting from reduced dimensions.

3.2.1.2. Quantum Confinement in One, Two, and Three dimensions

Other than spherical potential barriers, one-, two-, and three-dimensional (1-D, 2-D, and 3-D, respectively) quantum confinement can be achieved in "film-," "wire-," and "box-" like geometries, characterized by length scales L_x, L_y, and L_z in three separate directions. It has been customary in literature to call these nanostructures "quantum wells," "quantum wires," and "quantum dots," respectively. In the case of infinitely deep rectangular potential barriers in each restricted dimension, the energies of the confined carriers with respect to the band edges can be written in the form [23]

$$E_l = \frac{\pi^2\hbar^2 l}{2mL_x^2} + \frac{\hbar^2\left(k_y^2 + k_z^2\right)}{2m} \quad \text{(quantum wells)}, \tag{3.19a}$$

$$E_{l,m} = \frac{\pi^2\hbar^2}{2m}\left(\frac{l^2}{L_x^2} + \frac{m^2}{L_y^2}\right) + \frac{\hbar^2 k_z^2}{2m} \quad \text{(quantum wires)}, \tag{3.19b}$$

$$E_{l,m,n} = \frac{\pi^2\hbar^2}{2m}\left(\frac{l^2}{L_x^2} + \frac{m^2}{L_y^2} + \frac{n^2}{L_z^2}\right) \quad \text{(quantum dots)}, \tag{3.19c}$$

where, in the brackets, $l, m, n = 1, 2, 3\ldots$ are quantum numbers of the energy levels due to carrier confinement in the x, y, and z directions, respectively, $k_{x,y,z}$ are the wave vectors, and $\hbar^2 k_{x,y,z}^2/2m$ represents kinetic energies in the direction of unconfined dimensions. As discussed earlier, the bulk semiconductor can be regarded as an object in which carriers move freely in all three dimensions; the kinetic energy is thus given by

$$E = \frac{\hbar^2\left(k_x^2 + k_y^2 + k_z^2\right)}{2m} \quad \text{(bulk)}. \tag{3.19d}$$

As seen from the above equations, subbands exist in quantum wells or quantum wires in addition to the quantization energies, because the carriers are free to move in two or one dimensions. In a quantum dots, the carriers have no freedom of movement and only discrete (atomiclike) states are formed.

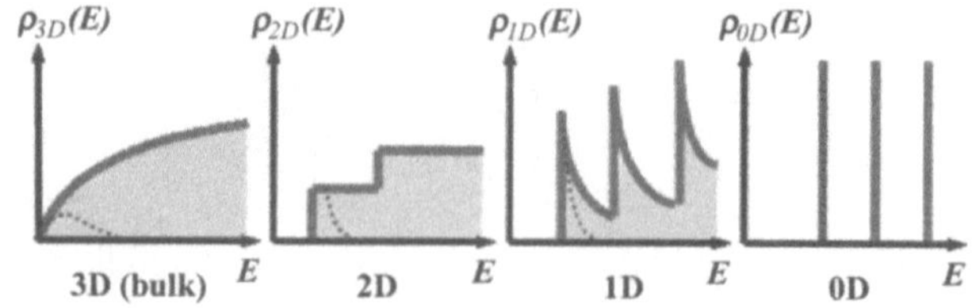

FIGURE 3.7. Schematic illustration of the density of states for bulk (3-D), quantum well (2-D), quantum wire (1-D), and quantum dot (0-D) nanostructures. The contour characterized by a dashed curve in each case represents the occupied states at similar carrier densities.

3.2.2. Higher Material Gain

Gain enhancement in semiconductor materials due to quantum confinement in one, two, and three directions has been investigated theoretically since 1982 [14]. Resulting from the narrowing of the carrier density of states, the laser gain profile of nanoscale semiconductor media was predicted to be concentrated into a much narrower spectral region than the corresponding bulk materials.

3.2.2.1. Density of States

One principal advantage of nanoscale semiconductor materials for laser applications originates from a noticeable increase of the density of states $\rho(E)$ for electrons and holes with the reduction of dimensions (Fig. 3.7). As demonstrated earlier, the density of states is a material property that quantifies the number of carriers that are permitted to occupy a given energy state of the semiconductor. Because most semiconductor lasers operate near the conduction band minimum, the density of states of a semiconductor material imposes an intrinsic limitation on the number of carriers that are allowed to contribute to the lasing performance at one time. The expressions of the density of states for electrons or holes in bulk semiconductors and in different nanostructures are given by [23]

$$\rho_{3D}(E) = \frac{(2m/\hbar^2)^{3/2}}{2\pi^2}\sqrt{E} \quad \text{(3-D)}, \tag{3.20a}$$

$$\rho_{2D}(E) = \frac{m}{\pi\hbar^2 L_x}\sum_l \Theta(E - E_l) \quad \text{(2-D)}, \tag{3.20b}$$

$$\rho_{1D}(E) = \frac{(2m)^{1/2}}{\pi\hbar L_x L_y}\sum_{l,m}(E - E_{l,m})^{-1/2} \quad \text{(1-D)}, \tag{3.20c}$$

$$\rho_{0D}(E) = \frac{2}{L_x L_y L_z}\sum_{l,m,n}\delta(E - E_{l,m,n}) \quad \text{(0-D)}, \tag{3.20d}$$

where Θ is the step function and δ is the delta function.

For 2-D quantum wells, the carriers are free to move in the plane of the film, which gives the total density of states a characteristic staircase shape, replacing the parabolic shape of the bulk density of states. In the 1-D quantum wire case, the carriers are confined also in a second direction and the carriers are free to move only along the wire axis. As a consequence of 1-D movement, the density of states has a $1/E^{1/2}$ dependence for each of the discrete pairs of states in the confined directions. Finally, if the movement of the carriers is limited in all three directions, one has a 0-D quantum dot for which the states are quantized in all directions and the density of states is a series of discrete, sharp states resembling that of an atom.

3.2.2.2. Material Gain

Considering the three possible routes of radiative band-to-band transitions occurred in a semiconductor as shown schematically in Fig. 3.3, the first process is spontaneous emission where a recombination of an electron–hole pair leads to the emission of a photon, random in direction, phase, and time. The second process is (stimulated) absorption; an electron–hole pair is generated as the result of the absorption of an incoming photon. The third process is stimulated emission; a recombination of an electron–hole pair is stimulated by a photon, with a second photon generated simultaneously, which has the same direction and phase as the first photon.

A net generation of coherent photons (positive gain) occurs if the rate of stimulated emission exceeds the rate of absorption. The optical gain of the material—the material gain, γ—can be defined as the proportional growth of the photon density (fractional increase of light intensity I_ω per unit length) as the light propagates along certain direction (e.g., z direction) inside the medium (Fig. 3.8):

$$\gamma(\omega) = \frac{1}{I_\omega}\frac{dI_\omega}{dz}. \quad (3.21)$$

For semiconductors with effective population inversion due to electrons and holes, the gain function for optical transition from E_2 in the conduction band to E_1 in the valence band can be written as [24, 25]

$$\gamma(\omega) = A_{21}\frac{\lambda^2}{8\pi n^2}\rho_{jnt}(\hbar\omega)\left[f_C(E_2) - f_V(E_1)\right], \quad (3.22)$$

where $\hbar\omega$ is the photon energy, λ is the wavelength, n is the refractive index, A_{21} is the spontaneous emission coefficient, f_C and f_V are the quasi-Fermi functions for the conduction and valance bands, respectively, and $\rho_{jnt}(\hbar\omega)$ is the joint density of states (defined using a reduce mass for electron–hole pairs) evaluated at $E = \hbar\omega$.

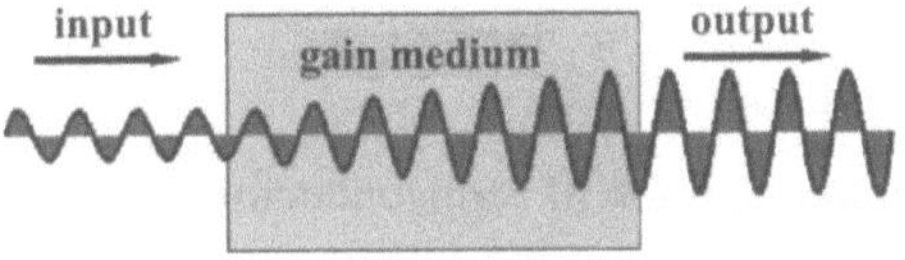

FIGURE 3.8. Optical gain (light amplification) in an active semiconductor medium.

In agreement with the earlier discussion, the expression of the material gain in semiconductors suggests that an incoming light wave of photon energy $\hbar\omega$ is amplified (positive γ) if the condition $f_C(E_2) > f_V(E_1)$ or, equivalently [12],

$$E_{FC} - E_{FV} > \hbar\omega > E_g \tag{3.23}$$

is satisfied. The gain material becomes transparent ($\gamma = 0$) when the separation in the quasi-Fermi levels (E_{FC} and E_{FV}) equals the energy gap between the highest and lowest available states for the electrons and holes. The carrier density required to provide this separation is known as the transparency carrier density, N_{tr}. Optical gain of the material is maintained when pumping the semiconductor creates a carrier density (N) larger than N_{tr}. The material gain can be approximated (near transparency) as

$$\gamma = \frac{\partial \gamma}{\partial N}(N - N_{tr}) = \gamma'(N - N_{tr}), \tag{3.24}$$

where γ' is the differential gain defined as the rate of gain increase as carriers are injected into the active medium. Notice that N_{tr} is not too different for nanostructures because it is determined by the relative locations of the quasi-Fermi levels at transparency, which are typically not too different between bulk and nanostructures.

The expression of the material gain (Eq. 3.22) indicates that γ is proportional to the products of the joint density of states of the semiconductor and the difference in the quasi-Fermi functions of the conduction and valence bands. The density of states has a more peaked structure with the decrease of the dimensionality, particularly for 0-D and 1-D nanostructures (Fig. 3.7). The major effect of the narrowing of the density of states with reducing dimensionality is to confine the carrier energy distribution to narrower spectral regions (Fig. 3.7). As a consequence, increasing quantum confinement yields narrowing of the gain spectrum (Fig. 3.9), thus higher optical gain at a given carrier density. The differential gain γ' increases correspondingly, due to the narrowing of the gain spectrum.

3.2.3. Lower Lasing Threshold

Threshold current density J_{th} is an important characteristic of a semiconductor laser device, at which gain overcomes overall (internal and external) losses, thus enabling lasing. Lasing threshold can be reduced by making the active volume smaller and by modifying the density of states for the carriers, for example, through the use of a quantum well in the active region containing a single layer of quantum dots [26–28].

3.2.3.1. Threshold Condition

To derive the threshold condition of semiconductor lasers, one needs to consider the total photon losses inside the optical cavity in addition to the optical gain of the active medium. A basic semiconductor laser structure consists of an active

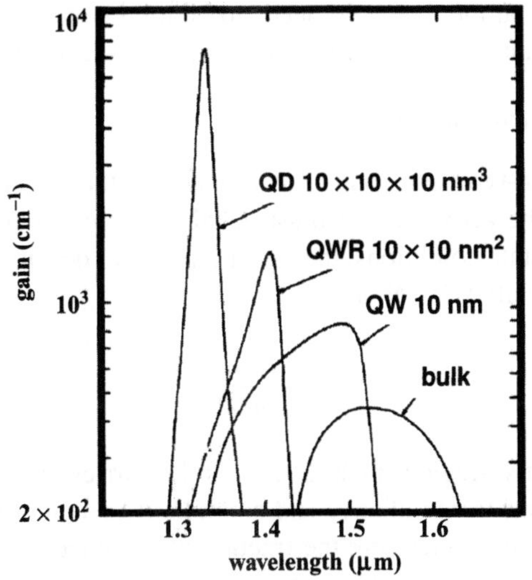

FIGURE 3.9. Calculated room-temperature gain spectra for GaInAs/InP quantum dot (QD), quantum wire (QWR), quantum well (QW), and bulk semiconductor active media. After Ref. 15.

gain medium enclosed inside an optical cavity bounded by two reflecting mirror planes (Fig. 3.1). The minimum material gain, γ_{th}, at which the device starts lasing operation can be estimated from the unity round-trip condition when the gain provided to the optical mode compensates the intrinsic absorption and the mirror losses for a round-trip:

$$R_1 R_2 \exp\{2(\Gamma\gamma_{th} - \alpha_i)L\} = 1, \tag{3.25}$$

where R_1 and R_2 are the mirror reflectivity, L is the laser cavity length, α_i is the internal loss, and Γ is the confinement factor characterizing the overlap of the lasing mode with the active region cross section. Since only a part of the intensity profile of the optical mode overlaps with the semiconductor active region, one has to distinguish between the gain of the active material γ and the gain (lower) of the optical mode $\Gamma\gamma$ reflecting light propagation in the entire device. Solving Eq. (3.25) for the threshold gain γ_{th},

$$\Gamma\gamma_{\text{th}} = \alpha_i + \frac{1}{L}\ln\left(\frac{1}{\sqrt{R_1R_2}}\right). \tag{3.26}$$

At the lasing threshold, the optical gain will compensate the sum of the intrinsic absorption and the mirror losses, which depend on the cavity length and the mirror reflectivity. Combining Eq. (3.26) and Eq. (3.24), the threshold carrier density N_{th} can be written as

$$N_{\text{th}} = N_{\text{tr}} + \frac{1}{\Gamma\gamma'}\left[\alpha_i + \frac{1}{L}\ln\left(\frac{1}{\sqrt{R_1R_2}}\right)\right]. \tag{3.27}$$

The carrier density at the threshold needs to be high enough to achieve material transparency ($N_{\rm tr}$) and overcome cavity losses simultaneously.

3.2.3.2. Threshold Current Density

For carrier-injection-type semiconductor lasers, since the rate at which the carriers are injected into the active region must equal the electron–hole recombination rate under steady-state conditions, the injection current density J is related to the injected carrier density [29], N, as

$$J = \frac{eNL_z}{\eta\tau_R}, \tag{3.28}$$

where L_z is the active-region thickness, τ_R is the carrier recombination time including contributions from radiative and nonradiative recombination, and η is the quantum efficiency characterizing the fraction of the injected carriers arriving in the active region. Using Eq. (3.27) for the expression of $N_{\rm th}$, the threshold current density can be written as

$$J_{\rm th} = \frac{eL_z}{\eta\tau_R}\left\{N_{\rm tr} + \frac{1}{\Gamma\gamma'}\left[\alpha_i + \frac{1}{L}\ln\left(\frac{1}{\sqrt{R_1R_2}}\right)\right]\right\}; \tag{3.29}$$

that is,

$$J_{\rm th} = J_{\rm th}^{\rm mat} + J_{\rm th}^{\rm cav} = \left\{\frac{eN_{\rm tr}L_z}{\eta\tau_R}\right\}_{\rm mat} + \left\{\frac{eL_z}{\eta\tau_R\Gamma\gamma'}\left[\alpha_i + \frac{1}{L}\ln\left(\frac{1}{\sqrt{R_1R_2}}\right)\right]\right\}_{\rm cav}. \tag{3.30}$$

The current necessary for lasing consists of two parts: $J_{\rm th}^{\rm mat}$, the current needed to maintain the carrier density at transparency in the semiconductor gain medium, and $J_{\rm th}^{\rm cav}$, the current needed to attain necessary gain to overcome losses inside the laser cavity. $J_{\rm th}^{\rm mat}$ represents a fundamental limit to achieving the lowest lasing threshold for semiconductor lasers, which is obtained with vanishing cavity losses.

In fact, part of the reduction of threshold current has its origin from a simple geometrical scaling of the physical size of the active region without invoking quantum confinement. As seen from Eq. (3.28), since the threshold carrier density does not change significantly between bulk and nanostructures, it is essential to reduce the thickness of the active region L_z in order to decrease the threshold current density.

The advantage of using nanoscale quantum structures for reducing threshold current density lies not only in their small active volume but also in the larger differential gainγ' resulted from the confinement of the carrier energy distribution [30–32], as seen from the expression for $J_{\rm th}^{\rm cav}$. Further reduction of the threshold current density can be achieved by using tight optical confinement (large Γ) as well as reducing cavity losses. It is informative to plot the evolution trend [33] of the minimum threshold current densities achieved since the invention of semiconductor laser in the early 1960s, as illustrated in Fig. 3.10.

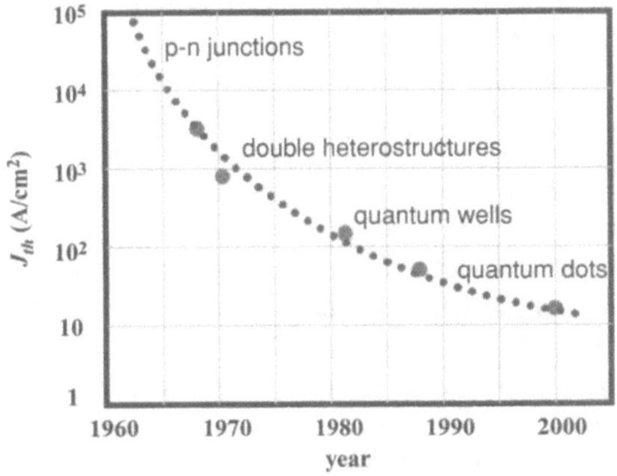

FIGURE 3.10. The evolution trend of the reduction of the semiconductor laser threshold.

The reduction of the threshold current density, or threshold current ($I_{\text{th}} \propto J_{\text{th}}$), can yield many improvements in the performance of semiconductor lasers. A need in a variety of laser applications is high-power conversion efficiencies, for instance. Because their threshold current can be very small, a semiconductor laser based on nanoscale semiconductor gain media is expected to have a large above-threshold power conversion efficiency, η_p, which reflects the efficiency of a semiconductor laser in converting electrical power input to optical power:

$$\eta_p = \frac{P_0}{V_b I} = \eta_d \frac{E_g}{V_b}\left(1 - \frac{I_{\text{th}}}{I}\right), \tag{3.31}$$

where V_b is the input bias voltage and η_d is the differential quantum efficiency. Neglecting the spontaneous component of optical emission, $P_0 = \eta_d E_g(I - I_{\text{th}})$, denoting the above threshold optical power output. Additionally, due to a reduced threshold current, the modulation bandwidth of semiconductor lasers with nanoscale gain media could be very large. The modulation bandwidth of a semiconductor laser can be written as [29]

$$f_{3\,\text{dB}} = 1.55 f_r = \frac{1.55}{2\pi \tau_R}\sqrt{\frac{\tau_R}{\tau_p}\left(\frac{I}{I_{\text{th}}} - 1\right)}, \tag{3.32}$$

where f_r denotes the relaxation frequency and τ_p is the photon lifetime. Clearly, high-speed modulation, essential for communication applications of semiconductor lasers that are transmitting the vast flux of information over optical fibers, is possible with a small I_{th} resulting from quantum confinement in nanoscale semiconductors.

The use of semiconductor quantum wells as the optical gain media represents the first-generation semiconductor laser technology implementing nanoscale materials. A quantum well is a thin semiconductor layer (a few nanometers thick in the z direction) sandwiched between layers of larger band-gap barriers. The band-gap difference causes electrons and holes to be trapped in the quantum well; the resulting confinement in the z direction leads to discrete energy levels greater than that of a bulk material layer. Quantum confinement in quantum wells results in a steplike density of electronic states that is nonzero at the band edge, enabling a higher concentration of carriers to contribute to the band-edge emission and leading to a reduced lasing threshold and improved characteristics. A further enhancement in the density of states at the band edge and the associated reduction in the lasing threshold can be achieved with quantum wires and quantum dots where quantum confinement occurs in two and three separate directions, respectively. Semiconductor lasers with quantum wells [34] are relatively mature and many of their characteristics have been realized in commercial applications. In contrast, quantum wire and quantum dot structures are still in their development stage. The focus of this chapter is on semiconductor lasers based on 1-D nanoscale quantum wires and 0-D nanoscale quantum dots. The progress of quantum wire [35] and quantum dot [33, 36–38] lasers has been discussed in several recent reviews.

3.3. SEMICONDUCTOR LASER CAVITY STRUCTURES: PHOTON CONFINEMENT

In addition to the requirement of an active semiconductor gain medium, an optical resonant cavity that confines the photons to create positive optical feedback is the other basic element necessary for realizing a semiconductor laser. The Fabry–Perot cavity (Fig. 3.1) is the most commonly adopted cavity structure for semiconductor as well as other types of laser. The cavity acts to determine a preferred spatial direction for stimulated emission and introduce a frequency-selective mechanism that defines the longitudinal cavity modes. Standing waves in a Fabry–Perot cavity are only possible for a discrete number of frequencies that define the longitudinal cavity modes. The frequency or wavelength mode spacing, or the free spectral range ($\Delta\nu$ or $\Delta\lambda$) of a laser cavity with length L, is

$$\Delta\nu = \frac{c}{2nL}, \tag{3.33a}$$

$$\Delta\lambda = \frac{\lambda_0^2}{2nL}, \tag{3.33b}$$

where λ_0 is the light wavelength and $\Delta\nu$ and $\Delta\lambda$ are the frequency and wavelength separation of the cavity modes, respectively. For an effective spectral width $\Delta\nu_g$ of a semiconductor gain medium, the ratio $\Delta\nu_g/\Delta\nu$ provides an estimate of the number of longitudinal modes that may become active. Figure 3.11 is a schematic illustration of the lasing spectrum as compared to the material gain spectra and position of cavity modes.

According their geometric features for photon confinement, one can distinguish semiconductor lasers with four basic types: edge-emitting laser, vertical

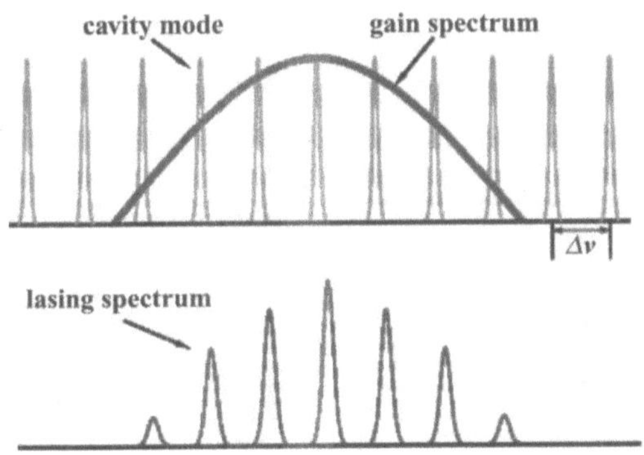

FIGURE 3.11. Schematic illustration of material gain spectrum and cavity modes (upper plot) and lasing spectrum (lower plot).

cavity surface-emitting laser, microdisk laser, and nanowire nanolaser, as schematically illustrated in Fig. 3.12. Edge-emitting, vertical cavity, and nanowire lasers all have a Fabry–Perot-type cavity structure.

3.3.1. Edge-Emitting Laser Cavity

Conventional edge-emitting semiconductor lasers have an in-plane optical cavity with a cleaved facet mirror (Fig. 3.12a). Light emission takes place from one side (edge) of the Fabry–Perot-type cavity typically a few hundred micrometers long (longitudinal) and 10 μm wide (lateral). Due to the relatively long optical cavity length, the longitudinal modes are very closely spaced, with frequency spacing of ~100 GHz; thus, a large number of longitudinal modes can participate in the laser action.

A proper design of an edge-emitting laser structure is essential to achieve low threshold currents. There are several disadvantages of the basic *p–n*-junction-type lasers (Fig. 3.4). Since the injected carriers are free to diffuse, the spatial distribution of recombination is diluted. There is also little guiding and confinement of the electromagnetic wave being amplified in the laser cavity. Double heterostructure,

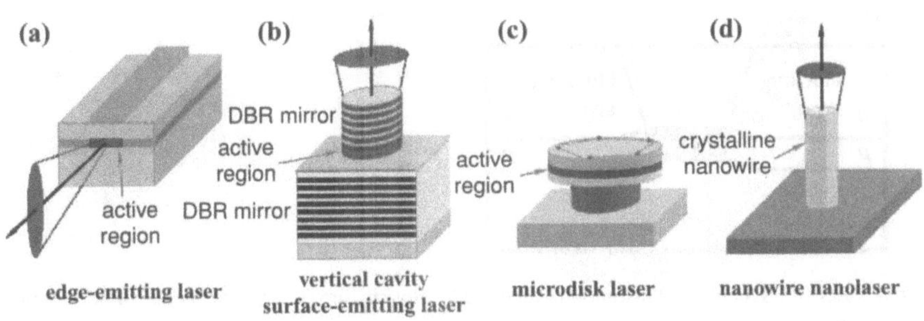

FIGURE 3.12. Four basic types of semiconductor laser cavity.

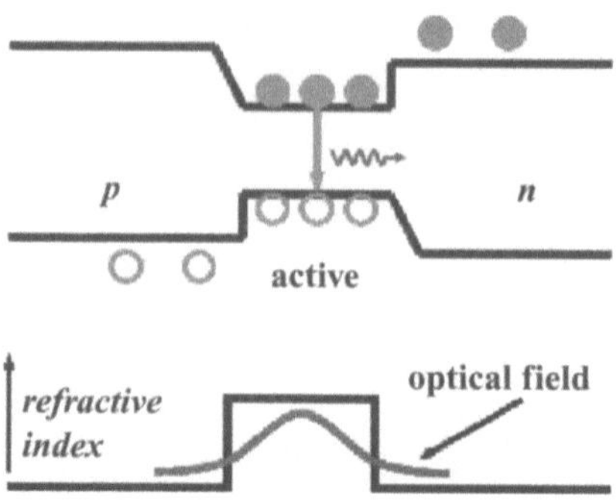

FIGURE 3.13. Carrier confinement (upper plot) and optical confinement (lower plot) of a double-heterojunction edge-emitting laser.

as proposed first in 1963 [6, 7], can produce efficient semiconductor lasers by offering both carrier and photon confinement.

Typically, a double-heterojunction edge-emitting laser involves an undoped material known as the active layer, sandwiched between p- and n-type layers (can be the same material system). As shown in Fig. 3.13, the active layer is usually a few hundred nanometers thick, having a lower band-gap energy but a higher refractive index compared to those of the p and n materials. The undoped active layer with higher refractive index provides optical confinement by forming a waveguide along its length. Additionally, the fact that the active layer has a lower energy gap than the surrounding layers enables effective carrier confinement by offering a potential well that makes it difficult for the majority carriers to diffuse out of the active region. Therefore, by confining simultaneously both majority carriers and photons in the active region, the double-heterojunction structure is able to facilitate more effective stimulated emission than p–n-junction-type lasers,

Advanced designs may further improve the performance of semiconductor lasers. The buried heterostructure (Fig. 3.14a) is a structure that offers better carrier and optical confinement [39–42]. The active area is reduced to a stripe that results in a lower threshold current. Burying the stripe inside lower refractive index materials

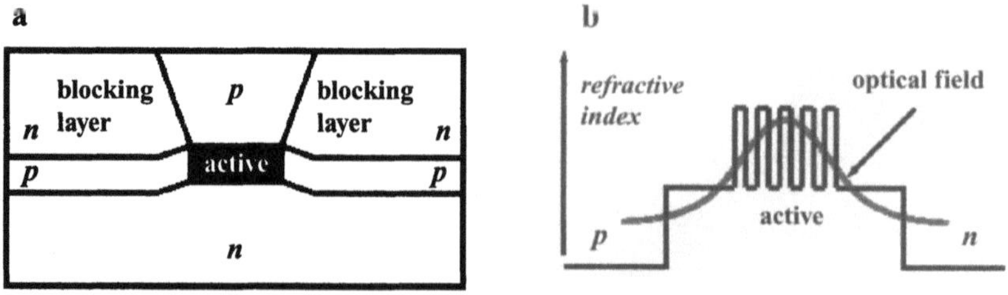

FIGURE 3.14. Schematic illustration of a buried heterostructure (a) and a separate confinement heterostructure (b).

enhances optical confinement. Further improvement of carrier confinement can be achieved by adding blocking layers configured to form a *p-n* junction the opposite way around to the main double heterojunction. Consequently, electric current passing through the junction will act as a reverse bias, thus prevents majority carriers crossing the junction.

A separate confinement heterostructure (Fig. 3.14b) can be used in conjunction with a buried heterostructure design to improve optical confinement, in particular for quantum well lasers [43–46]. Optical confinement factor of a laser device represents the fraction of light that is confined in the active region. A typical bulk double-heterojunction laser has a confinement factor on the order of 0.1. For a single quantum well of 10 nm width, the confinement factor is reduced significantly to approximately 0.001. The use of a separate confinement heterostructure, which usually consists of a material added at either side of the quantum wells and has a refractive index in between that of the wells and that of the surrounding material, can significantly improve the optical confinement in quantum well lasers by a factor of 10.

Although double-heterojunction edge-emitting lasers have been used for many practical applications, the combination of edge emission and active layer geometry has several disadvantages. The long optical cavity results in typically multimode operation, by which the spontaneous emission of the active gain material is distributed over a large number of nonlasing modes. Because the active region is transversely thin (versus. laterally wide) for carrier and photon confinement, edge-emitting lasers have highly elongated near and far fields. Additionally, edge-emitting semiconductor lasers are limited from being made smaller because they cannot be cleaved much shorter than 100 μm on a manufacturing basis.

3.3.2. Vertical Cavity Surface-Emitting Laser and Photonic Crystal Laser Cavity

The vertical cavity surface emitting laser [47] (VCSEL), originally demonstrated in 1979 [48–50], was one of the first semiconductor laser cavities with dimensions on the order of the wavelength of light. As shown in Fig. 3.12b, VCSEL structures, including mirrors and active regions, are defined by layered structures, unlike edge-emitting laser cavities where mirrors are implemented by cleaved facets (or by etched gratings). The very short cavity length makes VCSEL operation inherently single longitudinal mode, in contrast to edge-emitting lasers with typical cavity length several hundred times the lasing wavelength. Consequently, VCSELs exhibit a short gain region along with a large longitudinal mode spacing. The significant reduction in the gain length in VCSELs is compensated by a high-quality factor cavity; the quality factor represents the ratio of the amount of energy stored in an optical cavity in the form of a standing wave to the amount of energy lost during a round-trip inside the cavity. The development of distributed

Bragg reflectors (DBR) with reflectivity close to 100% [51] allows the active region to contain as few as a single quantum well [52].

The DBR is a mirror based on the constructive interference of light reflected from index steps in a layer stack of alternating low- and high-index materials [53]. High reflectivity is obtained by alternating quarter wavelength layers of high and low refractive index materials, so the multiple reflections add in phase to provide a high total reflectivity over a particular stop band. The resonant (Bragg) wavelength, λ_B, at which maximum reflectance occurs complies with the condition

$$\frac{\lambda_B}{4} = n_L d_L = n_H d_H, \tag{3.34}$$

where n_H and n_L are refractive indices of the high- index and low-index media and d_H and d_L are the physical thickness of the high- and low- refractive-index layer, respectively. The peak reflectivity of a DBR (Fig. 3.15) in the absence of optical loss, R_0, is given by [54]

$$R_0 = \left\{\left[1 - \frac{n_i}{n_e}\left(\frac{n_L}{n_H}\right)^m\right]\left[1 + \frac{n_i}{n_e}\left(\frac{n_L}{n_H}\right)^m\right]^{-1}\right\}^2, \tag{3.35}$$

where m is the number of mirror layers (not periods), n_i, n_e, n_H, and n_L are the refractive indices of the incident, exit, and high-index and low-index media, respectively, as illustrated in Fig. 3.15.

In addition to peak reflectivity, optical bandwidth, $\Delta\lambda$, is another important parameter of DBR. A large bandwidth allows for relaxed material growth tolerance. Also, the stop band must be sufficiently wide if multiple wavelength or tunable operation is desired. The relative bandwidth of a DBR mirror is related to the refractive indices by

$$\frac{\Delta\lambda}{\lambda_0} = \frac{4}{\pi}\sin^{-1}\left(\frac{1 - n_L/n_H}{1 + n_L/n_H}\right), \tag{3.36}$$

where λ_0 is the center wavelength of the DBR. $\Delta\lambda$ is the full width at half-maximum (FWHM) of the DBR stop band. Figure 3.16 shows an illustration of the reflectivity as a function of wavelength for a DBR.

FIGURE 3.15. Schematic illustration of refractive-index profile of a DBR structure ($m = 7$).

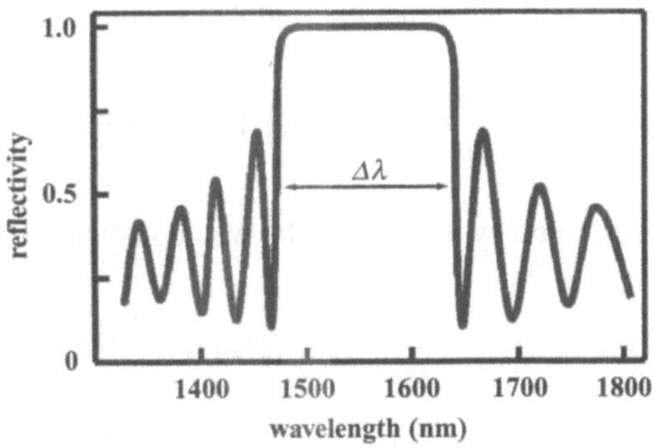

FIGURE 3.16. Reflectivity spectrum of a DBR. The center wavelength of this DBR is about 1550 nm.

With the laser emission from the wafer surface, it is possible to have a symmetric beam profile with a very small divergence. Surface emission also favors the fabrication of 2-D laser arrays of coherent light sources. Additionally, the epitaxial growth techniques for the VCSEL mirrors make wafer-scale device testing possible, enabling high-volume manufacturing. The development of VCSELs has resulted in devices that have characteristics and reliability suitable for various commercial applications. Low threshold current and high operating efficiency have been achieved by constructing smaller active areas (e.g., $\sim \mu m^2$) [55], band-gap engineering of the mirror stacks [56–58], and introduction of confinement apertures to reduce optical losses [59].

The DBR mirrors of VCSEL cavities are essentially 1-D photonic crystals. Alternatively, 2-d and 3-D photonic crystals have been applied to provide effective control of light on the microscale [60–62]. In general, a photonic crystal consists of a dielectric medium that is patterned into a regular structure with a length scale on the order of the optical wavelength. For a proper crystal design, light is Bragg scattered by the periodic structure and cannot propagate in any direction within the crystal.

In analogy to the electronic band gap in a semiconductor material, the range of optical frequencies that are excluded from the material is called the photonic band gap. It was not until 1987 that Yablonovitch suggested the existence of a complete photonic band gap. Electromagnetic waves in dielectric structures exhibiting 3-D periodicity will undergo multiple scattering if their wavelength is comparable to the lattice spacing. For some structures, due to coherent interference, there will be no propagating modes in any direction for a range of frequencies, giving rise to a photonic band gap. For the band-gap frequencies, all states are completely suppressed, resulting in modified radiative dynamics inside the structure.

There are many similarities between electromagnetic waves in a periodic dielectric structure and electron waves in the periodic ionic potential of a crystal.

An electron is described by the scalar Schrodinger equation,

$$\hat{H}_{\text{electron}}\Psi(\boldsymbol{r}) = \left(\frac{-\hbar^2}{2m}\nabla^2 + V(\boldsymbol{r})\right)\Psi(\boldsymbol{r}) = E\Psi(\boldsymbol{r}), \tag{3.37}$$

and the photon is described by the vector Maxwell equations [63],

$$\hat{H}_{\text{photon}}\vec{H}(\boldsymbol{r}) = \left(\nabla \times \frac{1}{\varepsilon(\boldsymbol{r})}\nabla\times\right)\vec{H}(\boldsymbol{r}) = \frac{\omega^2}{c^2}\vec{H}(\boldsymbol{r}). \tag{3.38}$$

With $\boldsymbol{R}$ representing the periodic lattice vectors, the periodicity is reflected in the potential $V(\boldsymbol{r}+\boldsymbol{R}) = V(\boldsymbol{R})$ for the electron and in the dielectric function $\varepsilon(\boldsymbol{r}+\boldsymbol{R}) = \varepsilon(\boldsymbol{R})$ for the photon. In the electron equation (3.37), E and Ψ are the electron energy eigenvalue and eigenfunction respectively; and in the photon equation (3.38), ω^2/c^2 and $\boldsymbol{H}$ are the photon frequency eigenvalue and the eigenmode respectively. Since both operators in (3.37) and (3.38) are linear and Hermitian [64], concepts of electrons in real crystals such as Brillouin zones and dispersion relations can be applied to the photon case. If the energies of the electrons in a crystal have energy bands separated by gaps, the same would be true for photons in a photonic crystal with corresponding periodic structures.

Like a semiconductor material, a defect in a photonic crystal will lead to a localized photon state within the band gap, which can serve as an optical cavity with a small mode volume. Notably, photonic crystal defect laser cavities with various designs and sizes have been realized [65–70]. By introducing a single-defect photonic crystal, a waveguide forms around the central defect region where the effective index is larger than the surrounding region. A thin semiconductor membrane suspended in air and perforated with air holes primarily has been used to provide optical confinement. Such 2-D slab photonic band-gap defect-mode lasers are easier to fabricate and more amenable to integration, compared to truly 3-D defect mode lasers. Figure 3.17 is a schematic illustration of a 2-D photonic crystal defect structure.

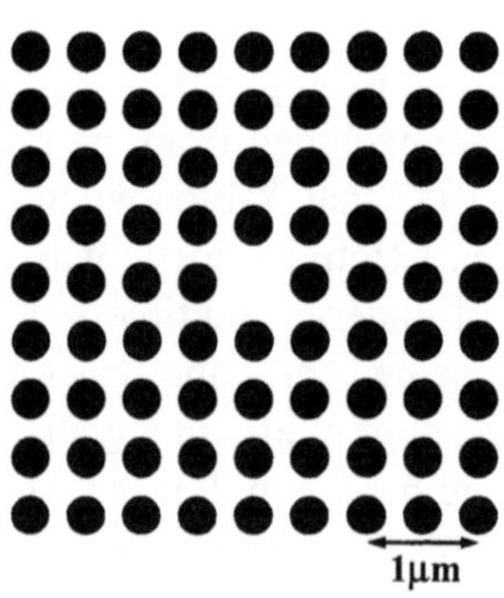

FIGURE 3.17. A schematic illustration of a 2-D photonic crystal defect structure.

3.3.3. Microdisk Laser Cavity

Different from the Fabry–Perot-type cavities, a microdisk laser [71, 72] uses total internal reflection at the edge of a high-refractive-index disk (Fig. 3.12c) to form low-loss whispering-gallery-type modes, in analogy to long-distance travel of whispers (sound wave) along a curved inner surface. The laser operates by confining the light through total internal reflection, reflected repeatedly from the boundary with the same angle of incidence.

By solving for the optical field in the 2-D transverse direction, for an ideal microdisk of radius r, the whispering-mode optical field takes the form [73, 74] $A_M J_M(rn_{\mathrm{eff}}\omega_{\mathrm{MN}}/c)e^{iM\theta}$, where J_M is the Bessel function of integer (M) order, A_M is a normalization constant, c is the speed of light in vacuum, and $n_{\mathrm{eff}}(\omega)$ is the (2-D) effective refractive index. The resonant frequencies may be calculated as $\omega_{M,N} = x_{M,N}c/n_{\mathrm{eff}}r$, where $x_{M,N}$ is the Nth zero of $J_M(x) = 0$, with $N = 1$ representing whispering-gallery modes.

As compared to Fabry–Perot laser cavities, microdisk cavities exhibit a higher (10^4) quality factor, leading to enhancement of the spontaneous emission [75]. However, it is usually difficult to couple the light into useful directions since the light in microdisk structures tends leak out in all directions due to scattering of surface imperfections.

3.3.4. Nanowire Single-Crystal Laser Cavity

Perhaps the smallest and simplest Fabry–Perot-type optical cavity that can effectively provide positive feedback of stimulated emission belongs to single-crystalline nanowires having a length on the order of light wavelength (Fig. 3.12d). Surface-emitting lasing action has been recently demonstrated in semiconductor nanowires [76, 77] in which two crystalline facets serve as reflecting mirrors of the cavity. The realization of the nanowire nanolasers may open up a new and plausible route for achieving simultaneous carrier and photon confinement in 1-D nanoscale cavities.

3.4. QUANTUM WELL LASERS

The thickness of the active layer in a conventional double-heterojunction laser is of the order of a fraction of 1 μm. When the thickness of the active layer is reduced to the order of the de Broglie wavelength of the carriers (~10 nm or smaller), the quantum effect emerges as carrier motion becomes quantized. Most significantly, the energy states seen for electrons and holes are no longer quasi-continuous, and discrete energy levels appear. The density of states is now the step function representing a 2-D system. Electrons and holes are now stimulated to recombine between discrete quantum well energy states.

3.4.1. *Quantum Well Fabrication Technologies*

Semiconductor lasers based on 2-D quantum well structures represent the first-generation nanoscale semiconductor laser technology. Their success is largely due to the breakthrough of two fabrication techniques: molecular-beam epitaxy (MBE) and metal–organic chemical vapor deposition (MOCVD), which allow for the deposition of semiconductor materials with atomic-layer precision. With nearly monolayer control available in MBE and MOCVD, layers of different materials can be reproducibly grown with dimensions small enough to display quantum effects. These quantum well fabrication techniques have triggered a wide spectrum of experimental and theoretical investigations resulting in not only the observation of a number of intriguing new phenomena due to electronic quantization but also the emergence of a new class of semiconductor laser devices.

There are many different materials that can be grown by MBE and MOCVD to form quantum well structures. AlAs and GaAs have almost identical lattice constants; that is, arbitrary structures can be grown with high quality in this materials system. Another commonly used system is InGaAs with InP; the proportions of In and Ga can be adjusted to yield a lattice constant for the ternary (three-component) InGaAs alloy that is equal to InP. Use of four-component (quaternary) alloys (e.g., InGaAsP) offers sufficient degrees of freedom to adjust both the lattice constant and the band-gap energy of the material. Up to a certain critical thickness, which depends on the degree of lattice constant mismatch, it is possible to grow structures with materials that naturally have different lattice constants. Any material can accommodate a certain amount of elastic strain without generating dislocations or defects. The strained-layer structure will be thermodynamically stable if its thickness is small enough to maintain the elastic strain energy below the energy of dislocation formation. Strained-layer structures can display new electronic and optical properties and represent a tool for band structure modification in a predictable fashion [78].

3.4.1.1. Molecular-Beam Epitaxy

Molecular-beam epitaxy was developed in the early 1970s as a means of growing high-purity epitaxial layers of compound semiconductors [79, 80]. In MBE, the constituent elements of a semiconductor in the form of "molecular beams" are impinged, in ultrahigh-vacuum environment, onto a heated crystalline substrate to form thin epitaxial layers. The molecular beams are typically from thermally evaporated elements heated in effusion cells. Atomic layer-by-layer deposition, with typical growth rates of a few angstroms per second, is achieved by using low beam flux, controlled by varying the source temperature. The atomic mean free path in the beams is usually larger than the distance between the source and the substrate; shutters in front of the individual sources are typically used to control growth time as well as the sequence of which material to grow. For example, with a shutter closed in front of the Al source, but open shutters in front of the Ga and As sources, GaAs layers will be grown. Opening the Al shutter will then grow the alloy AlGaAs, with the relative proportion of Ga and Al controlled by the temperatures of the shutters.

Growth of uniform layers can be obtained by rotating the substrate during the deposition process. To achieve low impurity levels of the deposited layer, very low background pressure (ultra-high vacuum) is typically required, which allows for in-situ monitoring of growth conditions and growth rate by techniques such as reflection high-energy electron diffraction (RHEED). Layer-by-layer epitaxy of lattice-matched materials proceeds within a range of growth rates and substrate temperatures, and the growth involves three basic steps: First, the atoms impinge on the surface and are adsorbed, followed by migration along the surface toward atomic steps where they are stabilized by the increased number of atomic bonds. Eventually the atoms migrate along the step edges to a kink site, where they are incorporated into the lattice. Epitaxy growth may involve the lateral motion of step edges or the growth of 2-D islands until an atomic layer is completed. MBE can produce high-quality layers with precise control of quantum well thickness, doping, and composition.

3.4.1.2. Metal–Organic Chemical Vapor Deposition

Metal–Organic CVD is one variation of CVD techniques that creates thin-film material through the use of chemical reactions [81, 82]. MOCVD is a gas-phase technique usually operated at low pressure. Modified MOCVD technique [83] using the gas sources of MOCVD in a high-vacuum molecular-beam system also exists, which is also known as chemical-beam epitaxy. A combination of organometallic compounds and hydrides (e.g., trimethylgallium and arsine for GaAs and trimethylgallium and ammonia for GaN) is typically used as the sources of elements for quantum wells.

A typical MOCVD deposition process takes place through a series of basic steps that include transport of the reactants to the substrate surface, surface chemical reaction, surface migration and lattice incorporation, and transport of reaction products outside of the deposition zone. MOCVD can be classified into three different growth regimes. The first is the thermodynamically limited regime that is characterized by a reduction of growth rate with increasing temperature. MOCVD is an exothermic process so high temperatures (typically $>800°C$) would limit the growth thermodynamically. The second regime is mass-transport limited at conventional growth temperatures between 600 and 800°C. The growth rate depends primarily on the input partial pressure of the metal-organic reactants, rather than the temperature. In this regime the surface reactions are relatively fast, and the limiting factor is the reagent arrival flux through a mass transport boundary layer. The third regime is characterized by reduced growth rate at lower temperatures ($<600°C$), for applications such as growing highly strained InGaAs/GaAs quantum wells and highly p-doped InGaAs layers. Nevertheless, surface-kinetics controlled growth under reduced temperature may cause incomplete pyrolysis of the precursor sources. Generally speaking, by rapidly changing the vapor phase composition, MOCVD technique enables the growth of quantum well structures with abrupt hetero-boundaries and steep doping profiles.

3.4.2. *Semiconductor Lasers Based on Quantum Wells*

There are several disadvantages in *p*–*n*-junction-type lasers. Since the injected carriers are free to diffuse, the spatial distribution of recombination is diluted. There is also little guiding and confinement of the light wave in the laser cavity. In double-heterostructure semiconductor lasers, charge carriers are concentrated and recombine in a narrow-band-gap layer between two slabs of wider-band-gap material (Fig. 3.13). The heterostructure also acts to confine the photons, which allows the stimulated emission to build up into a laser beam. By 1970, the first room-temperature continuously operating heterostructure (AlGaAs/GaAs) lasers were reported [9, 10]. In the process of double-heterostructure laser development, Alferov et al. [84] developed the idea of using AlGaAs/GaAs as the key heterostructure materials, which have close values of the lattice constants, thus little strain and small density of traps in the interface.

3.4.2.1. Quantum Well Lasers

Quantum well lasers represent the first generation of semiconductor lasers utilizing quantum size effects of a nanoscale dimension (in the direction of the quantum well thickness). In a quantum well structure, a series of energy levels and associated subbands are formed due to quantization of carriers in the direction of quantum well thickness. The carrier confinement and its resulting density of states in quantum wells promise more efficient lasing devices operating at a lower lasing threshold than those with "bulk" double-heterostructure active media. Shortly after the demonstration of quantum effects in the optical spectra of a quantum well heterostructure [85], lasing in quantum well structures was accomplished in 1975 [86], followed by the construction of a quantum well laser with parameters to match those of standard double-heterostructure lasers [87].

Figure 3.18 illustrates the key concept (energy levels) of a quantum well laser structure. Stimulated recombination of electron–hole pairs takes place in the quantum well region, where the confinement of carriers and the optical mode enhance the interaction between carriers and radiation. Quantum wells are usually used in a multiple configuration known as a multiple quantum well (MQW). The combination of several wells allows higher gains and more freedom in the design of devices. The addition of extra wells also improves the optical confinement of the device. There are many benefits in adopting the MQW design; for example, they yield greater gain per injected carrier, which results in lower threshold currents than for bulk devices. Also, by adjusting the width of the quantum wells, the exact

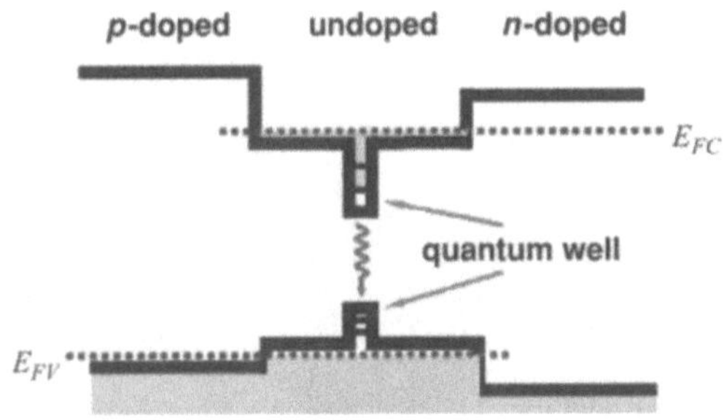

FIGURE 3.18. Energy levels of a quantum well laser with a single well.

heights of the energy bands can be tailored to give the adjustable band gap for the required emission wavelength.

A turning point in in the development of quantum well lasers is the demonstration of significant reduction in lasing threshold by incorporating a quantum well structure into the active region of a VCSEL structure [47, 88, 89]. Further performance improvements were reported by using graded heterojunction interfaces to the monolithic epitaxial distributed Bragg reflectors of the VCSEL to reduce the electrical resistance [90, 91]. Additionally, microdisk lasers with quantum wells as the active media have also been fabricated, and a low lasing threshold was achieved [92–96].

3.4.2.2. Quantum Cascade Lasers

A remarkable use of 2-D quantum well structures is the realization of lasing due to intersubband or interminiband transition, known as quantum cascade lasers [97, 98] originally proposed in 1971 [99]. What primarily distinguishes the quantum cascade lasers from conventional semiconductor lasers [100, 101] is the light generation scheme that allows one to achieve very high power by recycling many times (equal to the number of cascaded stages in a quantum cascade laser) the electrons making the optical transition. The original quantum cascade lasers [97] rely on transitions between quantized conduction band excited states (intersubband transitions) of double quantum wells. The population inversion between the states of laser transition is designed by reducing the final state lifetime using resonant optical phonon emission. Alternatively, the quantum cascade laser can be constructed based on optical transitions (interminiband transitions) between the energy bands of superlattices—periodic stacks of nanometer-thick quantum wells, in which superimposed potentials split the conduction and valence bands into a series of narrow minibands (∼100 meV) separated by energy gaps (minigaps) of comparable value [98].

A portion of the energy diagram of an interminiband cascade laser under operating bias is shown in Fig. 3.19, electrons are injected into the lowest lying

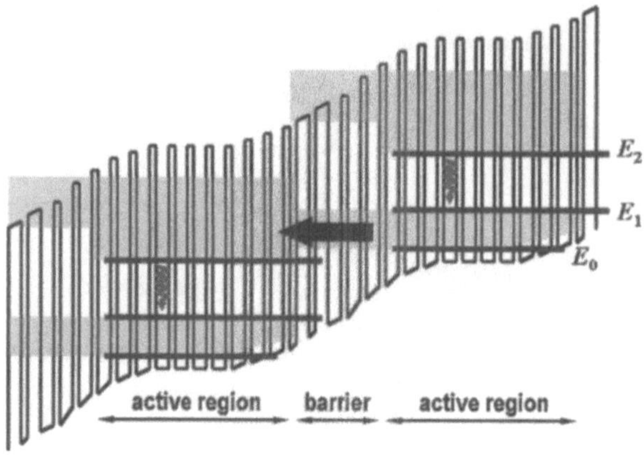

FIGURE 3.19. Schematic illustration of energy levels of a quantum cascade laser.

superlattice states of the upper miniband (E_2), where they emit a photon and relax to the highest lying superlattice states of the lower miniband (E_1). Electrons then quickly escape to the ground state (E_0) of the lower miniband by emitting optical phonons. The very short lifetime for phonon emission compared to the lifetime for interminiband transition makes it possible to maintain a population inversion. Eventually, electrons tunnel into the next active region, and typically 25 or more active regions are "cascaded" to provide enough gain to achieve efficient lasing action. In contrast to conventional quantum well lasers, the wavelength of quantum cascade lasers is essentially determined by quantum confinement (i.e., by the layers' thickness of the active region rather than by the band gap of the material). As such, the wavelength can be tailored over a very wide range using the same active material [100, 101].

3.5. QUANTUM WIRE LASERS

Beyond the accomplishments of the quantum well lasers that use 2-D nanostructures as the active gain media, the capability of growing 1-D quantum wires offers an additional degree of freedom in the design of semiconductor lasers. Both optically and electrically pumped semiconductor lasers have been achieved using nanoscale quantum wires as the active gain media. Since the initial demonstration of lasing from a V-groove quantum wire embedded in a micron-size edge-emitting optical cavity [102], semiconductor lasers based on nanowires that can act as both the gain medium and the optical cavity have been demonstrated [76]. Semiconductor lasers based on 1-D quantum wires fabricated using different growth techniques are discussed in this section, which includes lasing from lithographically defined, self-organized, selectively grown, and chemically synthesized nanoscale quantum wires.

3.5.1. Quantum Wire Fabrication Technologies

Nanoscale semiconductor structures in which carriers are confined in more than one direction (quantum wires or nanowires) have unique properties suitable for high-performance semiconductor laser applications. However, progress in the realization of practical quantum wire (and quantum dot) lasers has lagged behind advances in quantum well laser technology because the fabrication of nanoscale quantum wires with good crystalline characteristics has been a challenging task. Although a universal nanoscale quantum wire fabrication scheme for semiconductor laser applications is not available, optically or electrically pumped lasing has been achieved in nanoscale quantum wires produced by a variety of techniques.

Compared to the fabrication of quantum well structures, the realization of nanoscale quantum wires requires precise growth control in the lateral dimension. The nanoscale quantum wire fabrication techniques can be categorized as (1) nanoscale lithography, (2) self-organization, (3) selective growth, and (4) chemical synthesis.

3.5.1.1. Nanoscale Lithography

Taking the advantage of well-developed quantum well fabrication technologies (MBE and MOCVD), the most straightforward method to realize 1-D nanostructures is etching (and regrowth) through wire-defining masks placed above the quantum wells [103–109]. This nanoscale lithography scheme offers significant flexibility in the design of arrays of quantum wires that can be incorporated in semiconductor laser cavities. However, despite the fact that etching has been a standard process for realization of nanoscale transistor circuits, for optical applications, etching-induced lateral damage can degrade the light emission efficiency. These damages are nonradiative recombination centers that can dramatically reduce the optical gain of the quantum wires and, therefore, suppress the lasing action. Clearly, the primary challenge involved in optimizing nanoscale lithography of quantum wire fabrication is the elimination of damages introduced during the etching and regrowth steps.

Significant improvements of quantum wire structures have been demonstrated recently by careful control of etching and regrowth steps [110–112]. Figure 3.20 is a schematic illustration of quantum wire fabrication process involving two-step

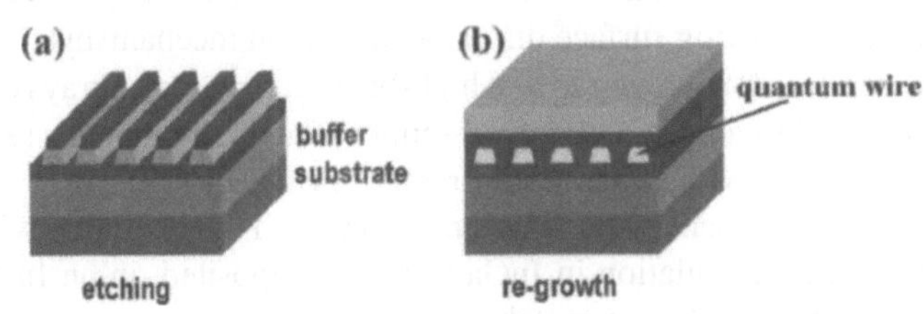

FIGURE 3.20. Quantum wire fabrication based on nanoscale etching and regrowth.

deposition (e.g., MOCVD) [111]. Fabrication starts by first depositing a single or multiple quantum wells on the top of the substrate. A 1-D wire pattern is then formed by etching the quantum well active region through an electron-beam written mask layer (Fig. 3.20a). After removing the mask layer, embedded growth of 1-D quantum wire heterostructures is performed by the second deposition process (Fig. 3.20b). In forming the wire pattern, wet chemical etching can yield the least damage [113]; however, the fabrication of quantum wires is usually limited to only one or two active layers because of their crystal orientation dependence. Dry etching such as the reactive ion-beam etching can be used to fabricate multiple layers of quantum wires, which are advantageous for low-threshold laser operation, followed by wet chemical etching to remove damages induced by the high-kinetic-energy beams of the dry etching step.

Alternative to etching through wire-defining masks, implantation and diffusion under masks have also been applied to realize 1-D nanostructures based on predeposited quantum wells [114, 115]. In these cases, the mask acts to prevent certain ion species from penetrating underneath, therefore generating a composition modulation in the form of nanoscale wires.

3.5.1.2. Self-organization

Different from nanoscale lithography that involves postgrowth wire patterning, *in situ* self-organized growth of 1-D nanostructures has been developed. Self-organized growth methods are promising for semiconductor laser applications since high-density nanometer-size quantum wires can be fabricated without introducing damages caused by etching. In addition, the size of self-organized quantum wires can be controlled simply by adjusting the growth and deposition conditions. Two primary self-organized quantum wire fabrication methods are strain-induced lateral ordering and epitaxial growth on vicinal surfaces.

Strain-Induced Lateral Ordering

One-dimensional quantum wire heterostructures can be self-organized by utilizing the strain-induced lateral ordering (SILO) process developed by Hsieh and co-workers in 1990 [116, 117]. Quantum wire formation by SILO is based on lateral composition modulation [i.e., the spontaneous formation of phase-separated, self-organized periodic structures during the epitaxial growth of a certain type of semiconductor thin films (e.g., short-period superlattices)]. The origin of lateral composition modulation is still under debate, but it is generally agreed that it is a kinetic process involving surface diffusion and strain mechanisms. Using SILO that leads to periodic low-band-gap and high-band-gap regions, arrays of quantum wires have been fabricated in a variety of semiconductor compounds (e.g., InGaP, InGaAs, and AlInAs) with either MBE or MOCVD [118–122].

Figure 3.21 is a schematic diagram showing 1-D quantum wire formation through lateral modulation in InGaAs alloys deposited on an InP substrate [123]. A buffer layer of lattice-matched InGaAs is first grown on the substrate (Fig. 3.21a), followed by the growth of a short-period superlattice (Fig. 3.21b)

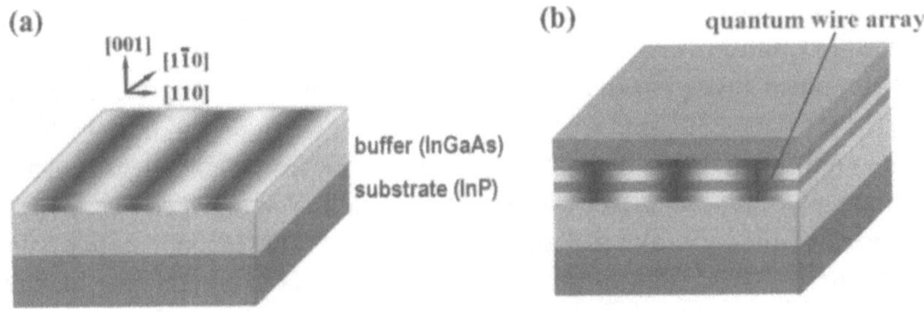

FIGURE 3.21. Formation of 1-D nanoscale quantum wires by SILO.

of $(InAs)_n/(GaAs)_m$ ($n, m \sim 2$), which is then capped with lattice-matched InGaAs. The resulting structure is vertical sheets of alternately Ga-rich and In-rich nanoscale wire structure, laterally modulated in the [110] direction, exhibiting 2-D quantum confinement.

Epitaxial Growth on Vicinal Surfaces

Epitaxial growth on vicinal substrates, as originally proposed by Petroff and co-workers [124], is another self-organization approach to create narrow quantum wire arrays without the use of lithography [125–130]. Quantum wires grown on vicinal surfaces are due to the existence of coherently aligned stepped surfaces and subsequent thickness modulation of the deposition layers at the multiatomic step edges. Figure 3.22 shows a schematic quantum wire array heterostructure fabricated using the vicinal growth technique. A typical vicinal growth process for GaAs quantum wires starts by growing a buffer layers (e.g., AlAs/GaAs superlattice) on a slightly misaligned GaAs substrate (Fig. 3.22a). Under proper deposition conditions, arrays of multiatomic steps form during superlattice growth on a misaligned substrate. The next step is depositing a single quantum well (e.g., AlAs/GaAs/AlGaAs) [127] or a fractional superlattice layer (e.g., $(AlAs)_{1/4}(GaAs)_{3/4}$) [130] to form a 1-D wire array (Fig. 3.22b). A substrate misalignment of 2° can yield lateral periodicity of approximately 8 nm. Since the

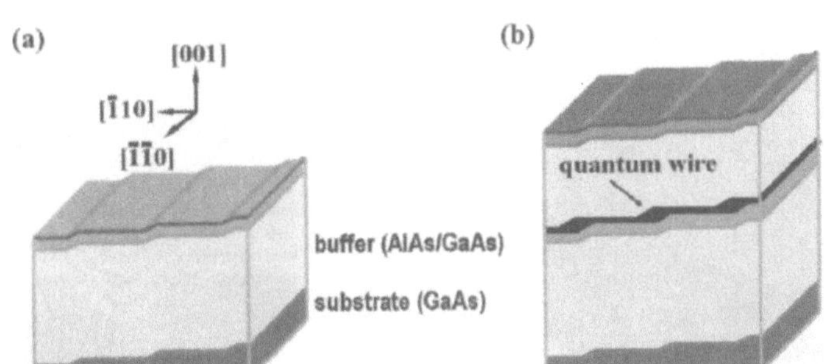

FIGURE 3.22. Growth of quantum wires on a vicinal surface with multiatomic steps.

thickness of the GaAs layer at the step edges is usually thicker than that at the terraces, GaAs quantum wires are naturally formed at these step edges. Deposition interruption during the growth of GaAs layer may help GaAs migrate to the step edges [131].

Although the vicinal growth technique is attractive for yielding dense quantum wire structures of high interface quality, its applications are limited due the strict requirement of preparing vicinal surfaces of multiatomic (or monatomic) steps with extremely high accuracy. Recent reports [128] of GaAs and InGaAs quantum wires grown on naturally corrugated high-index planes of GaAs substrates appear promising. For example, the surface of GaAs (or InGaAs) grown on a (775)B GaAs substrate above a substrate temperature of 640°C (or 620°C) was found to be corrugated; subsequent growth of self-organized vicinal quantum wires has been achieved. Room-temperature lasing has been reported [129] in semiconductor lasers using a quantum wire active media grown on high-index substrate planes.

3.5.1.3. Selective Growth on Prepatterned Substrates

As in the case of quantum wire growth on vicinal surfaces, flat or nonflat patterns generated using lithography can provide templates for direct growth of 1-D nanoscale structures. Epitaxial growth on prepatterned substrates provides more flexibility than growth on vicinal surfaces in the design of quantum wires. Selectively grown quantum wires have been frequently fabricated on prepatterned V-shape grooves and Λ-shape ridges, as well as T-shape cleaved edges of predeposited quantum well structures.

V-Groove Quantum Wires

Selective growth on prepatterned V-groove substrates, as originally developed by Kapon and co-workers [132, 133], has been a popular method to fabricate nanoscale quantum wire active medium beyond the lithographic resolution [102, 134–141]. *In situ* quantum wire growth on V-groove substrates generates minimum defect densities and allows the formation of vertical stacks of identical quantum wires. Figure 2.23 is a schematic illustration of V-groove GaAs quantum wires (Fig. 3.23b) fabricated by growing successive layers of GaAs and AlGaAs on V-grooves. The grooves are formed by wet etching into a (100)-oriented GaAs

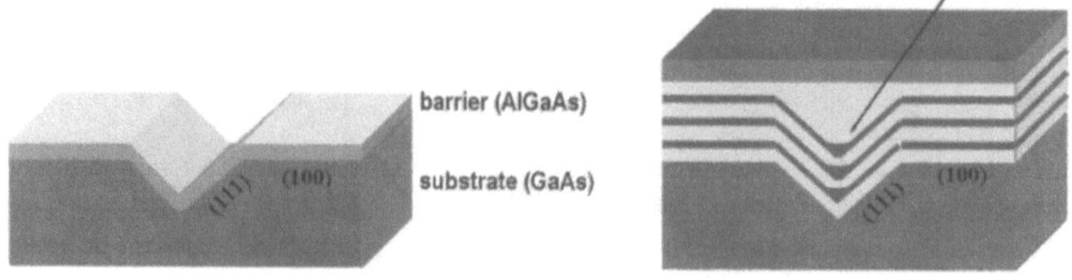

FIGURE 3.23. Selective growth of quantum wires on a prepatterned V-groove substrate.

substrate, a process that exposes (111)A sidewalls (Fig. 3.23a). The underlying mechanism of growing quantum wires on V-grooves is the use of differences in surface migration of reactive species (e.g., group III atoms) on different growth planes. For instance, the surface migration of a Ga atom is typically higher on the (111)A planes than on the (100) planes. This migration difference leads to a higher growth rate of GaAs on (100) planes as compared to (111)A planes, leading to a thin quantum well on the (111)A side walls and a significantly thicker quantum well at the bottom of the V-groove. Accumulation of GaAs occurs in the bottom of the groove and, hence, the formation of a crescent-shaped quantum wire along the direction of the V-groove. Additionally, the surface migration for Al on the (111)A planes is lower than that for Ga atom; therefore, AlGaAs growth is faster on the (111)A planes. The growth of stacked wires is therefore possible by alternating the AlGaAs and GaAs growth.

Λ-Ridge Quantum Wires

Crescent quantum wires can be fabricated not only in the bottoms of V-shape grooves but also on the ridge top by first growing a facet structure with a sharp ridge using lithography [142–145]. For example [143], GaAs quantum wires have been fabricated by growing successive layers a GaAs well on the top of the ridge structure as depicted in Fig. 3.24. The fabrication process starts by etching the substrate to form a mesa stripe, followed by epitaxial growth of GaAs on the top of the mesa stripe to form a narrow ridge with two (111)B adjacent surfaces (Fig. 3.24a). An AlAs barrier layer is deposited on the top of the ridge before a GaAs quantum well is grown to form a 1-D quantum wire structure (Fig. 3.24b).

T-Intersection Quantum Wires

Other than prepatterned nonplanar substrates, 1-D quantum wires can be formed at the T-intersection [146] of two parent quantum wells by cleaved-edge overgrowth, with two deposition (e.g., MBE) steps separated by *in situ* wafer cleaves. At the T-intersection of two quantum wells, quantum confinement of carriers forms a quantum wire. This type of T-shape quantum wires, as first proposed by Chang and co-workers [146], has been a useful nanoscale structure for the study of 1-D excitons [147]. The cleaved-edge overgrowth method, developed around 1990, enables one to combine two thin-film growth processes in two different directions

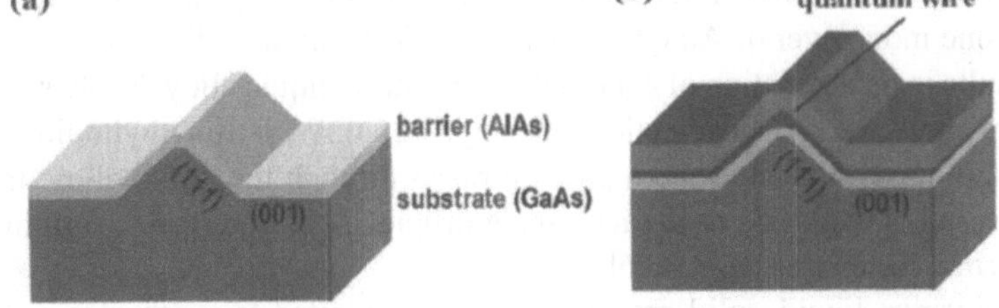

FIGURE 3.24. Selective growth of quantum wires on a prepatterned Λ-ridge substrate.

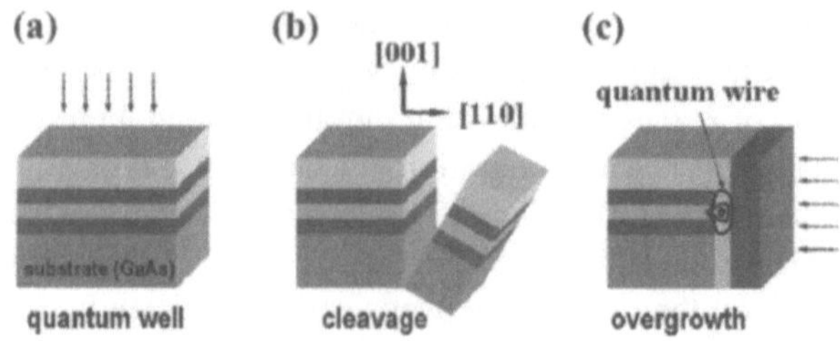

FIGURE 3.25. Formation of 1-D T-intersection quantum wire structure by cleaved-edge overgrowth.

to define quantum wires [148–152]. Lasing action in T-intersection quantum wires by optical pumping [153] and current injection [154] were both demonstrated. Figure. 3.25 illustrates schematic growth steps of a single T-intersection quantum wire structure. For quantum wires grown on a (001) GaAs substrate using MBE, for example, a quantum well sandwiched between two barriers is successively grown on the substrate in the first MBE process (Fig. 3.25a). Subsequently, the wafer was cleaved in the MBE chamber (Fig. 3.25b), and the second quantum well with barriers was grown on the newly exposed (110) edge (Fig. 3.25c).

3.5.1.4. Chemical (Bottom-up) Synthesis

Over the past decade, there has been tremendous progress in chemical synthesis techniques for growing nanoscale semiconductor quantum wires (1-D nanowires). As originally proposed by Wagner and Ellis in 1964 [155] for Au-catalyzed Si whisker growth, a unique gas-phase reaction technique based on the so-called vapor–liquid–solid process has resulted in high-quality lasing nanowires. In a typical vapor–liquid–solid process, 1-D anisotropic crystal growth is promoted by the presence of a liquid alloy–solid interface. A nanoscale catalyst liquid alloy droplet, which defines the diameter of the resulting 1-D nanostructure, serves as the preferential site for reactant adsorption and nucleation when supersaturated. The key feature of this vapor–liquid–solid process is that equilibrium-phase diagrams can be applied to help select catalysts and predict growth conditions. Ideally, the catalyst (e.g., Au) and the desired nanowire material (e.g., InAs) form eutectic alloys, and the growth temperature can be chosen between the eutectic point and the melting point of the nanowire material.

An example of the basic processing steps [156] for Au-catalyzed InAs nanowire growth is illustrated in Fig. 3.26. The first step involves the deposition of one monolayer of Au on an InAs (111)B substrate, followed by annealing that results in the formation of nanometer-size Au–In liquid alloy droplets at a temperature above the Au–In eutectic point (Fig. 3.26a). With trimethylindium (TMI) and arsine (AsH_3) as the source gases flowing through the liquid alloy droplets, InAs crystals precipitate at the nanoscale droplet–solid interfaces, forming long whiskerlike nanowires (Fig. 3.26b).

The vapor–liquid–solid process has been exploited to synthesize nanoscale semiconductor quantum wires of different compositions [157, 158], including the

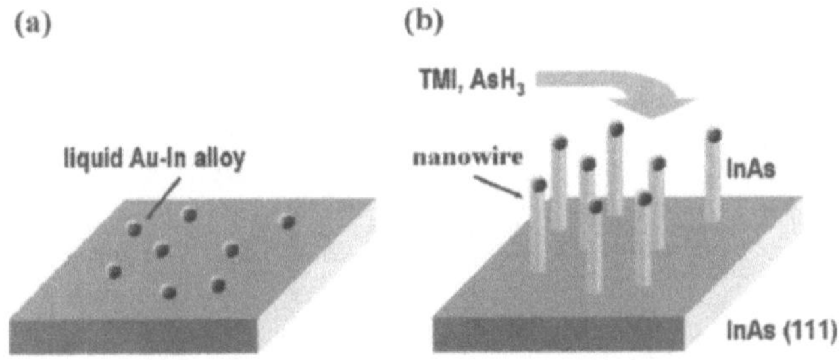

FIGURE 3.26. Catalyzed growth of nanowires through the vapor–liquid–solid process.

conventional III–V (InAs, GaAs, GaP, InP) and II–IV systems (ZnS, ZnSe, CdS, CdSe), oxides (ZnO, MgO, SiO_2), nitrides (GaN), as well as elemental semiconductors (Si and Ge). Different gas sources such as CVD [159] and laser evaporation [160–162] have been applied to supply material species during nanowire growth. In analogy to the vapor–liquid–solid process, a solution–liquid–solid mechanism for the growth of group III–V (InP, InAs, and GaAs) semiconductor nanowires has been developed [163] based on low-temperature solution-phase reactions. Other promising chemical synthetic methods that have been employed to produce semiconductor nanowires include template-directed synthesis [164, 165], aqueous precipitation [166], and solve-thermal chemistry [167].

3.5.2. Semiconductor Lasers Based on Quantum Wires

3.5.2.1. Lasers Based on Lithographically Defined Quantum Wires

Lasing from etched quantum wires showing 1-D carrier confinement in its spectrum [168] was reported around 1990 after years of efforts aimed at reducing the lateral dimensions of the etched wire structures [169]. Over the past 10 years, significant performance improvements [110–112, 170–173] of semiconductor lasers with the lithographically etched quantum wires as the active gain media have been achieved, for instance, by using a much slower growth rate than conventional deposition processes. In particular, low-damage etched and regrown interfaces have been fabricated by adopting a strain-compensated multiple quantum well structure as an initial wafer to suppress strain relaxation due to the large lattice mismatch along vertical structure [172]. Figure 3.27a is a schematic illustration of a low-damage GaInAsP quantum wire laser cavity grown on a *p*-type InP (100) substrate [110, 172]. The structure was fabricated by first growing a quantum well layer by MOCVD, then using reactive ion etching to form a wire pattern in the active region through an electron-beam written mask, followed by a second MOCVD regrowth step. The cavity mirrors were cleaved such that light propagated normal to the grating grooves. The laser structure consisted of a *p*-type InP buffer layer (2 μm), an undoped $Ga_{0.22}In_{0.78}As_{0.47}P_{0.53}$ lower optical confinement layer (170 nm, lattice matched to InP), five undoped $Ga_{0.22}In_{0.78}As_{0.82}P_{0.18}$ compressively strained quantum well layers (7 nm) sandwiched using six undoped

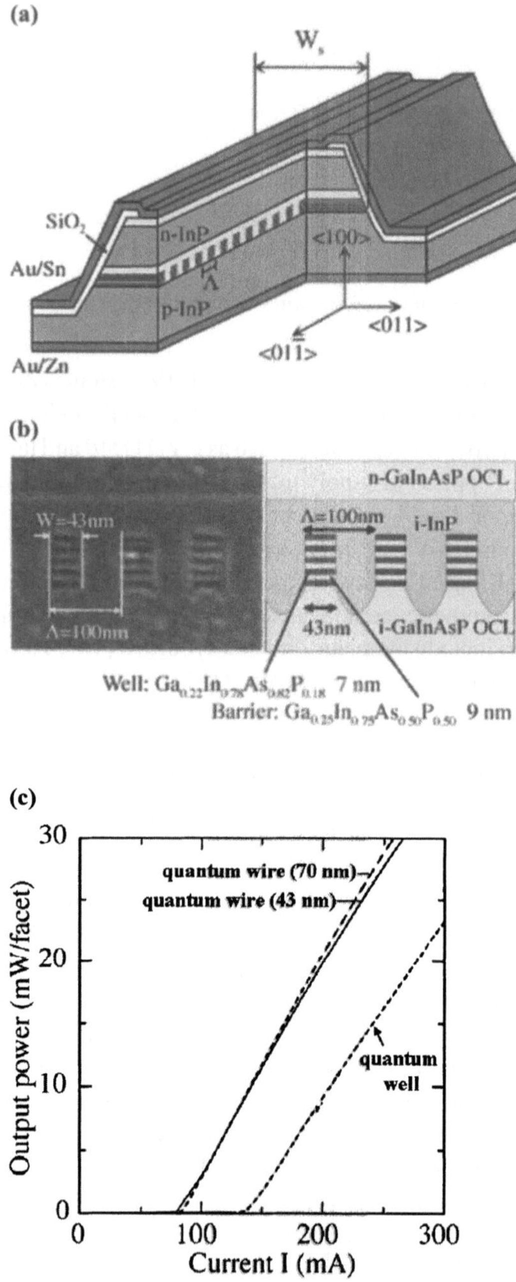

FIGURE 3.27. (a) Schematic illustration of an etched mesa quantum wire laser structures. (b) Cross-sectional SEM image of the active region of an etched quantum wire laser (43-nm-wide quantum wires). (c) Room-temperature light output versus current characteristics of two lithographically etched quantum wire lasers (with 43- and 70-nm-wide quantum wires) and of a quantum well laser prepared on the same wafer. After Refs. 110 and 172.

$Ga_{0.25}In_{0.75}As_{0.50}P_{0.50}$ tensile-strained barrier layers (9 nm thick), an upper optical confinement layer (45 nm) with the same composition as the lower one, and an InP cap layer (10 nm).

Shown in Fig. 3.27b is the cross-sectional scanning electron microscopic (SEM) view around the quantum wire active region, which has a period of about 100 nm and wire width of 43 nm. With respect to the lasing wavelength of quantum well laser (λ =1577 nm) fabricated on the same substrate, a 27-meV blue-shift (λ =1524 nm) was measured for the 43-nm quantum wire laser. As shown in Figure 3.27c, the room-temperature threshold current of the quantum wire laser was about 80 mA, corresponding to a current density of approximately 300 A/cm^2, lower than that for the quantum well laser of the similar cavity structure. Further reduction of threshold current density, as low as 94 A/cm^2, was accomplished for a 1.5-μm-wavelength GaInAsP distributed feedback laser structure consisting of two layers of a periodic quantum wire active region [174].

3.5.2.2. Lasers Based on Self-organized Quantum Wires

SILO Quantum Wire Lasers

Substitution of a quantum well with a short-period superlattice structure can create a self-organized quantum wire array with lateral composition modulation. Quantum wires grown by such a SILO scheme have been applied to serve as the active media for semiconductor lasers [119, 175–177].

Figure 3.28 shows a quantum wire laser based on SILO [119, 177] with a 60-μm-wide and 500-μm-long edge-emitting cavity fabricated by MBE. The GaInAs quantum wire laser structure (Fig. 3.28a), grown on a *n*-type InP (100) substrate, consists of 15 pairs of an *n*-type $Ga_{0.47}In_{0.53}As/Al_{0.48}In_{0.52}As$ (5 nm/10 nm) superlattice buffer layer, an *n*-type $Al_{0.48}In_{0.52}As$ cladding layer (1 μm), an undoped active region sandwiched between an undoped $Al_{0.24}Ga_{0.24}In_{0.52}As$ lower and upper wave-guiding layer (130 nm), a *p*-type $Al_{0.48}In_{0.52}As$ cladding layer (1 μm), and a *p*-type $Al_{0.47}In_{0.53}As$ cap layer. The active region consists of five quantum wells, each has eight pairs of $(GaAs)_2/(InAs)_2$ short period superlattice layers with a total thickness of about 10 nm. The barrier consists of a 7.5-nm $Al_{0.24}Ga_{0.24}In_{0.52}As$ layer. The SILO process occurs within the active region and a strong in-plane Ga/In lateral composition modulation is spontaneously formed in the [110] direction. The sandwiched In-rich regions by the higher-band-gap Ga-rich regions along the [110] direction combined with the barriers on top and bottom form an array of quantum wire heterostructure. Lasing spectra were measured at 77 and 300 K under pulsed excitation.

Figure 3.28b is the cross-sectional transmission electron microscopic (TEM) image of the active region; strong dark and bright fringes due to composition modulation can be seen in the cross section of the $(GaAs)_2/(InAs)_2$ short-period superlattice. The bright fringes correspond to the In-rich regions and the dark fringes correspond to the Ga-rich regions; both are 15 nm wide. The light-output versus current characteristics for two SILO quantum wire lasers with different contact strip orientations at 77 K temperature are shown in Fig. 3.28c. The threshold

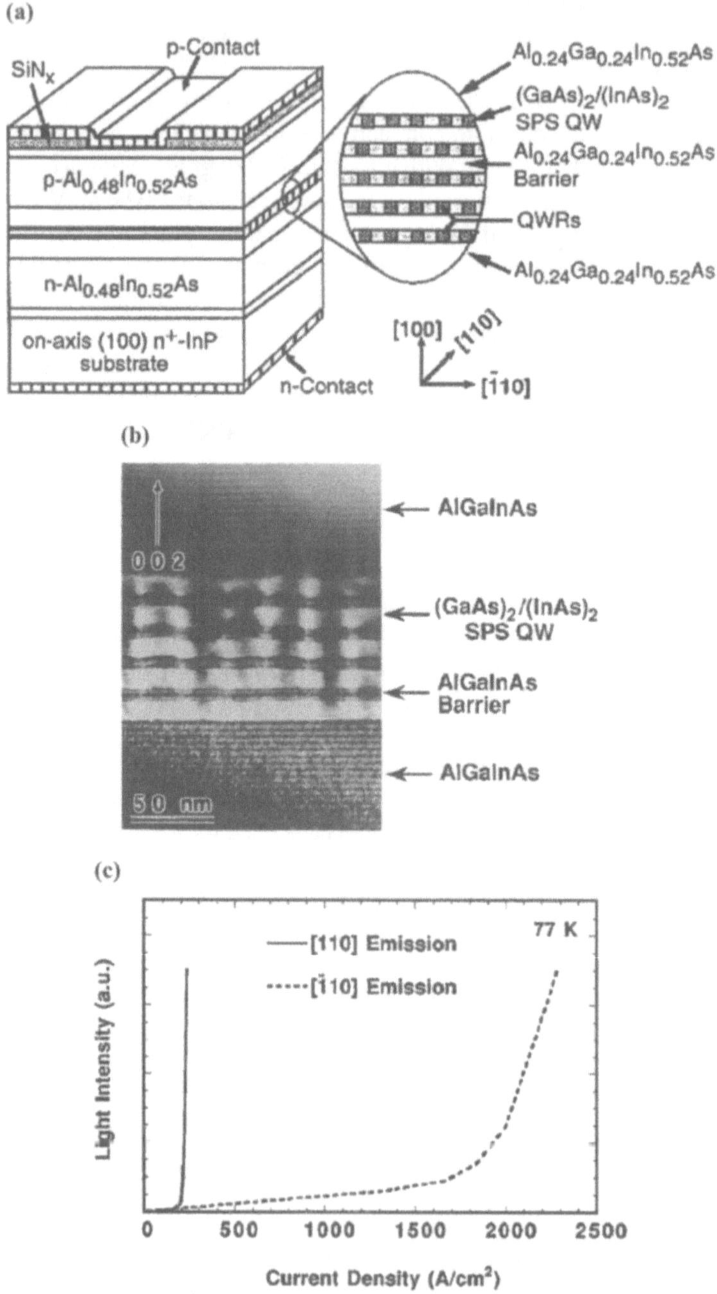

FIGURE 3.28. (a) Schematic illustration of an edge-emitting semiconductor laser with SILO quantum wires as the active gain media. (b) Cross-sectional TEM image of the active region of a GaInAs SILO quantum wire laser. (c) The light output versus current characteristics of two SILO quantum wire lasers operated at 77 K. The contact stripe was aligned either perpendicular ([110] emission) or parallel ([−110] emission) to the quantum wire array. After Refs. 119 and 177.

current density for laser emission in the [110] direction, which is perpendicular to the quantum wires, is significantly smaller than the case when laser emission is parallel to the quantum wires. The strong anisotropy observed in the threshold current density can be attributed to the direction of light propagation within the cavity with respect to the direction of the quantum wires. The effect of anisotropy of the electronic dipole moment in 1-D quantum wire lasers has been theoretically investigated.

Further investigations of quantum wire lasers based on SILO have shown temperature-stabilized lasing and gain spectra. Both the rates for lasing wavelength shift and for gain peak wavelength shift were found [177] stable to approximately 1 Å /K, a value much smaller than the quantum well case. This temperature-stability was attributed to a temperature-stabilized band gap in the quantum wires.

Vicinal Quantum Wire Lasers

Self-organization of quantum wires on vicinal multiatomic steps is an attractive technique since it does not involve postgrowth processing and therefore, can potentially yield quantum wire structures of high interface quality without lithography-induced damages. Semiconductor lasers using quantum wires self-organized on vicinal steps were demonstrated shortly after Kapon's report of the first V-groove quantum wire laser [178–182]. Figure 3.29a is a schematic illustration of a quantum wire edge-emitting laser structure using self-organized growth on vicinal GaAs multiatomic steps [183]. The InGaAs quantum wire active layer was grown on a vicinal GaAs (001) substrate, misoriented by 5° toward the [−110] direction. To form coherent GaAs multiatomic steps after the growth of an n-type AlGaAs cladding layer (320 nm), n-type $(GaAs)_2/(AlAs)_1$superlattice buffer layers with monoatomic steps were deposited prior to the growth of GaAs multiatomic steps. The average thickness of the InGaAs layers (0.15 In content) on the GaAs multiatomic steps was 3 nm, and the average step height and period were 6.1 and 70 nm, respectively. After the growth of the InGaAs quantum wires and an undoped GaAs and a p-type AlGaAs cladding layer on the top of the quantum wire layer, a standard Fabry–Perot cavity was formed with cleaved facet mirrors. The cavity length was 600 μm, and its direction was in either [110] or [−110] where the [110] direction is parallel to the quantum wires and the multiatomic steps and the [−110] direction is perpendicular to the quantum wires.

Figure 3.29b is a typical atomic force microscopy (AFM) image of the InGaAs quantum wire structures grown on multiatomic steps of a 5° misoriented GaAs (001) substrate. Figure 3.29c shows the lasing behaviors, through pulsed current injection at 77 K, of three lasers: two vicinal InGaAs quantum wire lasers emitting parallel and perpendicular to the quantum wires and an InGaAs quantum well laser fabricated under the same grown conditions but without substrate misorientation. Threshold currents of the quantum well laser and the quantum wire lasers with a cavity parallel and perpendicular to the wire array direction were approximately 150, 212, and 105 mA, respectively. These current values correspond to threshold current densities of 83, 118, and 58 A/cm^2, respectively. Clearly, minimum threshold current density was obtained for the quantum wires

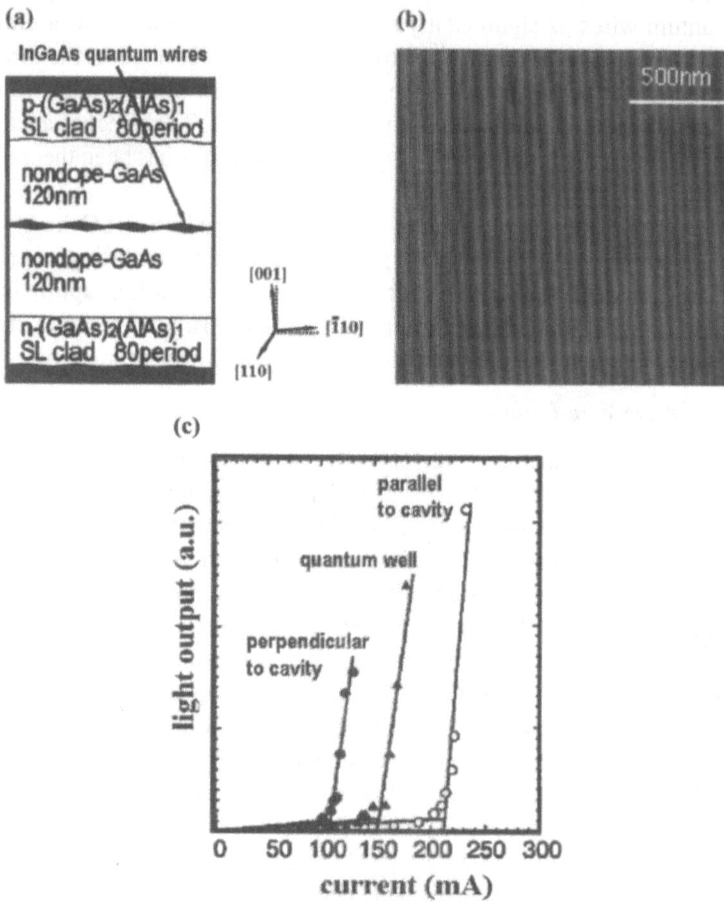

FIGURE 3.29. (a) Schematic illustration of a vicinal quantum wire laser structure, with InGaAs quantum wires grown on a GaAs vicinal substrate. (b) A typical AFM image of an InGaAs surface grown on GaAs multiatomic steps. (c) The 77-K lasing output versus current characteristics of two vicinal quantum wire lasers with emission parallel and perpendicular to the quantum wires and of a quantum well laser grown under similar fabrication conditions. After Ref. 183.

perpendicular to the light propagation (electric field vector parallel to the axis of the quantum wire). In terms of lasing wavelength, however, a red-shift of the emission wavelength (instead of blue-shift for quantum confinement) with respect to the reference quantum well laser was observed. This phenomenon was attributed to the change of composition of the locally thick InGaAs quantum wires at the edge of GaAs multiatomic steps.

Instead of simply growing a layer of quantum wires on a tilted substrate, quantum wire lasers with fractional superlattice layers grown on vicinal surfaces were also demonstrated [182]. This type of quantum wire laser may have better performance due to the benefit of the fractional superlattice layers that have additional compositional modulation perpendicular to the growth direction, thus less sensitive to small variations in deposition rates. Incorporation of fractional superlattice quantum wires, grown on vicinal misoriented GaAs (001) substrates, into microcavities (VCSEL) has resulted in room-temperature lasing under optical pumping [130, 184]. Additionally, room-temperature lasing from semiconductor lasers based on quantum wires grown on corrugated vicinal (775)B GaAs substrates and using a graded refractive-index separate confinement heterostructure [129] and a VCSEL [185] were reported.

3.5.2.3. Lasers Based on Selective Grown Quantum Wires

V-Groove Quantum Wire Lasers

Since Kapon et al. [102] reported the first observation of stimulated emission in quantum wires using V-groove structures, various types of semiconductor laser based on V-groove quantum wires have been reported [136, 137, 186–191], including lasers based on a single V-groove quantum wire such as single-wire AlGaAs/GaAs lasers [186] and single-wire InGaAsP/InP lasers [189]. A low operation threshold at room temperature has also been achieved by improving optical gain and minimizing electrical and optical losses using, for example, a short-period V-groove quantum wire array with submicron dielectric current blocking [137] or a V-groove quantum wire distributed feedback cavity structure [190], or strained V-groove quantum wires and large-index changes that enhance optical confinement factor and band-edge density of states [187].

Another advancement of V-groove quantum wire lasers is the demonstration of lasing from the ground 1-D subband transition [191, 192] in vertically stacked AlGaAs/GaAs multiple quantum wires. Because of the small optical confinement factor as well as structural imperfections, typical quantum wire lasers require a relatively high carrier density to reach threshold, resulting in lasing from the higher subbands or the excited states of the quantum wires rather than the ground-states. Ground state lasing is important in order to exploit the advantages of 1-D nanostructures, since quantum confinement is best manifested at the ground subbands as the excited subbands mix the motion of carriers in directions perpendicular to the axis of the wire.

Figure 3.30a shows a schematic structure of a MOCVD-grown V-groove quantum wire laser [141] with a SiO_2 current-blocking configuration. The V-grooves aligned along the [01-1] direction with a period of 4 μm were formed on a (001) GaAs substrate by wet chemical etching. The laser structure consists of an n-type GaAs buffer layer (0.3 μm),an n-type $Al_{0.5}Ga_{0.5}As$ lower cladding layer (1 μm), an undoped $Al_{0.2}Ga_{0.8}As$ guiding layer (0.2 μm), three GaAs quantum wires (8 nm) separated by two undoped $Al_{0.2}Ga_{0.8}As$ barrier layers (25 nm), an undoped $Al_{0.2}Ga_{0.8}As$ guiding layer (0.2 μm), a p-type $Al_{0.5}Ga_{0.5}As$ upper cladding layer (1 μm), and a p-type GaAs contact layer (0.2 μm). After the growth, the side wall and the top of the quantum wire laser structure were etched out in order

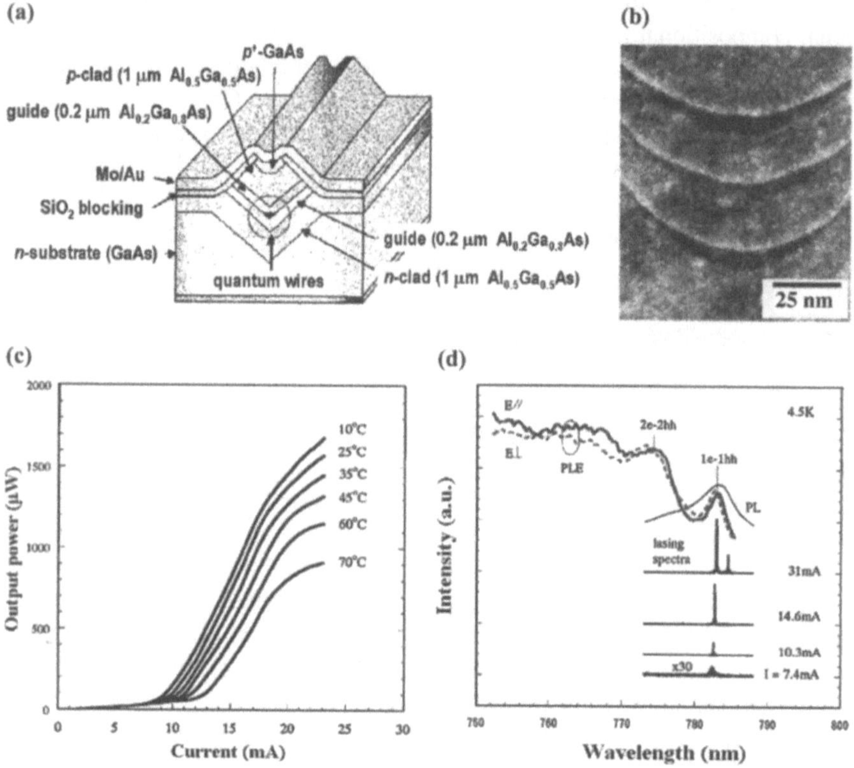

FIGURE 3.30. (a) Schematic illustration of a V-groove quantum wire laser. (b) A cross-sectional TEM image of three V-groove quantum wires. (c) Light output versus current characteristics for pulsed operation of an AlGaAs/GaAs V-groove quantum wire laser at temperatures between 10°C and 70°C. (d) Low-temperature photoluminescence (PL), photoluminescence excitation (PLE), and lasing spectra of a V-groove quantum wire laser. The photoluminescence peak positions are consistent with the lasing wavelengths. After Ref. 141.

to improve the carrier-injection efficiency. A 100-nm SiO_2 film was deposited using ion-beam sputtering on the etched surface for current confinement, leaving a 0.5-μm-wide opening for current feeding.

Figure 3.30b is the TEM cross-sectional image of three vertically stacked quantum wires. Each of the crescent-shaped V-groove quantum wire is 8 nm thick at the center. Figure 3.30c shows the pulsed output lasing power versus injection current at different temperatures for a V-groove quantum wire laser with a length of 300 μm. The rate of increase of the threshold current as a function of temperature is about 0.033 mA/K. Figure 3.30d is the low-temperature (4.5 K) photoluminescence (PL), photoluminescence excitation (PLE), and lasing spectra, with the PLE spectra measured by exciting the quantum wire laser structure normal to the wafer plane with two different linear polarization directions. The broad PL/PLE peak at 1.584 eV (782.8 nm) was attributed to the lowest-lying electron and heavy-hole-like states; the emission wavelength of the quantum wire laser virtually coincides with

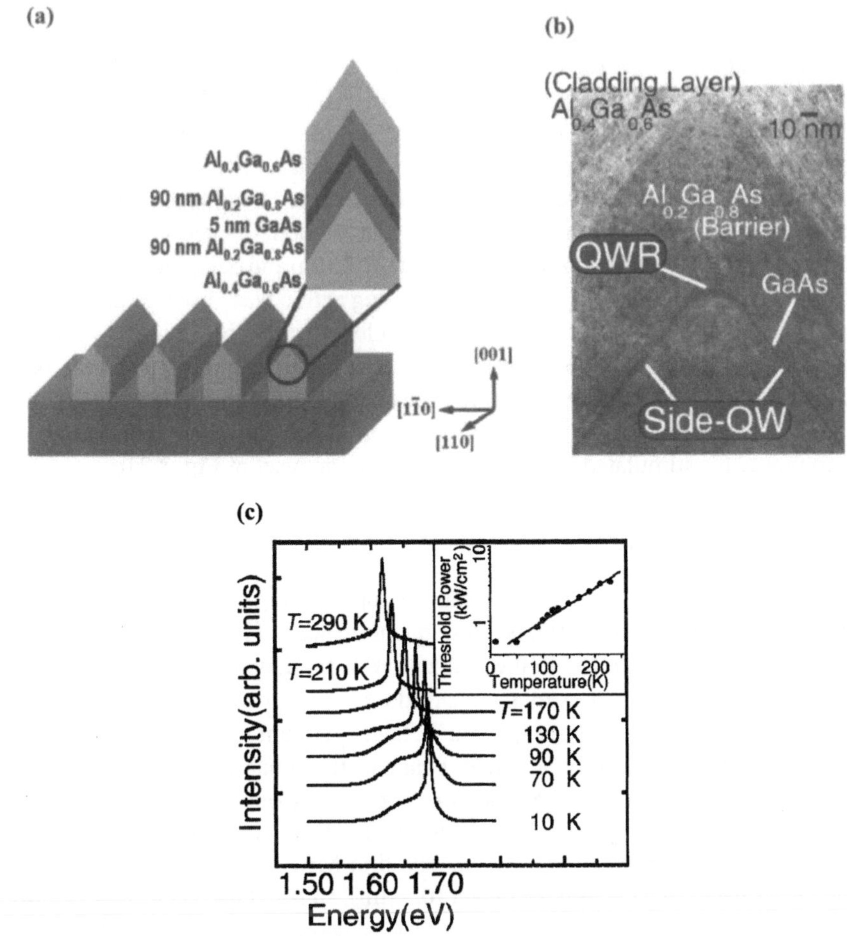

FIGURE 3.31. (a) Schematic illustration of a ridge quantum wire laser. (b) Cross-sectional TEM image of a ridge quantum wire laser active region. (c) Lasing spectra of a Λ-ridge quantum wire laser at 77 K for different optical pumping power. After Ref. 193.

this PL/PLE peak. This agreement between the lasing wavelength and the PL peak was found to hold at higher temperatures (up to 230 K) and can possibly be extended to room temperature.

Λ-Ridge Quantum Wire Lasers

Similar to V-groove quantum wire lasers, lasing action by optical pumping in Λ-ridge quantum wire structures has also been reported [193,194]. The ridge quantum wire structure may have the advantage of easily achieving large lateral carrier confinement at its ridge–air interface. Figure 3.31a shows a schematic structure [193] of a MBE-grownΛ-ridge quantum wire laser. The Λ-ridges with a period of 4 μm were formed on a patterned (001) GaAs substrate with reverse mesa stripes running along the [110] direction. The laser structure consists of an

$Al_{0.4}Ga_{0.6}As$ lower cladding layer, an $Al_{0.2}Ga_{0.8}As$ lower barrier layer (90 nm), a GaAs active quantum wire layer (5 nm), an $Al_{0.2}Ga_{0.8}As$ upper barrier layer (90 nm), and an $Al_{0.4}Ga_{0.6}As$ upper cladding layer. The curvature of the ridge top has a lateral width of approximately 10 nm. After the MBE growth, the sample was cleaved to 300 μm in length to form an optical cavity. Figure 3.31b is a cross-sectional TEM image of a ridge quantum wire laser active region.

Figure 3.31c shows the temperature dependence of lasing spectra. Optical excitation was performed by the second harmonics of mode-locked yttrium–lithium fluoride (YLF) laser pulses with a wavelength of 526 nm and pulse duration of 50 ps, incident on the top of the ridge structures. Lasing action was observed at temperatures between 4.7 and 290 K, and the peak wavelength corresponds to the transition from excited subbands of the quantum wires. The lasing energy became gradually lower at higher temperatures, which was considered to result from the temperature dependence of the band-gap energy. The temperature dependence of the threshold pump power for stimulated emission is shown in the inset of Fig. 3.31c.

T-Intersection Quantum Wire Lasers

Since the first demonstration of optically pumped lasing from ground quantum wire states using multiple T-intersection quantum wires [153], significant developments have been achieved in the realization of semiconductor lasers based on T-intersection quantum wires by the cleaved-edge overgrowth method [154, 195, 196]. Notably, the first current-injection T shape quantum wire laser was reported in 1994 [154], and, recently, a single T-intersection quantum wire was seen to exhibit lasing action from the ground-state subband by optical pumping [196].

A schematic view of a current-injection T-intersection quantum wire laser structure [154] is shown in Fig. 3.32a. The first MBE growth on a [001] GaAs substrate consists of a GaAs buffer layer (0.5 μm), an $Al_{0.5}Ga_{0.5}As$ lower cladding layer (1 μm), a 15-period *p*-type GaAs/$Al_{0.35}Ga_{0.65}As$ multiple quantum well structure (with well and barrier thickness of 7 and 58 nm), an $Al_{0.5}Ga_{0.5}As$ upper cladding layer (3 μm), and a GaAs cap layer (10 nm). The overgrowth along the [100] direction consists of an undoped GaAs quantum well (7 nm) followed by an $Al_{0.35}Ga_{0.65}As$ barrier (7 nm), an $Al_{0.1}Ga_{0.9}As$ wave setback layer (43 nm), an *n*-type $Al_{0.1}Ga_{0.9}As$ layer (124 nm), an *n*-type $Al_{0.5}Ga_{0.5}As$ cladding layer (1 μm), and an *n*-type GaAs cap layer (10 nm). The cladding layers (and the AlGaAs layer) serve as a T-shaped dielectric waveguide-confining optical mode in the vicinity of the quantum wire array. Figure 3.32b shows the light-current characteristics of two T-intersection quantum wire lasers with cavity lengths of 400 and 800 mm at 4.2 K. The 800-μm laser shows clear superlinear behavior with a threshold current of less than 0.6 mA, compared to the 400-μm laser, which has a threshold current of about 0.4 mA.

The evolution of emission from a T-intersection quantum wire laser with increasing current is shown in Fig. 3.32c. Increasing current-injection levels results in progressively narrowing laser emission spectra as well as clear individual Fabry–Perot peaks. Also seen in Fig. 3.32c, there is no appreciable shift of the

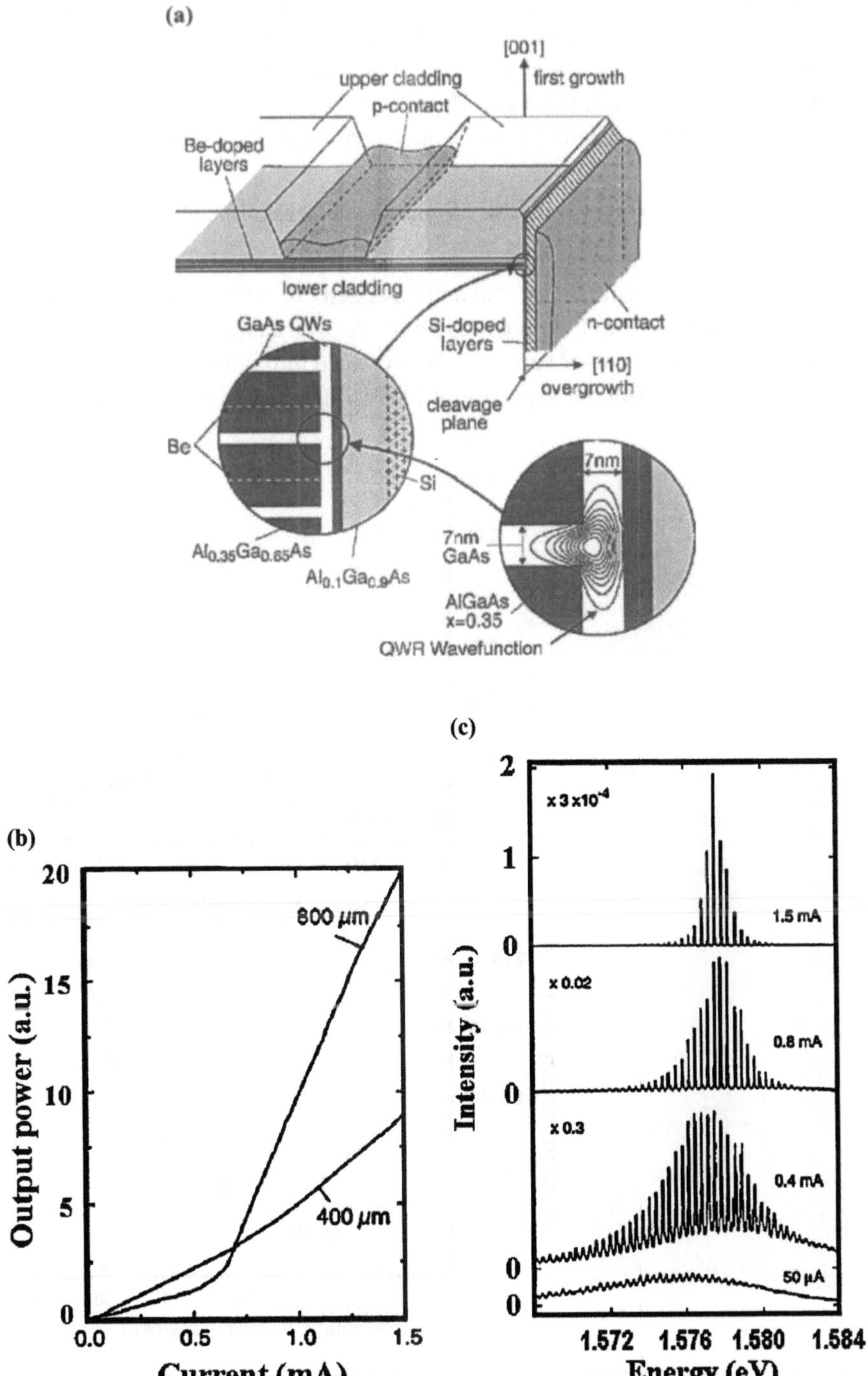

FIGURE 3.32. (a) Schematic illustration of a T-intersection quantum wire laser. (b) Light output versus current characteristics for two T-intersection quantum wire lasers. (c) Current-dependent lasing spectra of a T-intersection quantum wire laser. After Ref. 154.

T-intersection quantum wire laser emission energy over relatively large changes in excitation levels, in contrast to GaAs/AlGaAs quantum well lasers that typically display a pronounced red-shift with increasing injection current due to the carrier-density-dependent band-gap shrinkage. This observation was interpreted as the signature for excitonic gain in T-intersection quantum wires.

3.5.2.4. Lasers Based on Chemically Synthesized Crystalline Quantum Wires

Room-temperature lasing action from chemically synthesized nanoscale ZnO quantum wires (nanowires) was also demonstrated [76, 162]. The ZnO nanowires were grown with a vapor–liquid–solid process via catalyzed epitaxial crystal growth, where the precursor vapor can be produced through either thermal vaporization [76] or ultrafast laser vaporization [162]. Figure 3.33a shows a typical SEM

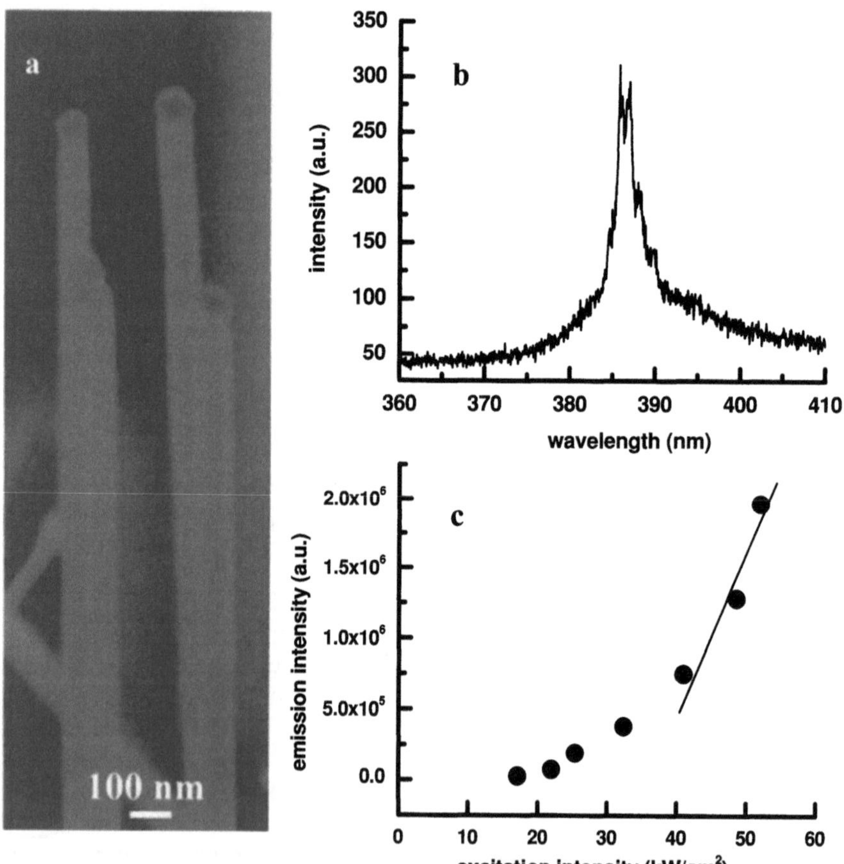

FIGURE 3.33. (a) A typical SEM image of ZnO nanowires with catalysts seen on the top of the wires. (b) A lasing spectrum of ZnO nanowires. (c) Optical (266 nm) pumping-energy-dependent light emission intensity from ZnO nanowires at room temperature.

image of ZnO nanowires, approximately 100 nm in diameter and with catalysts clearly seen on the top of the wire. When sapphire is used as the substrate for ZnO nanowire growth, due to the existence of a good epitaxial interface between the (0001) plane of the ZnO nanowire and the (110) plane of the substrate [197], ZnO nanowires tend to grow vertically from the substrate.

Figures 3.33b and 3.33c show respectively a light emission spectrum from ZnO nanowires and the dependence of emission intensity on pumping under room-temperature optical excitation. Excitons play a role in room-temperature light emission from ZnO nanowires [198, 199] due to the relatively large exciton binding energy in ZnO. The fourth harmonic of a solid-state Nd:yttrium–aluminum–garnet laser (Nd:YAG, 266 nm) was applied as the pumping source. As the excitation intensity exceeds a threshold at room temperature (Fig. 3.33c), lasing peaks emerge in the emission spectra. The lasing threshold is lower than that for lasing in disordered particles or thin films [200]. The observation of lasing action in these nanowire arrays without any fabricated mirror indicates that the crystalline well-faceted nanowires can act as natural optical cavities.

This surface-emitting nanolaser mechanism has been corroborated by near-field optical spectroscopy of single-nanowire lasing [201–203] and, more recently, electrically pumped lasing from CdS nanowires [204].

3.6. QUANTUM DOT LASERS

In view of a further scale-down of the semiconductor active gain media, impressive progress for lasers based on 0-D quantum dots [205] have been achieved recently, in parallel with the rapid development of quantum wire lasers. Additional stimulus toward the realization of quantum dot lasers comes from the prospect of quantum communication based on single-photon emission [206–208] from semiconductor quantum dots. The challenge in realizing quantum dot lasers with operation superior to quantum well lasers is the formation of high-quality, uniform quantum dots in the active layer. One critical aspect for semiconductor lasers based on quantum dots is the statistical nature of the size and location, which gives rise to the broadening in the emission spectra. Breakthroughs in defect-free quantum dot (quantum dots) growth techniques, especially the self-assembly method known as the Stranski–Krastanov process and the chemical synthetic routes based on stabilized replacement reactions, have led to a significant performance improvement of quantum dot lasers [209].

3.6.1. Quantum Dot Fabrication Technologies

Fabrication of high-quality quantum dots has lagged far behind fabrication of quantum wells for which two well-developed epitaxy techniques (MBE and MOCVD) allow for the deposition of semiconductor materials with atomic layer precision and very high purity. Early attempts at making semiconductor quantum dots focused on patterning quantum wells using nanoscale lithographic techniques, such as etching. A related technique is patterning the semiconductor substrate before epitaxial growth. However, damage introduced during the fabrication process generally reduced the optical efficiency of these dots, and limitations on the tolerances of microelectronic fabrication led to large variations in dot size. Deposition techniques that go beyond planar layer-by-layer growth were not developed until recently; uniform semiconductor quantum dots could be grown directly by a strain-induced process using MBE or MOCVD or by a variety of chemical approaches. These relatively new growth technologies have led to a recent rapid progress in the study of semiconductor quantum dots.

3.6.1.1. Nanoscale Lithography

As quantum wells fabrication is a mature technology, it is natural to start with a quantum well structure and impose additional confinement to achieve 3-D confinment. One approach (Fig. 3.34) is to physically etch away material everywhere until it becomes a small "dot." Both wet and dry etching technologies have been employed to define dots at nanometer scale using semiconductor quantum well materials [210–212], and I/V characteristics that can be attributed to carrier confinement have been observed. However, even though the etching processes are well understood and characterized, the optical properties of etched quantum dots are usually shadowed by the large-surface recombination and the etch-damaged defects on the side walls [213, 214]. Passivation has been used to reduce the surface recombination, but the recombination dynamics of etched quantum dots is still dominated by the defects. Additionally, with standard processing technology, the smallest quantum dot size of an etched pillar is usually not significantly smaller than 100 nm in diameter.

3.6.1.2. Self-organization

Growth of self-organized or self-assembled quantum dots involves the formation of defect-free 3-D islands during strained-layer epitaxy. The structure is a result of thermal equilibrium driven by epitaxial strain between the substrate and the

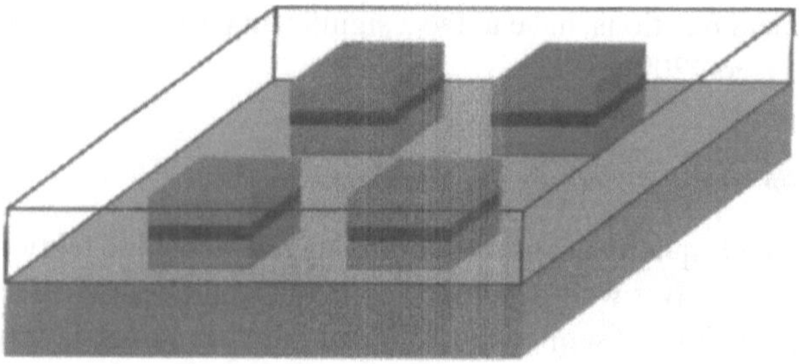

FIGURE 3.34. Schematic illustration of nanoscale quantum dots fabricated by lithography (etching).

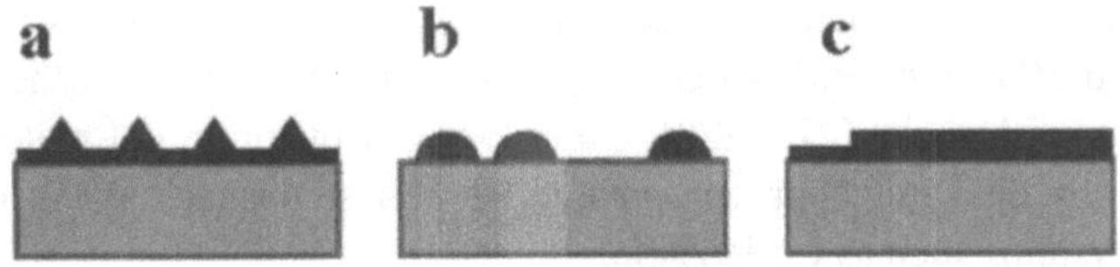

FIGURE 3.35. Schematic illustration of three different heteroepitaxial growth modes: (a) Stranski–Krastanov mode, (b) Volmer–Weber mode, and (c) Frank–van der Merwe mode.

deposited atoms. The islands have dimensions on the order of nanometers in all three directions, making them suitable as quantum dots as surrounding the islands with a wider-band-gap material can impose quantum confinement [215]. With no need of additional processing steps, the existing semiconductor fabrication processes can be easily adopted for quantum dot fabrication.

Self-organized growth occurs as a mechanism for strain relief for mismatches of approximately 5–10%. Growth initially begins with the mismatched atoms fitting onto the substrate undergoing biaxial strain [216]. The unit cells of the epitaxial material will be distorted in order to conform to the lattice constant of the substrate (Fig. 3.35), leading to strain in the epitaxial layer. However, The energy associated with this strain increases as the layer thickness increases, and the presence of the mismatch imposes a limit to the thickness of epitaxial layer, beyond which, dislocations form to relax the strain. As islanding competes with dislocation as a mechanism for relieving strain in epitaxial layers, if the energy lost by forming islands is greater than that lost by forming dislocations, strain can be relaxed without forming dislocations, but by laterally separated small islands on the surface. Deposition of laterally separated 3-D islands (quantum dots) is known as the Stranski–Krastanov growth mode [217], in which, after an initial 2D growth of a wetting layer, the built-in strain drives the system to a 3D growth for strain minimization.

Other than the Stranski–Krastanov growth mode, there are two heteroepitaxial crystal growth modes differentiated by the degrees of binding between the atoms and the surface of the substrate (Fig. 3.35). In the Volmer–Weber mode, the materials have high interface energy and the atoms are more strongly bound to each other than to the substrate. By minimizing the interface, material deposition results in small clusters formed directly on the substrate. In the Frank–van der Merwe mode, the situation is reversed, where the sum of the epitaxial surface energy and the interface energy is less than the substrate surface energy. The atoms are more strongly bound to the substrate rather than each other, thus sequential 2-D monolayers completely covering the surface can be achieved. This is the mode for growing high-optical-quality quantum well structures.

Research on the epitaxial growth using lattice-mismatch materials has provided new insights on the control of the quantum dot materials. The first observation of self-organized quantum dots was made in 1990 [218] for defect-free germanium islands deposited on silicon, followed by a demonstration [219] that GaInAs islands on GaAs exhibit quantum confinement and good optical properties. InAs and

InGaAs self-organized quantum dots grown in a GaAs matrix have by far been the most studied of all the semiconductor quantum dots fabricated by the Stranski–Krastanov growth mode [220–228]. Other types of quantum dot material [229–231] grown by the same approach include III–V systems such as (Ga)InP/GaAs and II–VI systems such as CdSe/ZnSe.

3.6.1.3. Chemical Synthesis

There are different methods to synthesize quantum dots or nanocrystals chemically, all of which involve nucleation and growth of nanocrystals in organic or aqueous solutions. Whereas fabrication of self-organized quantum dots based on heteroepitaxial process represents a natural extension of mature semiconductor film technology, chemical synthesis of quantum dots appears to be more versatile and easy to carry out, which typically involves heating a reactive mixture or adding reactants into a solution. Generally, the production of chemically synthesized quantum dots has a high yield, with control over composition, size, and shape. A unique advantage of the solution-phase chemical synthesis is that the solvent can cap the atoms on the surface to prevent quantum dots aggregation. As a consequence, quantum dots can be dispersed into different solutions or polymers, linked to solid surfaces, evaporated into thin films, or dried as powders.

Chemical synthesis of semiconductor quantum dots has been well developed for II–VI (CdSe, CdS, ZnSe, ZnS, CdTe) and III–V semiconductors (InP, InAs, GaAs). Growing monodisperse semiconductor quantum dots requires a separated nucleation period followed by controlled growth on the existing crystalline nuclei [232]. The supersaturation and subsequent nucleation can be triggered by rapid injection of metal–organic precursors into a coordinating solvent. If nanocrystal growth during the nucleation period were minimal compared to subsequent growth, quantum dots would be uniform over time [233]. The frequently used solvents include mixtures of long-chain alkylphosphines R_3P, alkylphosphine oxides R_3PO ($R =$ butyl or octyl), and alkylamines.

In the synthesis of II–VI quantum dots, metal alkyls (dimethylcadmium, diethylcadmium, diethylzinc, dibenzylmercury) are usually selected as the group II sources. The group VI sources are often organophosphine chalcogenides or bistrimethylsilyl chalcogenides TMS_2E ($E =$ S, Se, Te). CdSe quantum dots are probably the most extensively investigated chemically grown semiconductor nanocrystals. CdSe quantum dots prepared by the trioctylphosphine oxyde (TOPO) and trioctylphospine (TOP) route have shown good size distributions and crystalline structures [234, 235] Additional coordinating components such as hexadecylamine (HDA) have been introduced to reduce the nanocrystal size [236]. Figure 3.36 shows a TEM image of CdSe nanocrystals synthesized in HDA–TOPO–TOP mixture [236].

The chemical synthesis of III–V nanocrystals is more difficult than that of II–VI quantum dots. Because they are more covalent compounds and high temperatures are usually required for their synthesis [237–242] A step toward controllable synthesis of III–V quantum dots was the use of the dehalosylilation reaction [243] for InP quantum dots synthesis [237] and, later, InAs quantum dots [241]. Typically

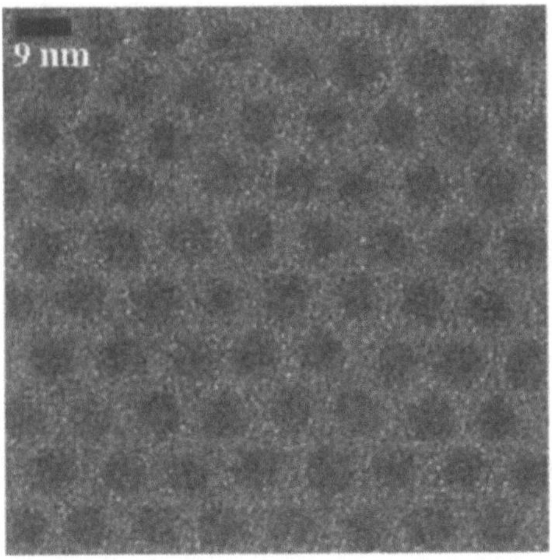

FIGURE 3.36. TEM image of CdSe nanocrystals synthesized in HDA–TOPO–TOP mixture. After Ref. 236.

$InCl(C_2O_4)$ is employed as an In source with TMS_3P or TMS_3As in R_3P or R_3PO solvents. The use of amines as stabilizing and size-regulating agent permits smaller quantum dots with a narrower size distribution.

Other chemical or electrochemistry synthesis techniques include nanocrystal growth in inverse micelles [244, 245] and electrodeposition [246, 247]. However, it is generally difficult to produce quantum dots with a very narrow size distribution (e.g., $<5\%$) using these methods.

3.6.2. Semiconductor Lasers Based on Quantum Dots

3.6.2.1. Lasers Based on Lithographically Defined Quantum Dots

Despite imperfections inherent to lithography, semiconductor lasers based on lithographically defined quantum dots were demonstrated in a MOCVD-grown GaInAs/GaInAsP/InP structure. In the first such demonstration [248] (Fig. 3.37), the starting device consisted of a 12-nm GaInAs quantum well deposited on a 200-nm *p*-GaInAsP optical confinement layer, grown on a *p*-InP substrate. Using polymethyl methacrylate (PMMA) as the resist, a pattern of quantum dots were defined by wet chemical etching with a 30-nm diameter and separated with a 70-nm pitch. A 200-nm *n*-GaInAsP top optical confinement layer was deposited after fabricating the quantum dots. The threshold current density of this quantum dot laser was 7.6 kA/cm^2 at 77 K, with a wavelength of approximately 1.26 μm, as shown in Fig. 3.37. This laser has a large threshold current due to a very small optical confinement of the quantum dots.

3.6.2.2. Lasers Based on Self-organized Quantum Dots

The first laser with an active single self-organized InGaAs quantum dot layer was grown by MBE [249, 250]. At 77 K, the lasing threshold current density was ~100

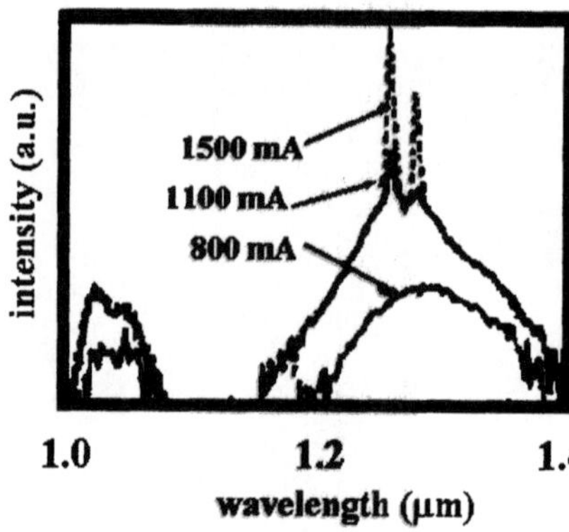

FIGURE 3.37. Lasing spectra of a semiconductor laser with lighographyically defined quantum dots as the gain medium. After Ref. 248.

A/cm^2 and remained essentially unaffected by temperature up to about 150 K. The threshold current was elevated at higher temperatures, due to thermally activated escape of carriers from quantum dots to the surrounding GaAs matrix. Therefore, in order for quantum dot lasers to operate at room temperature, the challenge is to reduce the thermally induced escape of carriers from the quantum dots. The approaches to improving quantum dot laser performance include [251–253] increasing quantum dot density, inserting a quantum dot in a quantum well, or using matrix material of greater band-gap energies. An example of lasing characteristics of a quantum dot double-heterojunction laser is shown in Fig. 3.38 [251].

With quantum dots surrounded by InGaAs layers, the threshold current density of room-temperature operating quantum dot lasers (26 A/cm^2 at 1.25 μm) was demonstrated to be lower than that of quantum well lasers [26]. Further

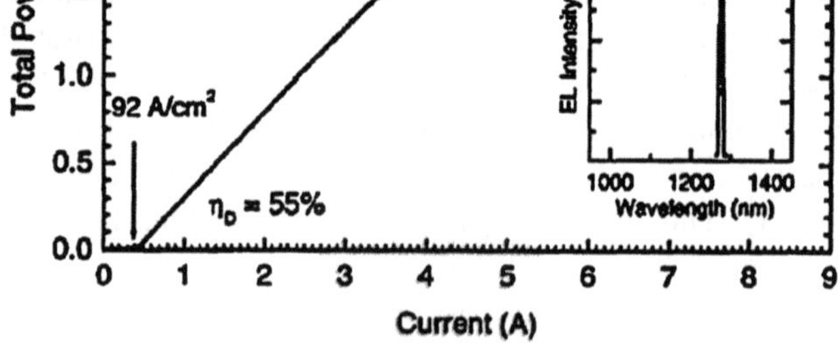

FIGURE 3.38. Lasing characteristics of a semiconductor quantum dot double-heterojunction laser. After Ref. 251.

improvement [254–256] of quantum dot laser structures yielded devices with the threshold current density in the vicinity of 10 A/cm^2.

Other than edge-emitting quantum dot lasers, the vertical cavity surface-emitting cavity structure [257, 258] is promising for achieving low-threshold quantum dot lasers. The incorporation of quantum dots in VCSELs has the benefit of suppressing the spreading of nonequilibrium carriers. Quantum dot lasers implemented in microdisk [259, 260] and photonic crystal [261] cavities have also been demonstrated. In addition, semiconductor lasers based on self-organized quantum dots other than GaAs material system, such as InGaN quantum dots [262] and CdSe quantum dots [230], was also achieved.

3.6.2.3. Lasers Based on Chemically Synthesized Quantum Dots

Cadmium selenide nanocrystals or quantum dots have been prime candidates for demonstrating the potential of chemically synthesized quantum dots in laser structures [263]. An optically pumped distributed feedback (DFB) laser based on chemically synthesized quantum dots was demonstrated recently [17]. Fabrication of this surface-emitting DFB laser structure is accomplished by spin-coating a CdSe nanocrystal/titania thin film on top of a DFB grating, which were patterned using interference lithography [264]. The refractive index of the film as determined by the volume fraction of nanocrystals was adjusted to match the Bragg condition of the grating with the emission peak of the nanocrystal/titania film. Photoluminescence measurements indicate a clear lasing threshold as a function of excitation pump pulse intensity.

The theoretically predicted temperature insensitivity of nanocrystal lasers in the strong confinement regime was confirmed from measurements. Figure 3.39 shows the above threshold emission profile of a CdSe nanocrystal/titania-based

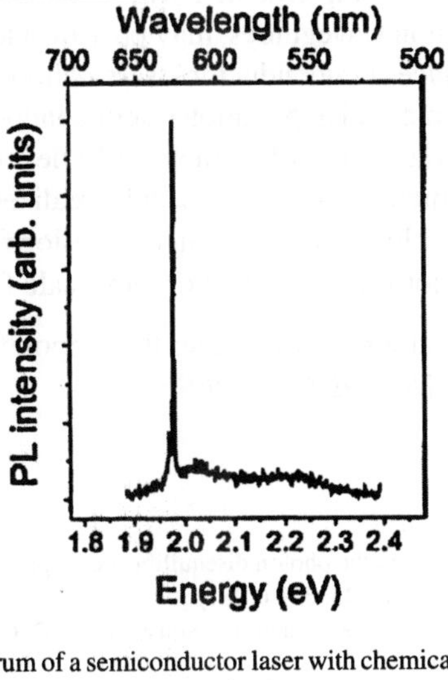

FIGURE 3.39. Lasing spectrum of a semiconductor laser with chemically synthesized CdSe nanocrystals (quantum dots) as the gain medium. After Ref. 17.

DFB laser with a grating period of 350 nm. The linewidth was observed not to vary when the device was cooled down to 80 K. A particular advantage of semiconductor lasers based on chemically synthesized nanocrystals is the size-dependent emission spectrum that can be achieved easily in a single-laser structure.

3.7. PERSPECTIVES

Miniaturization has been the subject of substantial research interest in semiconductor laser technology. One major direction is the realization of novel lasing devices that benefit from quantum confinement associated with low-dimensional semiconductor nanostructures. Dramatic progress in the development of nanoscale crystal growth and fabrication technologies has driven the miniaturization of semiconductor lasers, a trend also motivated by the desire to achieve greater color range, higher optical gain, and lower lasing threshold. This chapter provides a general picture of the current status of semiconductor laser technology based on 2-D, 1-D, and 0-D nanoscale quantum structures and offers an introduction to some of the essential characteristics of quantum confinement in nanoscale semiconductor materials. As discussed in the preceding sections, when the size of an active semiconductor gain material is reduced to the nanometer scale, quantization of the transition energies and narrowing of the density of states result in lasers with great color range, high material gain, and low lasing threshold. In the meantime, carrier confinement in nanoscale semiconductors can lead to an increase in the exciton-binding energy and the oscillator strength for radiative recombination [265–269]; photon confinement in miniature optical cavities can lead to an enhancement of the spontaneous emission rate and emission coupling to the fundamental cavity mode [270–273]. Still under intensive investigations, enhancement of the excitonic interaction and of the spontaneous emission rate in nanoscale quantum structures, in particular quantum wires and quantum dots implemented with nanostructured cavities such as photonic crystals or plasmon waveguides, may provide additional mechanisms for realizing better performance semiconductor lasers. Although quantum well lasers are maturing toward commercial applications, semiconductor lasers based on 1-D and 0-D nanoscale materials are still in their early development stage. The full promise of nanoscale semiconductor lasers will be realized based on a more complete understanding of the physics of stimulated emission from nanoscale quantum structures and the continuous development of nanoscale fabrication technology.

Acknowledgment. The author acknowledges the support from the US Department of Energy under contract No. DE-AC03-76SF00098.

REFERENCES

1. J. von Neumann, Notes on the photon disequilibrium amplification scheme, Sept 16, 1953, *IEEE J. Quantum Electron.* **23**, 659 (1987).
2. R.N. Hall, G.E. Fenner, J.D. Kingsley, T.J. Soltys, and R.O. Carlson, *Phys. Rev. Lett.* **9**, 366 (1962).

3. N. Holonyak, Jr. and S.F. Bevacqua, *Appl. Phys. Lett.* **1**, 82 (1962).
4. M.I. Nathan, W.P. Dumke, G. Burns, F.H. Dill, Jr., and G. Lasher, *Appl. Phys. Lett.* **1**, 62 (1962).
5. T.M. Quist, R.H. Rediker, R.J. Keyes, W.E. Krag, B. Lax, A.L. McWhorter, and H.J. Zeiger, *Appl. Phys. Lett.* **1**, 91 (1962).
6. H. Kroemer, *Proc. IEEE* **51**, 1782 (1963).
7. Z.I. Alferov and R.F. Kazarinov, U.S.S.R. Patent 181737 (1963).
8. J.M. Woodall, H. Rupprecht, and G.D. Petit, *IEEE Trans. Electron. Dev.* **14**, 630 (1967).
9. Z.I. Alferov, V.M. Andreev, D.Z. Garbuzov, Y.V. Zhilhayev, E.P. Morozov, E.L. Portnoi, V.G. Trofim, *Sov. Phys. Semicond.* **4**, 1573 (1971).
10. I. Hayashi, M.B. Panish, P.W. Foy, and S. Sumski, *Appl. Phys. Lett.* **17**, 109 (1970).
11. P. Yu and M. Cardona. *Fundamentals of Semiconductors: Physics and Material Properties*, Springer-Verlag, Berlin, 2001.
12. M.G.A. Bernard and G. Duraffourg, *Phys. Status Solidi* **1**, 699 (1961).
13. N.G. Basov, O.N. Krokhin, and Y.M. Popov, *JETP* **40**, 1320 (1961).
14. Y. Arakawa and H. Sakaki, *Appl. Phys. Lett.* **40**, 939 (1982).
15. M. Asada, Y. Miyamoto, and Y. Suematsu, *IEEE J. Quantum Electron.* **22**, 1915 (1986).
16. Y. Nambu and K. Asakawa, *Appl. Phys. Lett.* **67**, 1509 (1995).
17. H.J. Eisler, V.C. Sundar, M.G. Bawendi, M. Walsh, H.I. Smith, and V. Klimov, *Appl. Phys. Lett.* **80**, 4614 (2002).
18. J.-P. Laude, *DWDM Fundamentals, Components, and Applications*, Artech, London, 2002.
19. C.B. Murray, D.G. Norms, and M.G. Bawendi, *J. Am. Chem. Soc.* **115**, 8706 (1993).
20. A.L. Efros and M. Rosen, *Annu. Rev. Mater. Sci.* **30**, 475 (2000).
21. L.E. Brus, *J. Chem. Phys.* **80**, 4403 (1984).
22. L.E. Brus, *J. Chem. Phys.* **90**, 2555 (1986).
23. E. Kapon, *Proc. IEEE* **80**, 398 (1992).
24. J.T. Verdeyen, *Laser Electronics*, 3rd ed., Prentice-Hall, Englewood Cliffs, NJ, 1995.
25. A. Yariv, *Optical Electronics in Modern Communications*, 5th ed., Oxford University Press, New York, 1997.
26. G.T. Liu, A. Stintz, H. Li, K.J. Malloy, and L.F. Lester, *Electron. Lett.* **35**, 1163 (1999).
27. L.F. Lester, A. Stintz, H. Li, T.C. Newell, E.A. Pease, B.A. Fuchs, and K. Malloy, *IEEE Photon. Tech. Lett.* **11** 931 (1999).
28. P.G. Eliseev, H. Li, A. Stintz, G.T. Liu, T.C. Newell, K.J. Maloy, and L. Lester, *Appl. Phys. Lett.* **77**, 262 (2000).
29. K. Iga, *IEEE J. Select. Topics Quantum Electron.* **6**, 1201 (2000).
30. A. Yariv, *Appl. Phys. Lett.* **53**, 1033 (1988).
31. Y. Miyamoto, Y. Miyake, M. Asada, and Y. Suematsu, *IEEE J. Quantum Electron.* **25**, 2001 (1985).
32. M. Asada, Y. Miyamoto, and Y. Suematsu, *Jpn. J. Appl. Phys.* **24**, L95 (1985).
33. Z. Alferov, *IEEE J. Select. Topics Quantum Electron.* **6**, 832 (2000).
34. P.S. Zory, Jr., ed., *Quantum Well Lasers*, Academic, San Diego, CA, 1993.
35. S.S. Mao, *Int. J. Nanotechnol.* **1**, 42, 2004.
36. N.N. Ledentsov, *IEEE J. Select. Topics Quantum Electron.* **8**, 1015 (2002).
37. Y. Arakawa, *Trans. IEICE* **E85**, 37 (2002).
38. N.N. Ledentsov, *Semiconductors* **33**, 946 (1999).
39. S. Sugou, Y. Kato, H. Nishimoto, and K. Kasahara, *Electron. Lett.* **22**, 1214 (1986).
40. T. Ohtoshi and N. Chinone, *Electron. Lett.* **21**, 12 (1991).
41. J. Coleman, R.M. Lammert, M.L. Osowski, and A.M. Jones, *IEEE J. Select. Topics Quantum Electron.* **3**, 874 (1997).
42. Y. Yoshida, H. Watanabe, K. Shibata, A. Takemoto, and H. Higuchi, *IEEE J. Quantum Electron.* **35**, 1332 (1999).
43. D.Z. Garbuzov, N.Y. Antonishkis, A.D. Bondarev, A.B. Gulakov, S.Z. Zhigulin, N.I. Katsavets, A.V. Kochergin, and E.V. Rafailov, *IEEE J. Quantum Electron.* **27**, 1531 (1991).

44. M.C. Wu, Y.K. Chen, M. Hong, J.P. Mannaerts, M.A. Chin, and A.M. Sergent, *Appl. Phys. Lett.* **59**, 1046 (1991).
45. S.R. Chinn, *Appl. Opt.* **23**, 3508 (1984).
46. S. Ghosh, P. Bhattacharya, E. Stoner, H. Jiang, J. Singh, S. Nuttinck, and J. Laskar, *Appl. Phys. Lett.* **79**, 722 (2001).
47. J.L. Jewell, J.P. Harbison, A. Scherer, Y.H. Lee, and L.T. Florez, *IEEE J. Quantum Electron.* **27**, 1332 (1991).
48. H. Soda, K. Iga, C. Kitahara, and Y. Suematsu, *Jpn. J. Appl. Phys.* **18**, 2329 (1979).
49. K. Iga, S. Ishikawa, S. Ohkouchi, and T. Nishimura, *Appl. Phys. Lett.* **45**, 348 (1984).
50. F. Koyama, S, Kinoshita, and K. Iga, *Appl. Phys. Lett.* **55**, 221 (1989).
51. H. Kogelnik and C.V. Shanks, *J. Appl. Phys.* **43**, 2327 (1972).
52. Y.H. Lee, J.L. Jewell, A. Scherer, S.L. McCall, J.P. Harbison, and L.T. Florez, *Electron. Lett.* **25**, 1377 (1989).
53. L.A. Coldren and S.W. Corzine, *Diode Lasers and Photonic Integrated Circuits*, Wiley, New York, 1995.
54. D.I. Babic and S.W. Corzine, *IEEE J. Quantum Electron.* **28**, 514 (1992).
55. D.L. Huffaker, L.A. Graham, H. Deng, and D.G. Deppe, *IEEE Photon. Tech. Lett.* **8**, 974 (1996).
56. E.F. Schubert, L.W. Lu, G.J. Zydzik, R.F. Kopf, A. Benvenuti, and M.R. Pinto, *Appl. Phys. Lett.* **60**, 466 (1992).
57. S.A. Chalmers, K.L. Lear, and K.P. Killeen, *Appl. Phys. Lett.* **62**, 1585 (1993).
58. M.G. Peters, B.J. Thibeault, D.B. Young, J.W. Scott, F.H. Peters, A.C. Gossard, and L.A. Coldren, *Appl. Phys Lett.* **63**, 3411 (1993).
59. E.R. Hegblom, D.I. Babic, B.J. Thibeault, and L.A. Coldren, *IEEE J. Select. Topics Quantum Electron.* **3**, 379 (1997).
60. E. Yablonovitch, *Phys. Rev. Lett.* **58**, 2059 (1987).
61. S. John, *Phys. Rev. Lett.* **58**, 2486 (1987).
62. H.-Y. Ryu, H.-G. Park, and Y.-H. Lee, *IEEE J. Select. Topics Quantum Electron.* **8**, 891 (2002).
63. J.D. Jackson, *Classical Electrodynamics* Wiley, New York, 1962.
64. J.D. Joannopoulos, R.D. Meade, and J.N. Winn, *Photonic Crystals*, Princeton University Press, Princeton, NJ, 1995.
65. O.J. Painter, R.K. Lee, A. Scherer, A. Yariv, J.D. O'Brien, P.D. Dapkus, and I. Kim, *Science* **284**, 1819 (1999).
66. O.J. Painter, A. Husain, A. Scherer, J.D. O'Brien, I. Kim, and P.D. Dapkus, *J. Lightwave Tech.* **17**, 2082 (1999).
67. J.K. Hwang, H.Y. Ryu, D.S. Song, I.Y. Han, H.W. Song, H.K. Park, Y.H. Lee, and D.H. Jang, *Appl. Phys. Lett.* **76**, 2982 (2000).
68. J.K. Hwang, H.Y. Ryu, D.S. Song, I.Y. Han, H.K. Park, D.H. Jang, and Y.H. Lee, *IEEE Photon. Tech. Lett.* **12**, 1295 (2000).
69. D.S. Song, S.H. Kim, H.G. Park, C.K. Kim, and Y.H. Lee, *Appl. Phys. Lett.* **80**, 3901 (2002).
70. J. O'Brien, W. Kuang, P.-T. Lee, J.R. Cao, C. Kim, W. Kim, *Proc. SPIE* **4942**, 193 (2003).
71. S.L. McCall, A.F.J. Levi, R.E. Slusher, S.J. Pearton, and R.A. Logan, *Appl. Phys. Lett.* **60**, 289 (1992).
72. M. Fujita, A. Sakai, and T. Baba,*IEEE J. Select. Topics Quantum Electron.* **5**, 673 (1999).
73. N.C. Frateschi and A.F.J. Levi, *J. Appl. Phys.* **80**, 644 (1996).
74. A.F.J. Levi, *Solid State Electron.* **37**, 1297 (1994).
75. B. Gayral, J.M. Gerard, A. Lemaitre, C. Dupuis, L. Manin, and J. L. Pelouard, *Appl. Phys. Lett.* **75**, 1908 (1999).
76. M. Huang, S.S. Mao, H. Feick, H. Yan, Y. Wu, H. Kind, E.R. Weber, R.E. Russo, and P. Yang, *Science* **292**, 1897 (2001).
77. S.S. Mao, R.E. Russo, and P. Yang, *Proc. SPIE* **4608**, 225 (2001).
78. G.C. Osbourn, *J. Appl. Phys.* **53**, 1586 (1982).

79. A. Cho, *J. Vac. Sci. Technol.* **8**, S31 (1971).
80. A.Y. Cho, *Appl. Phys. Lett.* **19**, 467 (1971).
81. H.M. Manasevit, *Appl. Phys. Lett.* **12**, 156 (1968).
82. R.D. Dupuis and P.D. Dapkus, *Appl. Phys. Lett.* **31**, 466 (1977).
83. M.B. Panish, H. Temkin, and S. Sumski, *J. Vac. Sci. Technol.* **B3**, 657 (1985).
84. Z.I. Alferov, V.M. Andreev, V.I. Korolkov, D.N. Tretyakov, and V.M. Tuchkevich, *Sov. Phys. Semicond.* **1**, 1313 (1968).
85. R. Dingle, W. Wiegmann, and C.H. Henry, *Phys. Rev. Lett.* **33**, 827 (1974).
86. J.P. van der Ziel, R. Dingle, R.C. Miller, W. Wiegmann, and W.A. Nordland, Jr., *Appl. Phys. Lett.* **26**, 463 (1975).
87. R.D. Dupuis, P.D. Dapkus, N. Holonyak, Jr., E.A. Rezek, and R. Chin, *Appl. Phys. Lett.* **32**, 295 (1978).
88. W.W. Chow, K.D. Choquette, M.H. Crawford, K.L. Lear, and G.R. Hadley, *IEEE J. Quantum Electron.* **33**, 1810 (1997).
89. E. Towe, R.F. Leheny,and A. Yang, *IEEE J. Select. Topics Quantum Electron.* **6**, 1458 (2000).
90. R.S. Geels, S.W. Corzine, J.W. Scott, D.B. Young, and L.A. Coldren, *IEEE Photon. Tech. Lett.* **2**, 234 (1990).
91. R.S. Geels and L.A. Coldren, *Appl. Phys. Lett.* **57**, 1605 (1990).
92. A.F.J. Levi, R.E. Slusher, S.L. McCall, T. Tanbun-Ek, D.L. Coblentz, and S.J. Pearton, *Electron. Lett.* **28**, 1010 (1992).
93. A.F.J. Levi, R.E. Slusher, S.L. McCall, S.J. Pearton, and W.S. Hobson, *Appl. Phys. Lett.* **62**, 2021 (1993).
94. R.E. Slusher, A.F.J. Levi, U. Mohideen, S.L. McCall, S.J. Pearton, and R.A. Logan, *Appl. Phys. Lett.* **63**, 1310 (1993).
95. T. Baba, M. Fujita, A. Sakai, M. Kihara, and R. Watanabe, *IEEE Photon. Tech. Lett.* **9**, 878 (1997).
96. T. Baba, *IEEE J. Select. Topics Quantum Electron.* **3**, 808 (1997).
97. J. Faist, F. Capasso, D.L. Sivco, C. Sirtori, A.L. Hutchinson, and A.Y. Cho, *Science* **264**, 553 (1994).
98. G. Scamarcio, F. Capasso, C. Sirtori, J. Faist, A. L. Hutchinson, D. L. Sivco, and A.Y. Cho, *Science* **276**, 773 (1997).
99. R. Kazarinov and R.A. Suris, *Sov. Phys. Semicond.* **5**, 707 (1971).
100. F. Capasso, C. Gmachl, R. Paiella, A. Tredicucci, A. L. Hutchinson, D.L. Sivco, J.N. Baillargeon, A.Y. Cho, and H.C. Liu, *IEEE J. Select. Topics Quantum Electron.* **6**, 931 (2000).
101. F. Capasso, A. Tredicucci, C. Gmachl, D.L. Sivco, A.L. Hutchinson, A.Y. Cho, and G. Scamarcio, *IEEE J. Select. Topics Quantum Electron.* **5**, 792 (1999).
102. E. Kapon, D.M. Hwang, and R. Bhat, *Phys. Rev. Lett.* **63**, 430 (1989).
103. K. Kash, A. Scherer, I.M. Worlock, H.G. Craighead, and M.C. Tamargo, *Appl. Phys. Lett.* **49**, 1043 (1986).
104. D. Gershoni, H. Temkin, C.J. Dolan, J. Dunsmuir, S.N.G. Chu, and M.B. Panish, *Appl. Phys. Lett.* **53**, 995 (1988).
105. B.E. Maile, A. Forchel, R. Germann, D. Grützmacher, H.P. Meier, and J. P. Reithmaier, *J. Vac. Sci. Technol.* **B7**, 2030 (1989).
106. M. Kohl, D. Heitmann, P. Grambow, and K. Ploog, *Phys. Rev. Lett.* **63**, 2124 (1989).
107. E. M. Clausen, Jr., J.P. Harbison, L.T. Florez, and B.P. Van der Gaag, *J. Vac. Sci. Technol.* **B8**, 1960 (1990).
108. A. Izrael, B. Sermage, J.Y. Marzin, A. Ougazzaden, R. Azoulay, J. Etrillard, V. Thierry-Mieg, and L. Henry, *Appl. Phys. Lett.* **56**, 830 (1990).
109. M. Notomi, M. Naganuma, T. Nishida, T. Tamamura, H. Iwamura, S. Nojima, and M. Okamoto, *Appl. Phys. Lett.* **58**, 720 (1991).
110. N. Nunoya, M. Nakamura, H. Yasumoto, S. Tamura, and S. Arai, *Jpn. J. Appl. Phys.* **38**, L1323, (1999).

111. N. Nunoya, M. Nakamura, H. Yasumoto, S. Tamura, and S. Arai, *Jpn. J. Appl. Phys.* **39** 3410, (2000).
112. H. Yagi, K. Muranushi, N. Nunoya, T. Sano, S. Tamura, and S. Arai, *Appl. Phys. Lett.* **81**, 966 (2002).
113. P. Ils, M. Michel, A. Forchel, I. Gyuro, M. Klenk, and E. Zielinski, *Appl. Phys. Lett.* **64**, 496 (1994).
114. J. Cibert, P.M. Petroff, G.J. Dolan, S.J. Pearton, A.C. Gossard, and J.H. English, *Appl. Phys. Lett.* **49**, 1275 (1986).
115. H.A. Zarem, P.C. Sercel, M.E. Hoenk, J.A. Lebens, and K.J. Vahala, *Appl. Phys. Lett.* **54**, 2692 (1989).
116. K.C. Hsieh, J.N. Baillargeon, and K.Y. Cheng, *Appl. Phys. Lett.* **57**, 2244 (1990).
117. K.Y. Cheng, K.C. Hsieh, and J.N. Baillargeon, *Appl. Phys. Lett.* **60**, 2892 (1992).
118. S.T. Chou, K.C. Hsieh, K.Y. Cheng, and L.J. Chou, *J. Vac. Sci. Technol.* **B13**, 650 (1995).
119. S.T. Chou, K.Y. Cheng, L.J. Chou, and K.C. Hsieh, *Appl. Phys. Lett.* **17**, 2220 (1995).
120. S.T. Chou, K.Y. Cheng, L.J. Chou, and K.C. Hsieh, *J. Appl. Phys.* **78**, 6270 (1995).
121. S. Francoeur, M.C. Hanna, A.G. Norman, and A. Mascarenhas, *Appl. Phys. Lett.* **80**, 243, (2002).
122. C.M. Fetzer, R.T. Lee, S.W. Jun, G.B. Stringfellow, S.M. Lee, and T.Y. Seong, *Appl. Phys. Lett.* **78**, 1376 (2001).
123. R.D. Twesten, D.M. Follstaedt, S.R. Lee, E.D. Jones, J.L. Reno, J.M. Millunchick, A.G. Norman, S.P. Ahrenkiel, and A. Mascarenhas, *Phys. Rev.* **B60**, 13,619 (1999).
124. V.M. Petroff, A.C. Gossard, and W. Wiegmann, *Appl. Phys. Lett.* **45**, 620 (1984).
125. J.M. Gaines, P.M. Petroff, H. Kroemer, R.I. Simes, R.J. Geels and J.H. English, *J. Vac. Sci. Technol.* **B6**, 1378 (1988).
126. T. Fukui and H. Saito, *Jpn. J. Appl. Phys.* **29**, L731 (1990).
127. S. Hara, J. Ishizaki, J. Motohisa, T. Fukui, and H. Hasegawa, *J. Cryst. Growth* **145**, 692 (1994).
128. M. Yamamoto, M. Higashiwaki, S. Shimomura, N. Sano, and S. Hiyamizu, *Jpn. J. Appl. Phys.* **36**, 6285 (1997).
129. M. Higashiwaki, S. Ikawa, S. Shimomura, and S. Hiyamizu, *J. Cryst. Growth* **201**, 886 (1999).
130. A. Chavez-Pirson, H. Ando, H. Saito, and H. Kanbe, *Appl. Phys. Lett.* **62**, 3082 (1993).
131. H. Nakashima, T. Kato, K. Maehashi, T. Nishida, Y. Inoue, T. Takeuchi, K. Inoue, P. Fischer, J. Christen, M. Grundmann, and D. Bimberg, *Mater. Sci. Eng.* **B51**, 229 (1998).
132. E. Kapon, M.C. Tamargo, and D.M. Hwang, *Appl. Phys. Lett.* **50**, 347 (1987).
133. R. Bhat, E. Kapon, D.M. Hwang, M.A. Koza, and C.P. Yun, *J. Cryst. Growth* **93**, 850 (1988).
134. E. Kapon, K. Kash, E.M. Clausen, Jr., D.M. Hwang, and E. Colas, *Appl. Phys. Lett.* **60**, 477 (1992).
135. C. Percival, P.A. Houston, J. Woodhead, M. Al-Khafaji, G. Hill, J.S. Roberts, and A.P. Knights, *IEEE Trans. Electron. Dev.* **47**, 1769 (2000).
136. S. Simhony, E. Kapon, T. Colas, D.M. Hwang, N.G. Stoffel, and P. Worland, *Appl. Phys. Lett.* **59**, 2225 (1991).
137. T.G. Kim, K. Park, E.K. Kim, S. Min, and J. Park, *IEEE Photon. Tech. Lett.* **9**, 2 (1997).
138. T.G. Kim, Y. Suzuki, M. Shimiz, and M. Ogura, *Solid State Electron.* **43**, 2093 (1999).
139. T.G. Kim, X.-L. Wang, K. Komori, K. Hikosaka, and M. Ogura, *Electron. Lett.* **35**, 639 (1999).
140. T.G. Kim, Y. Suzuki, and M. Ogura, *IEEE Photon. Tech. Lett.* **12**, 104 (2000).
141. T.G. Kim, X.-L. Wang, Y. Suzuki, K. Komori, and M. Ogura, *IEEE J. Select. Topics Quantum Electron.* **6**, 511 (2000).
142. S. Koshiba, H. Noge, H. Akiyama, T. Inoshita, Y. Nakamura, A. Shimizu, Y. Nagamune, M. Tsuchiya, H. Kano, and H. Sakaki, *Appl. Phys. Lett.* **64**, 363 (1994).

143. H. Akiyama, S. Koshiba, T. Someya, K. Wada, H. Noge, Y. Nakamura, T. Inoshita, A. Shimizu, and H. Sakaki, *Phys. Rev. Lett.* **72**, 924 (1994).
144. C. Jiang, T. Muranaka, and H. Hasegawa, *Japan. J. Appl. Phys.* **40**, 3003 (2001).
145. C. Jiang and H. Hasegawa, *Jpn. J. Appl. Phys.* **41**, 972 (2002).
146. Y.C. Chang, L.L. Chang, and L. Esaki, *Appl. Phys. Lett.* **47**, 1324 (1985).
147. H. Akiyama, *J. Phys. Condens. Mat.* **10**, 3095 (1998).
148. L.N. Pfeiffer, K.W. West, H. L. Störmer, J. P. Eisenstein, K.W. Baldwin, D. Gershoni, and J. Spector, *Appl. Phys. Lett.* **56**, 1697 (1990).
149. L.N. Pfeiffer, H.L. Störmer, K.W. Baldwin, K.W. West, A.R. Goñi, A. Pinczuk, R.C. Ashoori, M.M. Dignam, and W. Wegscheider, *J. Cryst. Growth* **127**, 849 (1993).
150. D. Gershoni, J.S. Weiner, S.N.G. Chu, G.A. Baraff, J.M. Vandenberg, L. N. Pfeiffer, K.W. West, R.A. Logan, and T. Tanbun-Ek, *Phys. Rev. Lett.* **65**, 1631 (1990).
151. A.R. Goñi, L.N. Pfeiffer, K.W. West, A. Pinczuk, H.U. Baranger, and H.L. Störmer, *Appl. Phys. Lett.* **61**, 1956 (1992).
152. T. Someya, H. Akiyama, and H. Sakaki, *J. Appl. Phys.* **79**, 2522 (1996).
153. W. Wegscheider, L.N. Pfeiffer, M.M. Dignam, A. Pinczuk, K.W. West, S.L. McCall, and R. Hull, *Phys. Rev. Lett.* **71**, 4071 (1993).
154. W. Wegscheider, L.N. Pfeiffer, K.W. West, and R.E. Leibenguth, *Appl. Phys. Lett.* **65**, 2510 (1994).
155. R.S. Wagner and W.C. Ellis, *Appl. Phys. Lett.* **4**, 89 (1964).
156. K. Hiruma, M. Yazawa, T. Katsuyama, K. Ogawa, K. Haraguchi, M. Koguchi, and H. Kakibayashi, *J. Appl. Phys.* **77**, 447 (1995).
157. P. Yang, Y. Wu, and R. Fan, *Int. J. Nanosci.* **1**, 1 (2002).
158. J. Hu, T.W. Odom, and C.M. Lieber, *Acc. Chem. Res.* **32**, 435 (1999).
159. Y. Wu and P. Yang, *Chem. Mater.* **12**, 605 (2000).
160. A.M. Morales and C.M. Lieber, *Science* **279**, 208 (1998).
161. X.F. Duan, J.F. Wang, and C.M. Lieber, *Appl. Phys. Lett.* **76**, 1116 (2000).
162. Y. Zhang, R. Russo, and S.S. Mao, *Appl. Phys. Lett.*, **87**, 133115 (2005).
163. T.J. Trentler, K.M. Hickman, S.C. Geol, A.M. Viano, P.C. Gibbons, and W.E. Buhro, *Science* **270**, 1791 (1995).
164. C.R. Martin, *Science* **266**, 1961 (1994).
165. E. Braun, Y. Eichen, U. Sivan, and G. Ben-Yoseph, *Nature* **391**, 775 (1998).
166. L. Vayssieres, *Int. J. Nanotech.* **1**, 1 (2004).
167. Y. Xia, *Int. J. Nanotech.* **1**, 86 (2004).
168. M. Cao, Y. Miyake, S. Tamura, H. Hirayama, S. Arai, Y. Suematsu, and Y. Miyamoto, *Trans. IEICE* **E73**, 63 (1990).
169. M. Cao, P. Daste, M. Miyamoto, Y. Miyake, S. Nogiwa, S. Arai, K. Furuya, and Y. Suematsu, *Electron. Lett.* **24**, 824 (1988).
170. T. Kojima, M. Tamura, H. Nakaya, S. Tanaka, S. Tamura, and S. Arai, *Jpn. J. Appl. Phys.* **37**, 4792 (1998).
171. H. Yagi, K. Muranushi, N. Nunoya, T. Sano, S. Tamura, and S. Arai, *Jpn. J. Appl. Phys.* **41**, L186 (2002).
172. N. Nunoya, H. Yasumoto, H. Midorikawa, S. Tamura, and S. Arai, *Jpn. J. Appl. Phys.* **39**, L1042 (2000).
173. N. Nunoya, M. Nakamura, M. Morshed, S. Tamura, and S. Arai, *IEEE J. Select. Topics Quantum Electron.* **7**, 249 (2001).
174. M. Nakamura, N. Nunoya, H. Yasumoto, M. Morshed, K. Fukuda, S. Tamura, and S. Arai, *Electron. Lett.* **36**, 639 (2000).
175. J. Yoshida and K. Kishino, *IEEE Photon. Tech. Lett.* **7**, 241 (1995).
176. S.T. Chou, D.E. Wohlert, K.Y. Cheng, and K.C. Hsieh, *J. Appl. Phys.* **83**, 3469 (1998).
177. D.E. Wohlert, K.Y. Cheng, and S.T. Chou, *Appl. Phys. Lett.* **78**, 1047 (2001).
178. S. Hara, J. Motohisa, and T. Fukui, *Solid State Electron.* **42**, 1233 (1998).
179. M. Tsuchiya, P.M. Petroff, and L.A. Coldren, *IEEE Trans. Electron. Dev.* **36**, 2612 (1989).

180. S.Y. Hu, M.S. Miller, D.B. Young, J.C. Yi, D. Leonard, A.C. Gossard, P.M. Petroff, L.A. Coldren, and N. Dagli, *Appl. Phys. Lett.* **63**, 2015 (1993).
181. S.Y. Hu, J.C. Yi, M.S. Miller, D. Leonard, D.B. Young, A.C. Gossard, N. Dagli, P.M. Petroff, and L.A. Coldren, *IEEE J. Quantum Electron.* **31**, 1380 (1995).
182. H. Saito, K. Uwai, and N. Kobayashi, *Jpn. J. Appl. Phys.* **32**, 4440 (1993).
183. S. Hara, J. Motohisa, and T. Fukui, *Electron. Lett.* **34**, 894 (1998).
184. A. Chavez-Pirson, H. Ando, H. Saito, and H. Kanbe, *Appl. Phys. Lett.* **64**, 1759 (1994).
185. Y. Ohno, H. Kanamori, S. Shimomura, and S. Hiyamizu, *Physica* **E13**, 892 (2002).
186. E. Kapon, S. Simhony, R. Bhat, and D.M. Hwang, *Appl. Phys. Lett.* **55**, 2715 (1989).
187. S. Tiwari, G.D. Pettit, K.R. Milkove, F. Legoues, R.J. Davis, and J.M. Woodall, *Appl. Phys. Lett.* **64**, 3536 (1994).
188. Y. Qian, Z.T. Xu, J.M. Zhang, L.H. Chen, Q.M. Wang, L.X. Zheng, and X.W. Hu, *Electron. Lett.* **3**, 102 (1995).
189. D. Piester, P. Bönsch, T. Schrimpf, H.- H. Wehmann, and A. Schlachetzki, *IEEE J. Select. Topics Quantum Electron.* **6**, 522 (2000).
190. T.G. Kim, C.S. Son, and M. Ogura, *IEEE Photon. Tech. Lett.* **13**, 409 (2001).
191. L. Sirigu, D.Y. Oberli, L. Degiorgi, A. Rudra, and E. Kapon, *Phys. Rev.* **B61**, R10575 (2000).
192. T.G. Kim, X.-L. Wang, R. Kaji, and M. Ogura, *Physica* **E7**, 508 (2000).
193. S. Watanabe, S. Koshiba, M. Yoshita, H. Sakaki, M. Baba, and H. Akiyama, *Appl. Phys. Lett.* **73**, 511 (1998).
194. S. Watanabe, S. Koshiba, M. Yoshita, H. Sakaki, M. Baba, and H. Akiyama, *Appl. Phys. Lett.*, **75**, 2190 (1999).
195. L. Sorba, G. Schedelbeck, W. Wegscheider, M. Bichler, and G. Abstreiter, *Phys. Status Solidi* **A178**, 227 (2000).
196. Y. Hayamizu, M. Yoshita, S. Watanabe, H. Akiyama, L. N. Pfeiffer, and K. W. West, *Appl. Phys. Lett.* **81**, 4937 (2002).
197. P. Fons, K. Iwata, A. Yamada, K. Matsubara, S. Niki, K. Nakahara, T. Tanabe, and H. Takasu, *Appl. Phys. Lett.* **77**, 1801 (2000).
198. D.M. Bagnall, Y.F. Chen, Z. Zhu, T. Yao, S. Koyama, M.Y. Shen, and T. Goto, *Appl. Phys. Lett.* **70**, 2230 (1997).
199. P. Yu, Z.K. Tang, G.K.L. Wong, M. Kawasaki, A. Ohtomo, H. Koinuma, and Y. Segawa, *J. Cryst. Growth* **184**, 601 (1998).
200. H. Cao, J.Y. Xu, D.Z. Zhang, S.H. Chang, S.T. Ho, E.W. Seelig, X. Liu, and R.P.H. Chang, *Phys. Rev. Lett.* **84**, 5584 (2000).
201. P. Yang, H. Yan, S.S. Mao, R.E. Russo, J.C. Johnson, R.J. Saykally, N. Morris, J. Pham, R. He, and H.J. Choi, *Adv. Funct. Mater.* **12**, 323 (2002).
202. J.C. Johnson, H. Yan, R.D. Schaller, L.H. Haber, R.J. Saykally, and P. Yang, *J. Phys. Chem.* **B105**, 11,387 (2001).
203. J.C. Johnson, H.J. Choi, K.P. Knutsen, R.D. Schaller, P. Yang, and R.J. Saykally, *Nat. Mater.* **1**, 106 (2002).
204. X. Duan, Y. Huang, R. Agarwal, and C.M. Lieber, *Nature* **421**, 241 (2003).
205. V.M. Ustinov, A.E. Zhukov, A.Y. Egorov, N.A. Maleev, *Quantum Dot Lasers*, Oxford University Press, New York, 2003.
206. P. Michler, A. Imamoglu, M.D. Mason, P.J. Carson, G.F. Strouse, and S.K. Buratto, *Nature* **406**, 968 (2000).
207. P. Michler, A. Kiraz, C. Becher, W.V. Schoenfeld, P.M. Petroff, L. Zhang, E. Hu, and A. Imamoglu, *Science* **290**, 2282 (2000).
208. C. Santori, M. Pelton, G. Solomon, Y. Dale, and Y. Yamamoto, *Phys. Rev. Lett.* **86**, 1502 (2001).
209. D. Bimberg, M. Grundmann, and N.N. Ledentsov, *Quantum Dot Heterostructures*, Wiley, New York, 1999.

210. H.W. Lehmann, *J. Vac. Sci. Technol.* **B6**, 1881 (1988).
211. P. Grambow, T. Demel, D. Heitmann, M. Kohl, R. Schule, and K. Ploog, *Micro electron. Eng.* **9**, 357 (1989).
212. M.A. Reed, J.N. Randall, R.J. Aggarwal, R.J. Matyi, T.M. Moore, and A.E. Wetsel, *Phys. Rev. Lett.* **60**, 535 (1988).
213. R. Cheung, Y.H. Lee, C.M. Knowdler, K.Y. Lee, T.P. Smith, and D.P. Kern, *Appl. Phys. Lett.* **54**, 2130 (1989).
214. G. Mayer, B.E. Maile, R. Gernann, A. Forchel, P. Crambow, and H.P. Meier, *Appl. Phys. Lett.*, **56**, 2016 (1990).
215. M. Tabuchi, S. Noda, and A. Sasaki, in *Science and Technology of Mesoscopic Structures* edited by S. Namba, C. Hamaguchi, and T. Ando, Springer-Verlag, Tokyo, 1992.
216. V.A. Shchukin, N.N. Ledenstov, P.S. Kop'ev, and D. Bimberg. *Phys. Rev. Lett.* **75**, 2968 (1995).
217. E. Bauer and H. Poppa, *Thin Solid Films* **12**, 167 (1972).
218. D.J. Eaglesham and M. Cerullo, *Phys. Rev. Lett.* **64**, 1943 (1990).
219. D. Leonard, M. Krishnamurthy, C.M. Reaves, S.P. DenBaars, and P.M Petroff, *Appl. Phys. Lett.* **63**, 3203 (1993).
220. J.M. Moison, F. Housay, F. Barthe, L. Leprince, E. Andre, and O. Vatel, *Appl. Phys. Lett.* **64**, 196 (1994).
221. D. Leonard, K. Pond, and P.M. Petroff, *Phys. Rev.* **B50**, 11687 (1994).
222. H. Drexler, D. Leonard, W. Hansen, J.P. Kotthaus, and P.M. Petroff, *Phys. Rev. Lett.* **73**, 2252 (1994).
223. G. Medeiros-Ribeiro, D. Leonard, and P.M. Petroff, *Appl. Phys. Lett.* **66**, 1767 (1995).
224. M. Grundmann, O. Stier, and D. Bimberg, *Phys. Rev.* **B62**, 1969 (1995).
225. R. Heitz, M. Grundmann, N.N. Ledentsov, L. Eckey, M. Veit, D. Bimberg, V.M. Ustinov, A. Yu. Egorov, A.E. Zhukov, P.S. Kop'ev, and Z.I. Alferov, *Appl. Phys. Lett.* **68**, 361 (1996).
226. M. Grundmann, N.N. Ledentsov, O. Stier, J. Bohrer, D. Bimberg, V.M. Ustinov, P.S. Kop'ev, and Z.I. Alferov, *Phys. Rev.* **B53**, R10,509 (1996).
227. M.A. Cusack, P.R. Briddon, and M. Jaros, *Phys. Rev.* **B54**, R2300 (1996).
228. G. Medeiros-Ribeiro, F.G. Pikus, P.M. Petroff, and A.L. Efros, *Phys. Rev.* **B55**, 1568 (1997).
229. G. Walter, N. Holonyak, Jr., J.H. Ryou, and R.D. Dupuis, *Appl. Phys. Lett.* **79**, 3215 (2001).
230. M. Klude, T. Passow, R. Kroger, and D. Hommel, *Electron. Lett.* **37**, 1119 (2001).
231. B. Daudin, F. Widmann, G. Feuillet, Y. Samson, M. Arlery, and J. L. Rouviere, *Phys. Rev.* **B56**, R7069 (1997).
232. C.B. Murray, C.R. Kagan, and M.G. Bawendi, *Annu. Rev. Mater. Sci.* **30**, 545 (2000).
233. H. Reiss, *J. Chem. Phys.* **19**, 482 (1951).
234. C.B. Murray, D.J. Norris, and M.G. Bawendi, *J. Am. Chem. Soc.* **115**, 8706 (1993).
235. X. Peng, J. Wickham, and A.P. Alivisatos, *J. Am. Chem. Soc.* **120**, 5343 (1998).
236. D.V. Talapin, E.V. Shevchenko, A. Kornowski, N. Gaponik, M. Haase, A.L. Rogach, and H. Weller. *Adv. Mater.* **13**, 1868 (2001).
237. O.I. Micic, C.J. Curtis, K.M. Jones, J.R. Sprague, and A.J. Nozik, *J. Phys. Chem.* **98**, 4966 (1994).
238. A.A.Guzelian, J.E.B. Katari, A.V. Kadavanich, U. Banin, K. Hamad, E. Juban, A.P.Alivisatos, R.H. Wolters, C.C. Arnold, and J.R. Heath, *J. Phys. Chem.* **100**, 7212 (1996).
239. O.I. Micic, H.F. Cheong, A. Zunger, J.R. Sprague, A. Mascarenhas, and A.J. Nozik, *J. Phys. Chem.* **B101**, 4904 (1997).
240. O.I. Micic, J.R. Sprague, C.J. Curtis, K.M. Jones, J.L. Machol, A.J. Nozik, H. Giessen, B. Fluegel, G. Mohs, and N. Peyghambarian, *J. Phys. Chem.* **99**, 7754 (1995).
241. A.A. Guzelian, U. Banin, A.V. Kadavanich, X. Peng, and A.P. Alivisatos, *Appl. Phys. Lett.* **69**, 1432 (1996).
242. U. Banin, C.J. Lee, A.A. Guzelian, A.V. Kadavanich, A.P. Alivisatos, W. Jakolski, G.W. Bryant, A.L. Efros, and M. Rosen, *J. Chem. Phys.* **109**, 2306 (1998).

243. R.L. Wells, C.G. Pitt, A.T. McPhail, A.P. Purdy, S. Schafieezad, and R. B. Hallock, *Chem. Mater.* **1**, 4 (1989).
244. M.L. Steigerwald, A.P. Alivisatos, J.M. Gibson, T.D. Harris, R. Kortan, A.J. Muller, A.M. Thayer, T.M. Duncan, L.E. Douglass, and L.E. Brus, *J. Am. Chem. Soc.* **110**, 3046 (1990).
245. A.R. Kortan, R. Hull, R.L. Oplia, M.G. Bawendi, M.L. Steigerwald, P.J. Carroll, and L.E. Brus, *J. Am. Chem. Soc.* **112**, 1322 (1990).
246. Y. Golan, L. Margulis, I. Rubinstein, and G. Hodes, *Langmuir* **8**, 749 (1992).
247. B. Alperson, S. Cohen, I. Rubinstein, and G. Hodes, *Phys. Rev.* **B52**, 17,017 (1995).
248. H. Hirayama, K. Matsunaga, M. Asada, and Y. Suematsu, *Electron. Lett.* **30**, 142 (1994).
249. N. Kirstaedter, N.N. Ledentsov, M. Grundmann, D. Bimberg, V.M. Ustinov, S.S. Ruvimov, M.V. Maximov, P.S. Kop'ev, Z.I. Alferov, U. Richter, P. Werner, U.Gösele, and J. Heydenreich, *Electron. Lett.* **30**, 1416 (1994).
250. N.N. Ledentsov, V.M. Ustinov, A. Yu. Egorov, A.E. Zhukov, M.V. Maximov, I.G. Tabatadze, and P.S. Kop'ev, *Semiconductors* **28**, 832 (1994).
251. N.N. Ledentsov, M. Grundmann, F. Heinrichsdorff, D. Bimberg, V.M. Ustinov, A.E. Zhukov, M.V. Maximov, Z.I. Alferov, and J.A. Lott, *IEEE J. Select. Topics Quantum Electron.* **6**, 439 (2000).
252. M.V. Maximov, I.V. Kochnev, Yu.M. Shernyakov, S.V. Zaitsev, N.Yu. Gordeev, A.F. Tsatsul'nikov, A.V. Sakharov, I.L. Krestnikov, P.S. Kop'ev, Z.I. Alferov, N.N. Ledentsov, D. Bimberg, A.O. Kosogov, P. Werner, and U. Gösele, *Jpn. J. Appl. Phys.* **36**, 4221 (1997).
253. N.N. Ledentsov, V.A. Shchukin, M. Grundmann, N. Kirstaedter, J. Böhrer, O. Schmidt, D. Bimberg, S.V. Zaitsev, V.M. Ustinov, A.E. Zhukov, P.S. Kop'ev, Z.I. Alferov, A.O. Kosogov, S.S. Ruvimov, P. Werner, U. Gösele, and J. Heydenreich, *Phys. Rev.* **B54**, 8743 (1996).
254. C. Ribbat, R. Sellin, M. Grundmann, and D. Bimberg, *Phys. Status Solidi* **B224**, 819 (2001).
255. G.T. Liu, A. Stintz, H. Li, T.C. Newell, A.L. Gray, P.M. Varangis, K.J. Malloy, and L.F. Lester, *IEEE J. Quantum Electron.* **36**, 1272 (2000).
256. R.L. Sellin, Ch. Ribbat, M. Grundmann, N.N. Ledentsov, and D. Bimberg, *Appl. Phys. Lett.* **78**, 1207 (2001).
257. D.L. Huffaker, O. Baklenov, L.A. Graham, B.G. Streetman, and D.G. Deppe, *Appl. Phys. Lett.* **70**, 2356 (1997).
258. J.A. Lott, N.N. Ledentsov, V.M. Ustinov, A.Y. Egorov, A.E. Zhukov, P.S. Kop'ev, Z.I. Alferov, and D. Bimberg, *Electron. Lett.* **33**, 1150 (1997).
259. M. Arzberger, G. Bohm, M.-C. Amann, and G. Abstreiter, *Appl. Phys. Lett.* **79**, 1766 (2001).
260. D.K. Young, L. Zhang, D.D. Awschalom, and E.L. Hu, *Phys. Rev.* **B66** 081307 (2002).
261. T. Yoshie, O.B. Shchekin, H. Chen, D.G. Deppe, and A. Scherer, *Electron. Lett.* **38**, 967 (2002).
262. K. Tachibana, Someya, Y. Arakawa, R. Werner, and A. Forchel, *Appl. Phys. Lett.* **75**, 2605 (1999).
263. V.I. Klimov, A.A. Mikhailovsky, S. Xu, A. Malko, J.A. Hollingsworth, C.A. Leatherdale, H.-J. Eisler, and M.G. Bawendi, *Science* **290**, 314 (2000).
264. M. Farhoud, J. Fererra, A.J. Lochtefeld, T.E. Murphy, M.L. Schattenburg, J. Carter, C.A. Ross, and H.I. Smith, *J. Vac. Sci. Technol.* **B17**, 3182 (1999).
265. T. Ogawa and T. Takagahara, *Phys. Rev.* **B44**, 8138 (1991).
266. S. Nojima, *Phys. Rev.* **B50**, 2306 (1994).
267. M. Grundmann and D. Bimberg, *Phys. Rev.* **B55**, 4054 (1997).
268. F. Rossi, G. Goldoni, and E. Molinari, *Phys. Rev. Lett.* **78**, 3527 (1997).
269. S.D. Sarma and D.W. Wang, *Phys. Rev. Lett.* **84**, 2010 (2000).
270. A. Kuther, M. Bayer, T. Gutbrod, A. Forchel, P.A. Knipp, T.L. Reinecke, and R. Werner, *Phys. Rev.* **B58**, 15744 (1998).

271. C. Constantin, E. Martinet, D.Y. Oberli, E. Kapon, B. Gayral, and J.-M. Gerard, *Phys. Rev.* **B66**, 165306 (2002).
272. J. Vuckovic, M. Pelton, A. Scherer, and Y. Yamamoto, *Phys. Rev.* **A66**, 23,808 (2002).
273. Y. Yamamoto, F. Tassone, and H. Cao, *Semiconductor Cavity Quantum Electrodynamics*, Springer-Verlag, Berlin, 2000.

4

Silicon Nanocrystal Nonvolatile Memory

R. A. Rao,* M. A. Sadd, R. F. Steimle, C. T. Swift, H. Gasquet, and M. Stoker

4.1. INTRODUCTION

Silicon nanocrystal memory devices [1, 2] such as shown in Fig. 4.1, offer the potential to solve the challenging problem of scaling nonvolatile memories. Scaling of floating-gate (FG) nonvolatile memory cells has been limited to bottom oxide thicknesses in the range of 80–110 Å primarily because of the vulnerability to charge loss from the conducting FG through isolated defects in the tunnel oxide that arise after repeated write/erase operations. As a result the FG, operating voltages are in the range of 16–20 V required for erasing the memory cell by Fowler–Nordheim tunneling of carriers from the FG to the channel. This voltage is sometimes split as ± 8 to ± 10 V using fully isolated wells. Silicon nanocrystal memory cells that store charge in isolated centers inside a gate dielectric are less susceptible to charge loss through isolated defect paths in the tunnel oxide due to their discontinuous nature of charge storage. In other words, an underlying oxide defect leads to charge loss only from charge storage sites in its immediate proximity. Once the impact of defect-mediated charge loss is mitigated, charge loss is primarily due to tunneling and the tunnel oxide in these devices can be scaled down to about 50–60 Å based on retention-time requirements. The scaling of the tunnel oxide results in embedded memory modules that can operate with a maximum on-chip voltage of ± 6 V, allowing reduction of the memory module size by up to a factor of 2 at the 90-nm technology node, as shown in Fig. 4.2 [3]. Furthermore, this reduction in operating voltage enables sharing of logic I/O device implants with the high-voltage periphery devices, which are used to charge and discharge the

* Technology Solutions Organization, Freescale Semiconductor, Inc., Austin, TX 78721
rajesh.rao@freescale.com

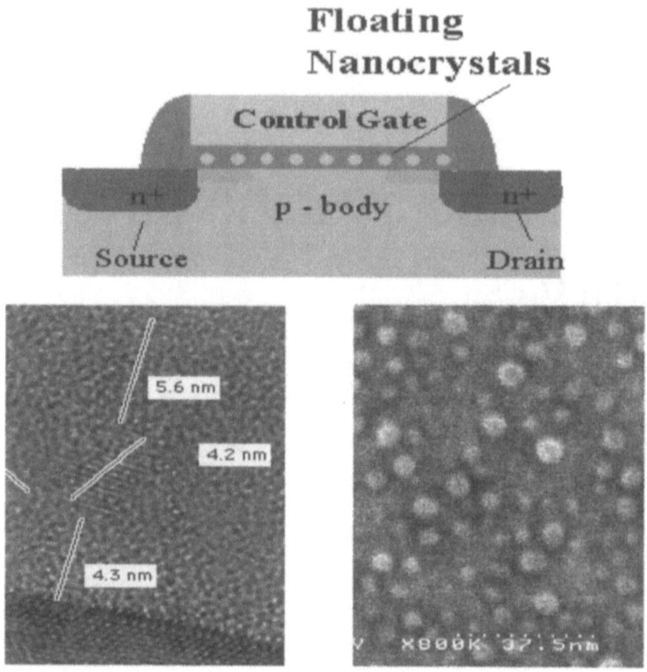

FIGURE 4.1. Silicon nanocrystal nonvolatile memory bitcell showing the floating silicon nanocrystals used for isolated charge storage. A cross-section transmission electron microscopic image through the gate stack of a bitcell and a plan view scanning electron microscopic image of the nanocrystals is also shown.

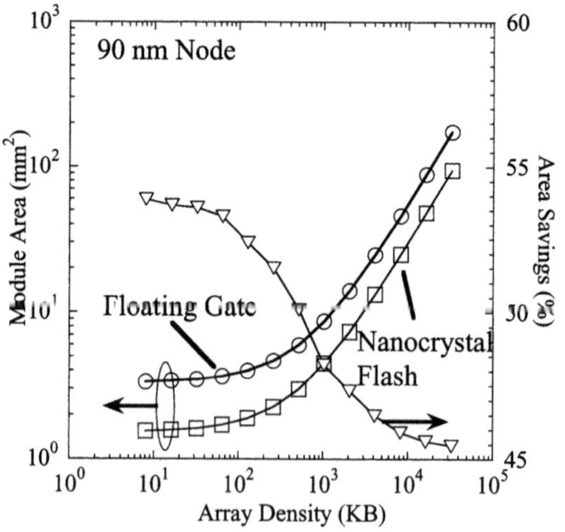

FIGURE 4.2. Memory module size for both conventional FG nonvolatile memory (NVM) and nanocrystal-based NVM showing the approximate factor-of-2 reduction in memory module size for nanocrystal-based NVM.

Process Step	Logic	Floating Gate	Nano-crystal
Isolation Formation	●	●	●
High Voltage Wells		2 masks	
NVM Array Well (1 mask)		1 mask	1 mask
Tunnel Oxidation		●	
Floating Gate Deposition/Patterning		1 mask	
ONO/Nanocrystal Dep./Patteming		1 mask	1 mask
Low Voltage Wells	●	●	●
DGO Wells	●	●	●
High Voltage Oxidation/Paterning		1 mask	1 mask
DGO Oxidation/Patterning	●	●	●
Low Voltage Oxide Growth	●	●	●
Gate Depostion	●	●	●
NVM Stack Patterning		1 mask	
NVM Source Halo Implant		1 mask	
NVM Drain Implant		1 mask	1 mask
Gate Patterning	●	●	●
High Voltage LDD Implants		2 masks	
DGO LDD Implants	●	●	●
S/D and Backend Processing	●	●	●
Masking Step Adder		**+11**	**+4**

FIGURE 4.3. Mask adders required to integrate nanocrystal-based NVM into a standard CMOS flow. Substantial reductions in mask count can be achieved over conventional NVM.

memory bitcells in the array. As a result, the integration of the nanocrystal-based bitcell into a conventional complementary metal–oxide–semiconductor (CMOS) flow can be accomplished with the addition of only four noncritical masks over the baseline logic process, as depicted in Fig. 4.3 [4].

New aspects of nanocrystal memories over conventional FG memories include (1) charge confinement effects in small nanocrystals of a size in the range 30–100 Å, (2) tunneling transport of charge to and from nanocrystals, (3) nanocrystal formation by nonlithographic processes, and (4) nanocrystal passivation or protection from the impact of processing subsequent to deposition. In this article, we discuss in detail nonvolatile memory (NVM) devices with silicon nanocrystals emphasizing the new aspects alluded to earlier. The outline of this chapter is as follows: In Section 4.2, we discuss the key physics that govern bitcell characteristics and contrast it with the well-studied Semiconductor-Oxide-Nitride-Oxide-Semiconductor (SONOS) memory [5, 6] case; in Section 4.3, we review the deposition of silicon nanocrystals by chemical vapor deposition (CVD) and their subsequent passivation; in Section 4.4, we discuss memory bitcell data characteristics; in Section 4.5, we highlight key results from 4-Mb nanocrystal memory arrays, and in Section 4.6, we summarize the chapter.

4.2. NANOCRYSTAL MEMORY DEVICE PHYSICS

In this section, we summarize some of the salient aspects of nanocrystal memory devices, such as that shown in Fig. 4.1, in which the nanocrystals are embedded in a dielectric stack between the channel and the gate of a metal–oxide–semiconductor field-effect transistor (MOSFET). Typically, the devices are n-channel MOSFETs due to faster charge transport to and from the nanocrystals with electrons compared to holes. The threshold voltage shift due to electron storage is given by

$$\Delta V_t = \frac{npq}{\varepsilon_{ox}}\left(t_{cntrl} + \frac{\varepsilon_{ox} d}{2\varepsilon_{Si}}\right), \tag{4.1}$$

where n is the nanocrystal number density, p is the average number of electrons stored per nanocrystal, q is the electronic charge, ε_{ox} and ε_{Si} represent the permittivity of oxide and silicon and respectively, d is the nanocrystal diameter, and t_{cntrl} represents the control oxide thickness. The threshold voltage shift arises because the electrons stored in the nanocrystals screen part of the gate bias and this shift can be deduced from the electric field of a sheet of electronic charge situated at the center of spherical nanocrystals [1]. Clearly, for storing a given number density of electrons, np, the susceptibility to isolated defects in the tunnel oxide is mitigated by having a higher density, n, of nanocrystals and minimizing the number, p, of electrons stored per nanocrystal. For nanocrystals to be sufficiently electrically isolated with respect to lateral tunneling transport, their typical separation must be greater than about 40 Å from one another. Having the minimal area fraction of nanocrystals mitigates the probability of defects underlying any nanocrystal. However, charge confinement effects or Coulomb blockade effects to be briefly discussed increase the energy levels when multiple electrons are stored and this becomes pronounced below a nanocrystal size of about 30 Å. The increase of energy levels may be approximately computed from the reversible work needed to charge the nanocrystal with the additional electron and may be computed from the self-capacitance of the nanocrystal, which is given by

$$C = 2\pi\varepsilon d, \tag{4.2}$$

where ε is the dielectric constant of the surrounding medium. The increase in energy for a nanocrystal on addition of the nth electron is given by

$$\Delta E_{n,n-1} = \frac{q^2}{2C}(n^2 - (n-1)^2) = \frac{nq^2}{C} - \frac{q^2}{2C}. \tag{4.3}$$

This implies that the electrochemical potential change due to the addition of each electron is given by

$$\Delta\mu = \Delta E_{n,n-1} - \Delta E_{n-1,n-2} = \frac{q^2}{C}. \tag{4.4}$$

Since the self-capacitance of nanocrystals in the size range of interest is in attofarads (10^{-18} F) or smaller, this change in energy level is significant. For silicon nanocrystals in the size range of a few nanometers, the energy level increase is of the order of a few tenths of an electron volt. In a typical nanocrystal memory, the capacitance of a nanocrystal is higher due to capacitive coupling to other nanocrystals and the substrate. As such, the energy level for a five-electron nanocrystal is about 0.5 eV above the ground state. This increase in energy levels adversely impacts tunneling mediated charge loss through the bottom oxide during retention and through the top oxide during read operations, and thus, limits the number of electrons that can be stored or the memory window. It should be noted, however, that the increase in energy levels facilitates erasure of the memory, as it is easier to remove the charges by Fowler–Nordheim tunneling.

We now present an argument to show that there is an optimum in the size of nanocrystals. Let us assume that the nanocrystals of diameter d are arranged in a square lattice with a fixed edge-to-edge separation of s. Clearly, s is set to be large enough that the tunneling transport from one nanocrystal to another does not occur in the timescale of interest. Since the nanocrystal density is proportional to $(d+s)^{-2}$, the memory window or threshold voltage shift becomes

$$\Delta V_t \propto \frac{p}{(d+s)^2}. \tag{4.5}$$

However due to Coulomb blockade effects discussed earlier, there is a penalty to be paid for multiple-electron storage in a nanocrystal. Since only a fixed increase in energy levels can be tolerated based on data retention and read disturb criteria,

$$p \sim (\Delta\mu)^{-1} \sim d. \tag{4.6}$$

Thus, the memory window or threshold voltage shift becomes

$$\Delta V_t \propto \frac{d}{(d+s)^2}, \tag{4.7}$$

which for a given value of s has a distinct maximum at d equal to s. Since a nanocrystal separation of about 50 Å is deemed necessary to mitigate charge transport from one nanocrystal to another, the optimum nanocrystal diameter is 50 Å and, hence, the optimal nanocrystal density is about 10^{12} cm^{-2}.

We now discuss the Fowler–Nordheim (FN) erase mechanism of nanocrystal memories and compare it with the erase behavior of SONOS. In the SONOS device,

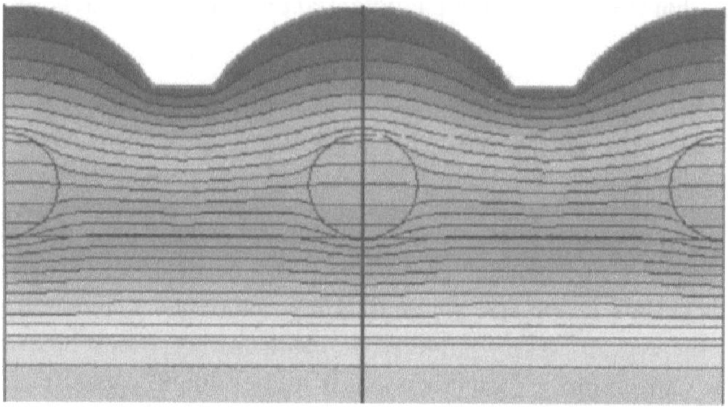

FIGURE 4.4. Simulation of the electrostatic potential contours corresponding to the device in Fig. 4.1.

the FN erase is limited by the saturation of the threshold voltage with gate bias and tunnel oxide thickness [7]. The nanocrystal device, however, does not show this limitation on the tunnel erase, and it continues to have acceptable erase performance with a tunnel oxide sufficiently thick for reliable operation. A simple model for the erase operation illustrates how this is achieved in the nanocrystal device.

To model the erase process, both the charge leaving the nanocrystal and the charge injected from the substrate or from the gate are calculated based on a standard Wentzel-Kramers-Brillouin (WKB) current model [8]. As the nanocrystals discharge, a portion of the back-injected current is captured by the nanocrystals, causing competition with the discharge process. To calculate the amount of charge in the nanocrystal during erase, the Poisson equation is solved at each point in time, and the current continuity equations are integrated forward in time.

For the nanocrystal device, the model must correctly treat the geometry of the nanocrystal embedded inside the gate stack. The conformal deposition of the top oxide around the nanocrystal introduces contours in the polysilicon gate. The calculation of the top and bottom fields accounts for this geometry, as shown in Fig. 4.4. The charge distribution within the nanocrystal results from an approximate solution for the wavefunction inside its potential well, as shown in Fig. 4.5. The

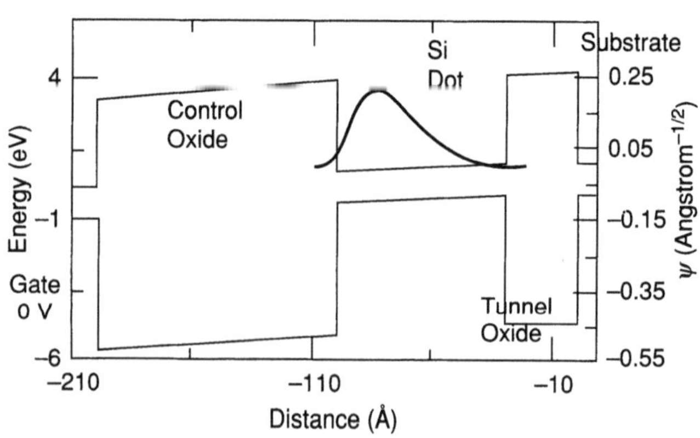

FIGURE 4.5. Energy band diagram of the device in Fig. 4.1 under zero bias.

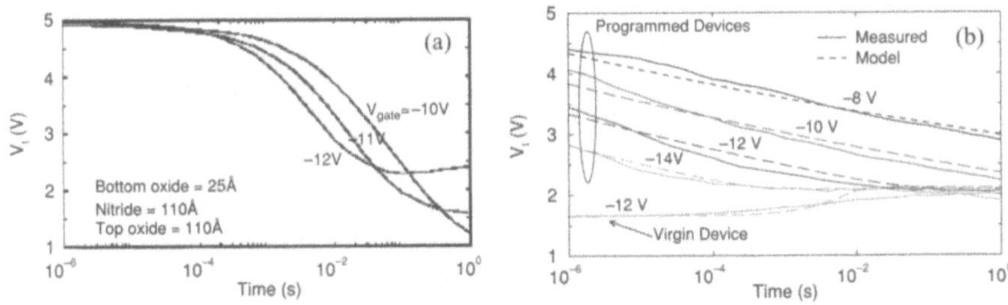

FIGURE 4.6. Erase transients for (a) SONOS memory and (b) a nanocrystal device with a 50-Å bottom oxide, 70-Å top oxide, and nanocrystals with a mean diameter of approximately 60 Å.

Coulomb confinement effect reduces the effective tunnel barrier, as discussed earlier.

Figure 4.6 shows a characteristic set of measured and modeled erase transients as a function of gate voltage for SONOS and nanocrystal memory. The saturation of the erase V_t does not show the bias dependence that is seen in SONOS. In addition, a virgin device increases its threshold voltage upon application of a negative voltage to a point that corresponds to the same saturation V_t as for the erase curves. Another feature of the nanocrystal memory device is that it will erase with a positive gate bias, as shown in Fig. 4.7.

At the point of erase saturation, two competing currents through the bottom and top oxides balance each other, leading to a steady-state condition. The V_t difference between the uncharged and steady-state condition corresponds to approximately one electron stored per nanocrystal. In the case of the nanocrystal memory, both currents result from tunneling from the silicon into the oxide, so that their electric field dependences are similar. If the gate bias is increased after the steady-state condition is met, then the steady state will be maintained, because both the current through the top and bottom oxides are increased by a similar factor. In the case of SONOS, however, the current through the bottom results from a combination

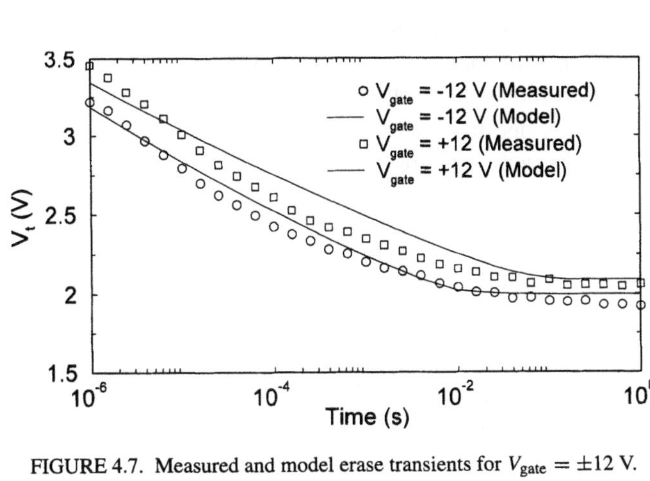

FIGURE 4.7. Measured and model erase transients for $V_{gate} = \pm 12$ V.

of hole injection and electron detrapping from Si_3N_4, both of which respond less to an increase in the field than does tunneling from the gate. In this case, the amount of stored charge and, therefore, the difference between the bottom and top electric fields, must increase as the gate bias increases to maintain the steady-state condition.

Likewise, the nanocrystal device will continue to show acceptable erase performance as the stack thickness increases, as long as field is held constant. In a SONOS device, the FN erase performance degrades with the oxide thickness such that it is not practical to use FN erase in a SONOS device at thicknesses beyond 30 Å.

The slight increase in the speed and depth of erase with positive compared to negative polarity is modeled by degradation of the barrier of the top, deposited oxide relative to that of the grown, tunnel oxide [9], which reduces the amount of back injection from the channel. The effect of this barrier reduction is partly countered by the geometry in Fig. 4.4, which tends to concentrate the field in the bottom oxide.

As discussed earlier, the distributed nature of the charge storage in the nanocrystal memory limits the impact of defects on the device reliability so that the dielectric film may be thinned to the limit implied by tunneling alone. Therefore, an important test of this hypothesis is to compare the observed rate of charge loss with that calculated in a tunneling model. To obtain accurate results, the Schrödinger equation is solved within the effective mass approximation in the region of the nanocrystal, while allowing the evanescent portion of the wavefunction to penetrate into the surrounding silicon dioxide. Since the dominant contribution to the currents derive from the points closest to the substrate or gate, a one-dimensional (1-D) treatment taken along the axis of the nanocrystal describes the device data well. A plot of the resulting lowest-energy wavefunction with the gate and substrate grounded is shown in Fig. 4.5.

The lifetime of the resulting quasi-bound states determines the rate of charge loss [10]. In the weak coupling regime of a thick tunneling oxide, the evanescent portion of the wavefunction, Ψ, may be treated as an incident wave impinging on the interface between the silicon dioxide and substrate. The boundary condition is then solved at this interface, assuming a reflected and transmitted wave. Mathematically, the lifetime of each state results from the introduction of an imaginary part to the energy, which is the well-known approach to treat a quasi-bound level. Physically, however, one may interpret this result as integrating the probability current, which leaks to the substrate,

$$\Gamma = \frac{\hbar^2}{2m^*} \int (\psi^* \nabla \psi - \psi^* \nabla \psi) \, d\vec{S}, \tag{4.8}$$

so the lifetime, τ, is equal to $\hbar/\Gamma$. To obtain the final lifetime for an N-electron system, the rates from each state are thermally averaged.

A kinetic model [11] may then be used to calculate the time evolution of charge loss. The probability, f_N, that the nanocrystal is occupied by N electrons

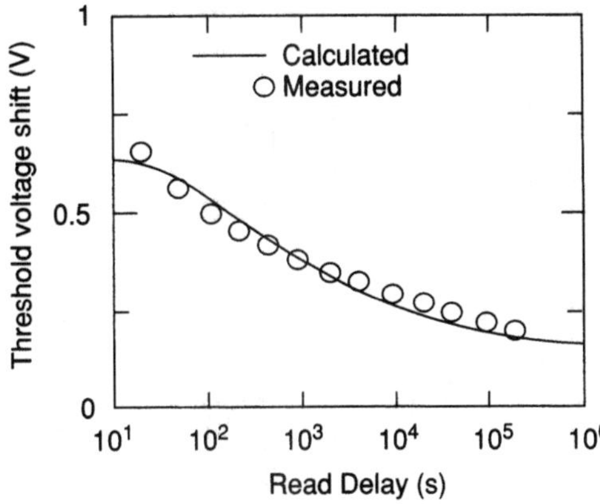

FIGURE 4.8. Comparison between the rate of threshold voltage decay measured in the device described in the text and that calculated using the present model.

obeys a set of kinetic equations,

$$\frac{\partial f_N}{\partial t} = \frac{f_{N+1}}{\tau_{N+1}} - \frac{f_N}{\tau_N}, \tag{4.9}$$

which can be integrated in time to obtain the average charge in the nanocrystal, $\sum N f_N$. Experimental results [12] suggest that charge may occupy interface states associated with the nanocrystal. A kinetic model may account for these states as an additional reservoir of charge that excites into the conduction band.

Figure 4.8 shows the threshold voltage decay of a nanocrystal bitcell at zero bias compared to that predicted by this tunneling mode. The model is seen to describe the reliability data well. The device on which these measurements were conducted had a 30-Å tunnel oxide and a control dielectric with equivalent oxide thickness of 106 Å. The model predicts that the tunnel oxide must be thicker to meet the required reliability for NVM applications [13]. The voltage dependence of charge loss under an accelerating bias provides an additional test for the model. Figure 4.9 shows the time evolution of the V_t under such bias acceleration with measured results on a bitcell with a 37-Å tunnel oxide.

The combined effect of Coulomb confinement energy on the difference between the end-of-life programmed V_t and the saturated erase V_t is estimated in Fig. 4.10. As the nanocrystal becomes small, the energy due to Coulomb confinement increases inversely to the diameter. This added energy has the effect of decreasing the barrier height, reducing the threshold voltage of the programmed state at the end-of-life of the device (taken here as 10 years). Since a single electron is not effected by Coulomb confinement, the saturated erase state remains relatively unaffected. (Note that for very small nanocrystals less than about 30 Å in diameter, increasing the ground state can impact erase also.) The difference between the programmed and erased states (memory window), therefore, narrows as the nanocrystal size approaches traplike, and the resulting window for memory

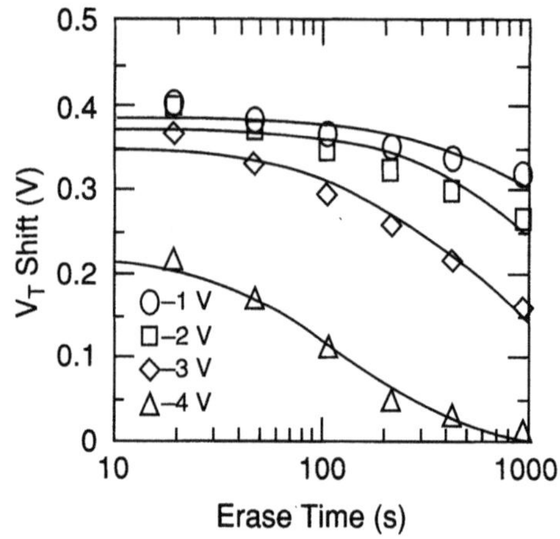

FIGURE 4.9. Comparison between measured discharge rates and model predictions under the indicated erase biases for a nanocrystal memory with 37-Å tunnel oxide.

operation becomes more limited. This simple result illustrates the importance of the stored Coulomb energy of the programmed state to aid in the erase process and to distinguish the programmed state from the erased state when substrate hole injection is limited by a thick bottom oxide. This observation partly explains why the erase in a tunnel erased memory based on silicon-rich oxide [14] loses its memory window as the bottom oxide thickness is increased, just as in the SONOS device, despite the similarity in the trap depth to the nanocrystal device.

Thus, in summary, nanocrystal memory devices offer advantages over SONOS at the expense of processing complexity to be discussed later. The advantages include the ability to use Fowler–Nordheim erasure at tunnel oxide thicknesses needed to mitigate READ disturb. The erase saturation characteristics of SONOS imply that with thicker tunnel oxides, one has to use hot-hole injection [6] for erasure, a process inherently demanding a more robust tunnel oxide. Further, the larger trap depth of nanocrystals compared to nitride traps of SONOS implies better

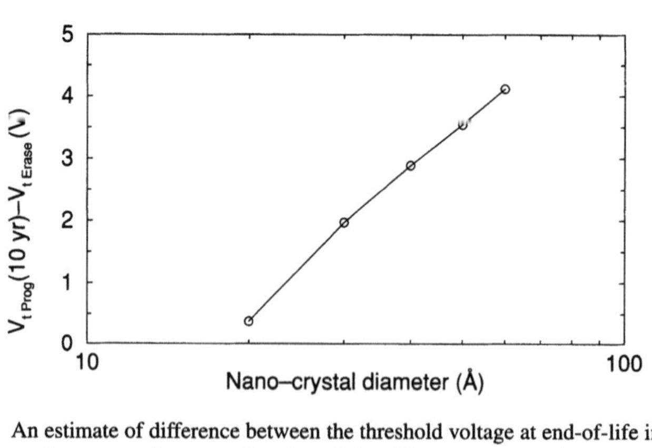

FIGURE 4.10. An estimate of difference between the threshold voltage at end-of-life in the program state and the saturation erase voltage as a function of size. The area coverage is constant at 0.23 Å.

reliability of the programmed state. The ability to physically observe nanocrystal traps (e.g., by Energy Filtered Transmission Electron Microscopy (EFTEM)) facilitates their engineering.

4.3. NANOCRYSTAL ENGINEERING

As has been discussed in the previous section, crucial to making a nanocrystal memory is the ability to form nanocrystals at densities around 10^{12} cm^{-2} and a mean size around 50 Å. Numerous efforts have focused on obtaining a high density of nanocrystals through a variety of techniques, including aerosol, ion implantation, direct CVD, and recrystallization anneal of amorphous Si. In this chapter, we focus on CVD, as the other methods have either not demonstrated required densities in a single layer of nanocrystals (e.g., recrystallization) or result in multilayered nanocrystals without a fixed tunnel oxide thickness (e.g., aerosol deposition). Further, nucleation and growth by CVD provides appropriate simpler processing controls to manipulate the size and density of nanocrystals. Si nanocrystals with number densities between 10^{11} and 10^{12} cm^{-2} have been deposited on various dielectrics such as SiO_2, Si_3N_4, and Al_2O_3 using CVD [15–17]. Si island growth during CVD on amorphous dielectrics such as SiO_2 and Si_3N_4 is believed to proceed by atomistic nucleation, with a critical size of between 1 and 4 atoms. Figure 4.11a shows a typical nucleation and growth curve for Si nanocrystal formation during CVD along with scanning electron microscopic (SEM) images of the surface during various phases of the nucleation and growth curve. During the initial incubation phase, there are not enough adatoms formed on the surface for nucleation to occur and the surface adatom concentration increases with time. Once a sufficient surface concentration of adatoms is attained, nucleation occurs as adatoms diffusively encounter each other to form small clusters. During this nucleation phase, the number of nanocrystals increases rapidly as fresh nuclei are formed. The nanocrystals formed by nucleation grow by adatom attachment

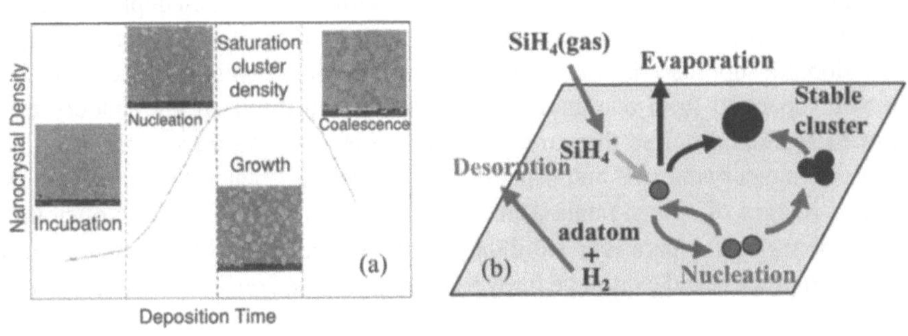

FIGURE 4.11. (a) Typical nucleation and growth curve along with SEM images showing the evolution of nanocrystals on SiO_2 surface during CVD of Si and (b) major processes, including SiH_4 adsorption, adatom reevaporation, and diffusion, as well as H2 desorption, during atomistic nucleation of Si nanocrystals.

through surface diffusion and direct epitaxy. Initially, the growth phase overlaps the nucleation phase. However, once a certain saturation nanocrystal density is attained, fresh nucleation is shut off, as all incoming adatoms are captured by existing nanocrystals. Eventually, the growing clusters merge with adjacent ones by coalescence and the nanocrystal density decreases as a continuous network of clusters is formed. Figure 4.11b depicts the major processes occurring on the dielectric surface during atomistic nucleation of Si nanocrystals by SiH_4 CVD. An incoming SiH_4 molecule from the gas phase is adsorbed on the surface at a physisorption site of surface energy minima and dissociates to form a Si adatom accompanied with H_2 desorption. The formed Si adatom then either contributes to the formation a fresh nanocrystal through atomistic nucleation or is consumed by an existing nanocrystal through surface diffusion. To obtain a high nanocrystal density, a high adatom flux and low surface diffusion is preferred. In addition, fast surface H_2 desorption and low adatom evaporation are also desired. Therefore, by controlling the partial pressure of precursor and surface temperature, Si nanocrystals of the desired size and density can be deposited. In a molecular-beam-epitaxy (MBE)-type deposition, the saturation nanocrystal density, N_s, is given by Venables [18] for initially incomplete condensation and three dimensional island growth as

$$N_s \sim R^{2i/5} \exp\left(\frac{2}{5}\frac{E_i + i E_a}{kT}\right), \tag{4.10}$$

where R is the flux of adatoms, i is the critical size, and E_i and E_a are respectively the binding energy of a cluster of i atoms and the adsorption energy. In general, the saturation density increases with adatom flux and decreasing temperature. In a CVD process, the presence of chemical reaction determines the adatom flux and, as such, the dependencies on temperature can be more complicated.

Another key aspect of nanocrystal formation by CVD is the partial self-ordering of nanocrystals on the surface, as demonstrated by Lombardo et al. [19]. This self-assembly characteristic arises from the diffusive concentration gradient of adatoms around stable nanocrystals. As adatoms are lost to stable clusters, they are depleted in the vicinity of nanocrystals, and within this adatom depletion zone, nucleation events that are at least second order with respect to nanocrystal concentration are suppressed. This necessarily implies that neighboring stable nuclei are well separated from one another at the outset and the resulting nanocrystals are well separated unless they are allowed to grow to coalescence. Figure 4.12 shows the nanocrystal size and separation distributions obtained from plan-view EFTEM images of nanocrystals deposited on SiO_2 using CVD [19]. Notice that the distribution of separations exhibits a peak indicating a spatially nonrandom nucleation process and cannot be fit by a simulation based on a random nucleation model.

We now summarize the salient aspects of a nucleation and growth model presented by Stoker et al. [20] that accounts for atomistic nucleation, growth of nanocrystals by adatom diffusion and epitaxy, as well as coalescence. In order to simplify the model, we assume that all nanocrystals are hemispherical in shape

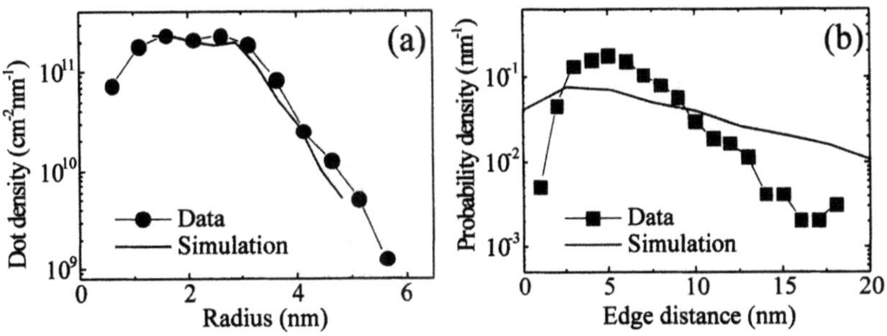

FIGURE 4.12. Measured nanocrystal size distribution (a) and edge-to-edge separation extracted from experimental data. The solid curve in both parts represents simulation based on random nucleation model. After Ref. 18.

and that stable nanocrystals are immobile. Precursor adsorption and dissociation are assumed to occur in a two-step process:

$$\begin{aligned} SiH_4(g) + * &\rightarrow SiH_2(s) + H_2(g), \\ SiH_2(s) &\rightarrow Si(s) + H_2(g), \end{aligned} \tag{4.11}$$

where the asterisk indicates a vacant adsorption site on the SiO_2 surface. Both processes are assumed to be first order with respect to the concentration of the reacting surface species.

The critical size is assumed to be 1 atom and, as such, the nucleation rate is described by the simple rate equation

$$\frac{dN}{dt} = k_n \theta_{Si}^2, \tag{4.12}$$

where θ_{Si} corresponds to the silicon adatom coverage and k_n is a rate constant for the nucleation process.

As discussed earlier, nanocrystals grow by surface adatom diffusion and direct epitaxy. The dotted line in Fig. 4.13a shows the expected evolution of nanocrystal size with time based on solution of the steady-state diffusion equation for the adatom flux to the nanocrystal and including the contribution from direct epitaxy. In order to simplify the calculations, our model approximates this evolution as two linear growth regimes. These are shown as solid lines in Fig. 4.13a along with experimental data. Nanocrystal coalescence, the merging of two adjacent nanocrystals into a single nanocrystal, becomes increasingly important as the nanocrystal surface coverage increases. Coalescence is assumed to occur whenever growth causes two nanocrystals to intersect and the size of the resulting nanocrystal is determined by mass conservation.

Nucleation exclusion zones around nanocrystals that result from concentration gradients and lead to partial self-organized growth discussed earlier are included by solving the adatom diffusion equation around a nanocrystal including appropriate source (surface adatom generation) and sink terms (attachment to other clusters).

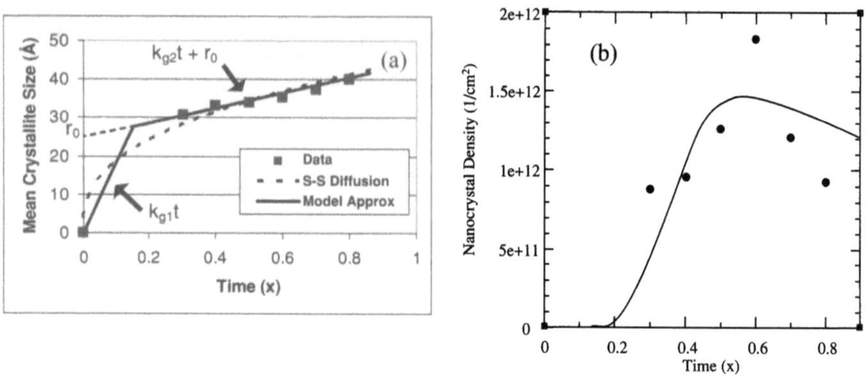

FIGURE 4.13. (a) Evolution of the nanocrystal size with time. Squares indicate experimental data, the dashed line indicates the results corresponding to solution of the steady-state diffusion equation, and solid lines indicate the approximation used by our model. (b) Nanocrystal density versus deposition time for nanocrystals >2.5 nm. Solid dots indicate experimental measurements and the black line indicates the model predictions.

Mathematically, the resulting expression for the exclusion zone radius, r_{ex}, is

$$r_{ex}^2 = r_k^2 + 2r_k\sqrt{D\tau}\frac{K_1\left(r_k/\sqrt{D\tau}\right)}{K_0\left(r_k/\sqrt{D\tau}\right)} - \frac{8D\tau}{\pi^2} \times \int_0^\infty \frac{e^{-(m^2+1)/\tau}\,dm}{m(m^2+1)\left[J_0^2\left(mr_k/\sqrt{D\tau}\right) + Y_0^2\left(mr_k/\sqrt{D\tau}\right)\right]}, \quad (4.13)$$

where r_k is the nanocrystal radius, D is the adatom diffusivity, τ is the time constant of adatom attachment to other nuclei, desorption, and other processes, and J_0, Y_0, K_0, and K_1 are Bessel functions. The third term on the right-hand side of Eq. (4.13) corresponds to the transient solution and is important for short times (relative to τ).

Model predictions and experimental measurements of the nanocrystal density as a function of deposition time are shown in Fig. 4.13b. Both the model and experimental data indicate that the density initially increases very rapidly to a maximum value and then decays. The maximum in the model prediction occurs at roughly the same time as the peak in the experimental data, although the model underestimates the maximum density by about 20%.

Another issue is the preservation of nanocrystals during subsequent processing involved in making a memory. In particular, protection from oxidizing ambients is needed. At an 850°C anneal in NO ambient, the nanocrystals are passivated by forming a thin shell of SiO_xN_y around the Si nanocrystals. This helps to prevent oxidation of nanocrystals during subsequent processing steps [21] and leads to a narrower size distribution. Figure 4.14 shows the ability of passivated nanocrystals to withstand aggressive oxidizing ambients (1000°C, O_2, 30 min). Although the

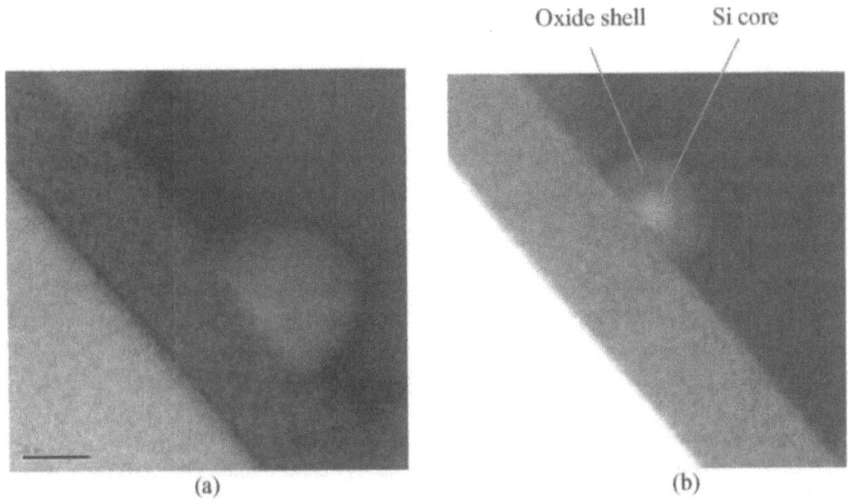

FIGURE 4.14. Cross-section EFTEM images of (a) passivated and (b) unpassivated nanocrystals subjected to severe oxidizing ambient (1000°C, O_2, 30 min). The unpassivated nanocrystals oxidize, partially forming a thick oxide shell around a Si core, whereas the passivated nanocrystals do not show any signs of oxidation.

unpassivated nanocrystals oxidize partially forming a thick oxide shell around a Si core, the passivated nanocrystals do not show any signs of oxidation.

4.4. NVM BITCELL CHARACTERISTICS

All devices discussed in this chapter have been fabricated using 130-nm and 90-nm CMOS technology platforms. For NVM applications in a one transistor per bitcell architecture, read disturb and data retention criteria limit bottom oxide thickness to the range of 50–70 Å to suppress defect-site-mediated charge exchange with the channel. After tunnel oxide growth, the silicon nanocrystals were deposited at required densities using optimized CVD processes and passivated as described in the previous section. Subsequently, a 100 Å high-temperature control oxide was deposited on the nanocrystals. A key aspect of the bitcell integration is the gate stack etch for removal of nanocrystals from the source/drain regions, without causing any silicon recess in the source/drain.

The nanocrystal characteristics and stack balance (tunnel and control oxide thickness) are key features affecting the performance of nanocrystal memories. As described earlier, there is an optimal value of nanocrystal number density, size, and spacing for maximum threshold voltage shift per electron stored. Similarly, the optimal control and tunnel oxide thickness are determined by a trade-off between read disturb at lower thicknesses ($T_{bottom} \sim 40$ Å, $T_{top} \sim 80$ Å) and erase voltage at larger thicknesses ($T_{bottom} \sim 70$ Å, $T_{top} \sim 140$ Å). Depending on the tunnel oxide thickness and the applied bias conditions, the nanocrystal memory device can be

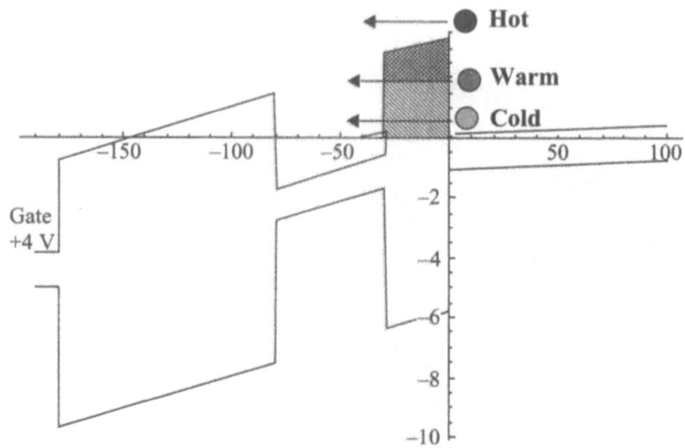

FIGURE 4.15. Energy band diagram of nanocrystal memory under positive gate bias of 4 V showing cold, warm, and hot electron injection.

programmed by direct tunneling, warm carrier programming, or by hot-carrier injection, as shown in Fig. 4.15. Si-nanocrystal memory cells with thin bottom oxides (~45 Å) can be programmed and erased by direct quantum mechanical tunneling of charge carriers from the channel into the nanocrystals. However, programming the device takes longer time at moderate gate biases (6–10 V), as shown in Fig. 4.16a. Furthermore, due to the thin tunnel oxide (<40 Å), the direct tunneling memories show read disturb of the erased state. Figure 4.16b shows the V_t shift of a 0.15-μm device with a 38-Å tunnel oxide in an erased state, which is subjected to typical read conditions of $V_g = 3$ V and $V_d = 0.7$ V. The device gains a significant amount of charge in less than 100 s. Thus, design solutions are needed to mitigate READ disturb in direct tunneling memory devices or multitransistor per bitcell architectures need to be employed.

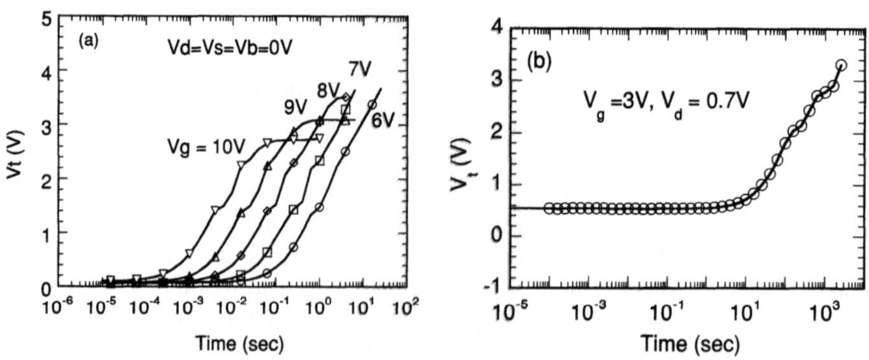

FIGURE 4.16. (a) Program speed at different gate biases and (b) READ disturb of erased state of a direct tunneling nanocrystal memory with a 38-Å tunnel oxide and a 100-Å control oxide.

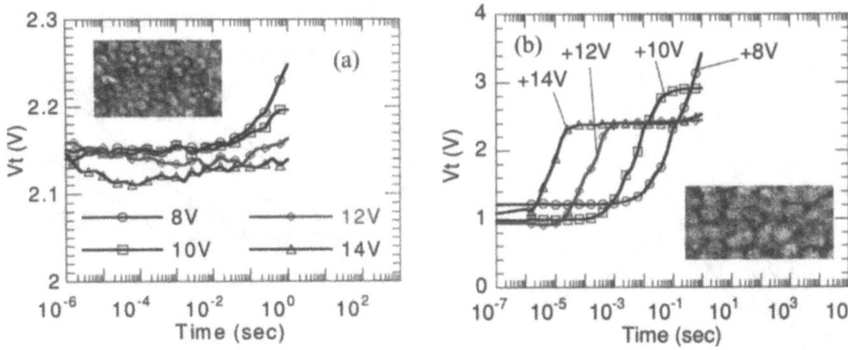

FIGURE 4.17. Gate bias dependence of FN program speed for memory bitcell with a 38-Å tunnel oxide and a 100-Å control oxide and with (a) 15% area coverage of nanocrystals, as shown in the inset, and (b) 50% area coverage of nanocrystals, as shown in the inset.

As the thickness of the tunnel oxide is increased to greater than 40 Å, programming by direct tunneling is impractical and one has to employ FN tunneling or hot-carrier injection (HCI) to program the nanocrystals. The FN programming of the devices shows interesting characteristics depending on the area coverage of nanocrystals. Figure 4.17a shows the tunneling program curves for a bitcell with 15% area coverage of nanocrystals. At low gate voltages (8–10 V), the device is in the direct tunneling regime and limited programming, albeit slowly, by direct tunneling is possible. At higher gate bias (12–14 V), the device is in the FN tunneling regime. However, at a low area coverage (~15%) of nanocrystals, when electrons are injected into the nanocrystal from the substrate, the local control oxide field in the vicinity of the nanocrystal becomes greater than the tunnel oxide field. This facilitates removal of electrons from the nanocrystal to the gate, thus preventing any significant programming of memory cells. Figure 4.17b shows the tunneling program curves for a bitcell with 50% area coverage of nanocrystals and with the same bottom and top oxide thicknesses. When the nanocrystal area coverage is large enough to form a network of nanocrystals, the local field in the floating gate is smaller than at lower nanocrystal area fractions for the same average charge density. Further, since electrons can rapidly tunnel from one nanocrystal to another, making the gate resemble a continuous floating gate, the capacitive coupling of the control gate to the floating gate is stronger. Both of these effects lead to the FN programming seen in Fig. 4.17b and is not observed in the case of low areal coverage (Fig. 4.17a). However, saturation sets in when currents from the channel to the nanocrystal and from the nanocrystal to the poly-Si gate are in balance.

Programming the nanocrystals in a thick tunnel oxide device is possible by local charge injection using hot carriers. For HCI, a drain bias of greater than 3 V is applied to impart sufficient energy to electrons near the drain to overcome the oxide barrier, as shown in Fig. 4.15. A portion of the generated hot carriers are drawn by the gate field ($V_g \geq 6$ V) and are injected into the nanocrystals near the

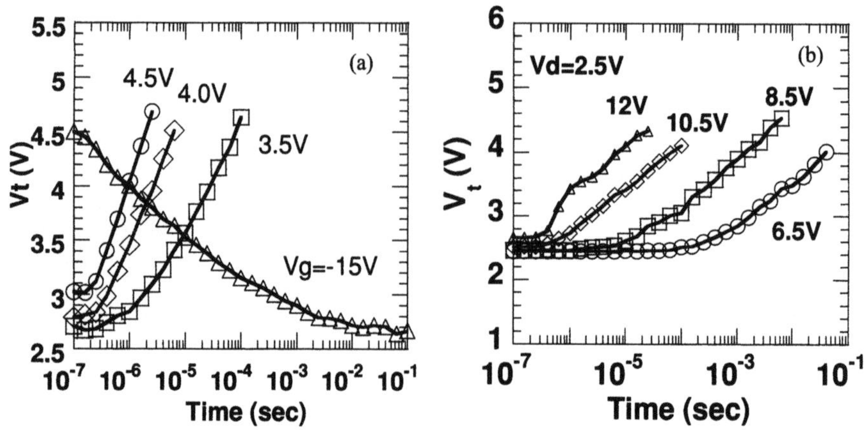

FIGURE 4.18. The effect of (a) increasing drain bias at $V_g = 10$ V (fixed) and (b) increasing gate bias at $V_d = 2.5$ V on HCI programming speed for a bitcell with a 50-Å tunnel oxide and a 100-Å control oxide. An FN erase curve from the same bit is superimposed in (a).

drain. Figure 4.18 shows the effect of drain and gate bias on HCI program speed for a 0.24-μm (W) × 0.24-μm (L) device with a 50-Å bottom oxide and a 100-Å top oxide. Increasing the drain bias increases the lateral field, which is responsible for hot-carrier generation. As a result, the program speed improves even for small changes in drain bias (Fig. 4.18a). Increasing the gate bias increases the number of carriers in the channel and the gate field, which attracts the hot electrons to the gate. However, the gate bias has a lesser impact on HCI program speed than the drain bias (Fig. 4.18b). It is evident that a 2-V threshold voltage shift can be obtained with sub-10-μs programming time.

The memory device can be erased by FN tunneling of charges from the nanocrystals either to the gate or to the substrate by applying either positive or negative gate bias, as described earlier (Fig. 4.7).

For a given set of device biasing conditions, the program and erase behavior of the nanocrystal memory depends on the nanocrystal number density and size distribution. This is illustrated in Fig. 4.19, which compares the program and erase operations of devices with different nanocrystal characteristics. Three devices with identical bottom oxide (45 Å SiO_2) and top oxide (120 Å SiO_2) but different nanocrystal depositions, as shown in Fig. 4.19a were used to obtain the data. All three devices were programmed using HCI as shown in Fig. 4.19b. In all cases, the programmed state V_t eventually saturates when the rate of charge injection into the nanocrystals due to HCI is balanced by the rate of charge removal from nanocrystals due to FN tunneling through the control dielectric. By comparing the program curves for the samples with similar nanocrystal density, it can be seen that as the mean nanocrystal size is reduced, the program speed and saturation V_t decrease. This is due to a reduction in capture cross section and increased Coulomb blockade effects as the nanocrystal size is reduced. At very low nanocrystal densities (2×10^{11} cm^{-2}), approximately 15 electrons have to be stored in every nanocrystal to achieve the same memory effect. This results in a high field across the top oxide

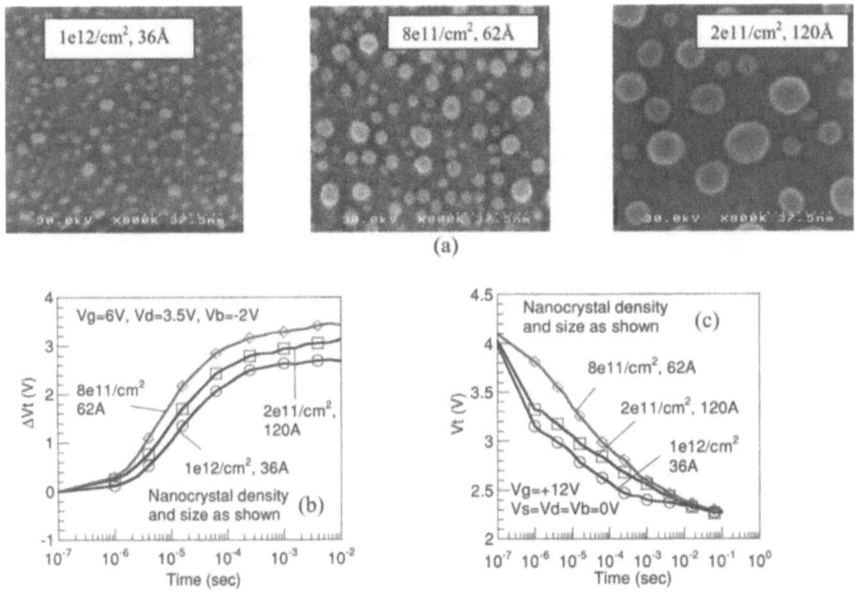

FIGURE 4.19. (a) SEM images of nanocrystals deposited on SiO_2 surface at three different conditions to produce nanocrystals with different number density and size distribution, as mentioned below the images. (b) HCI program speed and (c) FN erase speed for memory devices with a 45-Å tunnel oxide and a 120-Å control oxide, with different nanocrystal depositions, as shown in (a).

during the program operation, which, in turn, causes a lower program speed and saturation V_t during the program compared to the case with nanocrystal density of 8×10^{11} cm^{-2} and mean size of 62 Å.

The nanocrystal characteristics influence the erase speed as well through the Coulomb blockade effect. In small nanocrystals and in nanocrystals with a large number of stored electrons, the energy levels in the nanocrystal conduction band are raised due to the Coulomb blockade. This results in a faster erase, as shown in Fig. 4.19c. However, the Coulomb blockade effects decrease as charge is erased from the nanocrystals. As a result, although the device with the smallest nanocrystals erases fast initially, the erase saturates at the same value as the other devices.

Figure 4.20 shows the HCI program and FN erase cycling endurance of the cell with a 50′ tunnel oxide and 100′ control oxide. The program and erase threshold voltages drift up by approximately 1 V after 10000 cycles due to charge trapping in the control oxide. Unlike a floating gate device that captures all hot electrons injected towards the gate, the nanocrystal layer has lower capture cross section due to its small area fraction (~20%) and hence only a portion of the injected electrons are captured by nanocrystals, while the rest are either trapped in the control oxide or reach the gate. Improvement in endurance behavior can be obtained by optimizing the control dielectric to minimize charge trapping.

Figure 4.21 shows data retention at 150°C as well as READ disturb for a memory cell with 50 Å tunnel oxide and 85 Å control oxide. The data retention

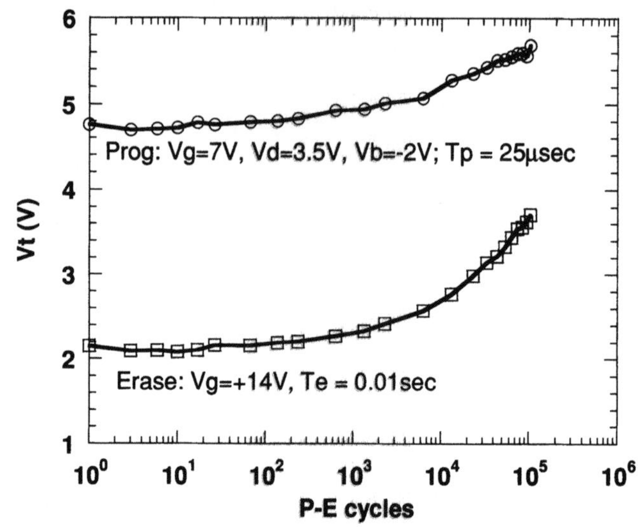

FIGURE 4.20. HCI program and FN erase cycling endurance behavior of a memory cell with a 50-Å tunnel oxide and a 100-Å control oxide.

was measured before and after program–erase cycling under conditions described in Fig. 4.20. Whereas the uncycled device shows good charge retention in the high-V_t state, the cycled device shows some initial charge loss, presumably due to detrapping of oxide charges. The READ disturb measurements were conducted with a gate stress of 1 V over the erase V_t state and with $V_d = 1$ V. There is practically no READ disturb of the erase state even after 10,000 HCI–FN cycles, due to the thick and high-quality thermally grown tunnel oxide. However, READ disturb of the programmed state is observed due to tunneling from the control dielectric, which can reduce the end-of-life memory window by approximately 1

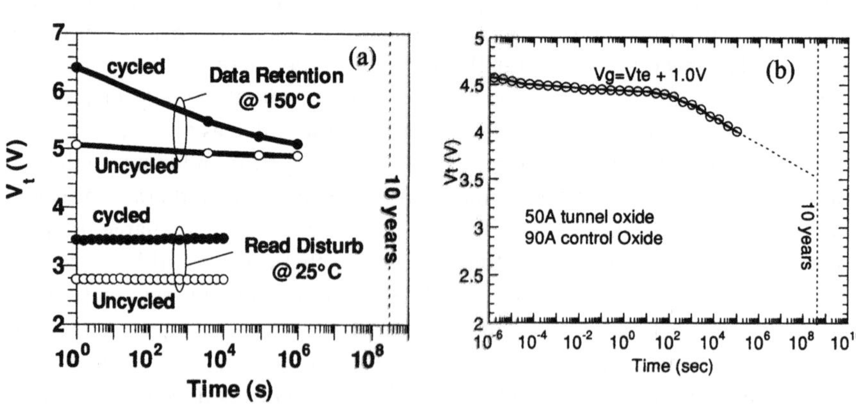

FIGURE 4.21. (a) 150°C data retention of high-V_t state and 25C READ disturb before and after 10,000 HCI–FN cycles and (b) READ disturb of the high-V_t state for a bitcell with a 50-Å tunnel oxide and a 90-Å control oxide.

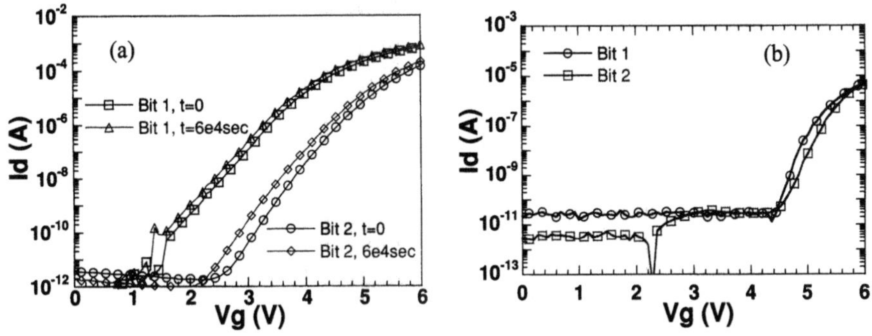

FIGURE 4.22. (a) Forward and reverse READ operations at $t = 0$ and $t = 65$ h showing data retention of 2 bits at 150°C after localized HCI programming for a memory bitcell with ~20% areal coverage of nanocrystals. (b) Memory bitcell with 50% area coverage of nanocrystals showing no significant difference in the forward and reverse READ operations at room temperature after localized HCI programming.

V. Modeling of the charge loss using tunneling calculations shows that this can be substantially mitigated by a slight increase in control oxide thickness.

The isolated nature of charge storage centers (low area fraction of nanocrystals) and the energy well (3.1 eV to silicon dioxide) makes the nanocrystal memory cell a good candidate for local charge storage such as in 2-bit/cell applications. By using HCI programming, charge can be injected locally near the source or drain region. The injected charge can be sensed by reading the device using reverse read [22]. Figure 4.22a shows the asymmetric threshold voltages obtained by forward and reverse reading operations in a 0.13-μm channel length device with nanocrystal area coverage of 20%, which has been programmed by HCI near the drain. It is seen that at such a low area coverage, excellent local storage and data retention are observed. However, as the nanocrystal area coverage is increased to 50%, lateral tunneling transport between nanocrystals prevents local charge storage and the forward and reverse read operations do not show any asymmetry (Fig. 4.22b).

4.5. MEMORY ARRAY FABRICATION AND CHARACTERIZATION

For NOR applications, the bitcells in the memory array share a common p-well and a common source that is tied to a ground. The word line, which connects the control gate of all bitcells along a row, and the bit line, which connects the drain terminals of all bitcells along a column, can be biased so as to select an individual bit for programming, as shown in Fig. 4.23. The entire array is erased using a block erase by biasing all word lines and the p-well.

The 4-Mb NVM arrays were fabricated using a 90-nm and a 130-nm CMOS process flow. Key aspects of the integration scheme include the ability to prevent nanocrystal oxidation and tunnel oxide thickness increase during subsequent processing as well as the ability to completely remove nanocrystals from undesired

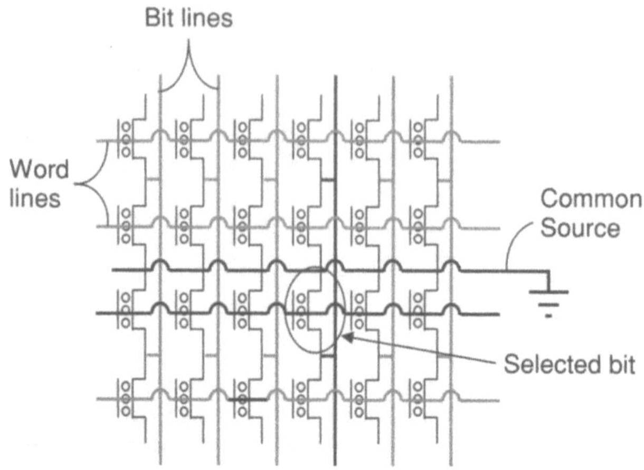

FIGURE 4.23. NOR architecture of nanocrystal NVM array. The biased word and bit lines are shown in bold and the unbiased word and bit lines are shown in light gray.

areas such as source and drain regions of the array and logic device regions. The first is particularly important, as nanocrystals can be oxidized and tunnel oxide thickness can increase during the growth of high-voltage-device oxides.

The integration process flow is shown in Fig. 4.24. A 50-Å thermally grown SiO_2 layer was used as the tunnel dielectric. The nanocrystal process was varied to produce arrays with a range of nanocrystal densities (5×10^{11} cm^{-2} to 1×10^{12} cm^{-2}) and sizes (50–90 Å). After nanocrystal deposition and passivation, a 100-Å SiO_2 layer was deposited using CVD as a control dielectric. Using an ONO control dielectric instead of SiO_2 mitigates nanocrystal oxidation and tunnel oxide thickness increase during growth of high-voltage-device oxides [23]. A combination of wet and dry etches have been developed to remove the control oxide/nanocrystals/tunnel oxide stack from undesired regions. Figure 4.25 shows a

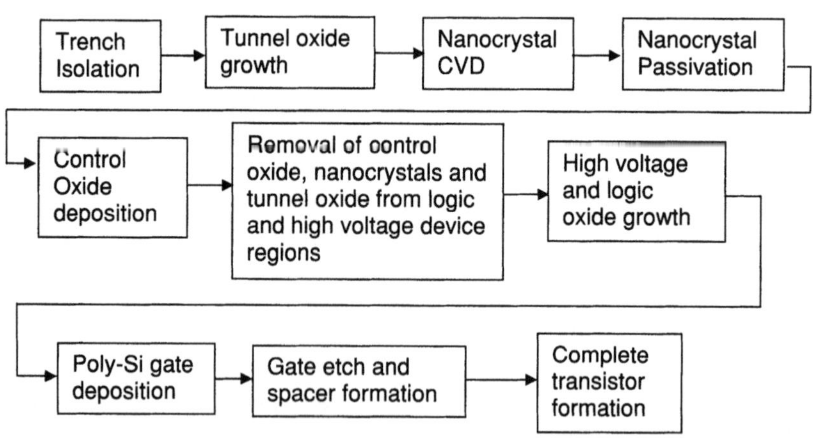

FIGURE 4.24. Process flow for integrating a nanocrystal memory with standard CMOS technology.

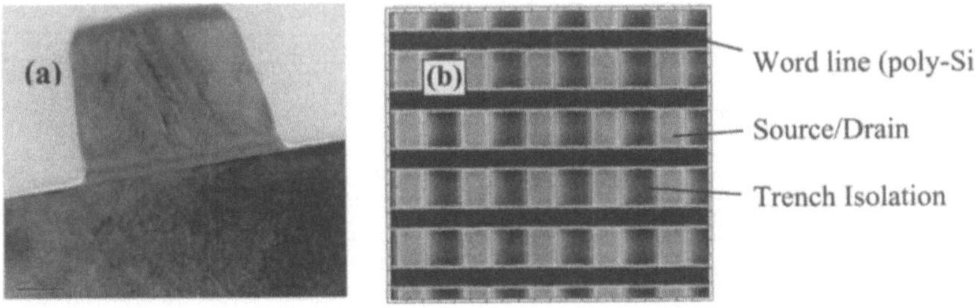

FIGURE 4.25. (a) TEM image of a nanocrystal NVM bitcell after gate etch and (b) top down SEM image of the memory array at the same stage. Removal of silicon nanocrystals from the source, drain, and field regions of the memory array is critical to embedding the memory technology.

cross-sectional transmission electron microscope (TEM) and a plan-view SEM after the nanocrystals have been successfully removed from the source drain regions of the array.

Figure 4.26 shows the threshold voltage distributions for programmed and erased states of one sector from a 4-Mb array containing of 512 K bits. A very tight erased state distribution of approximately 0.8 V is obtained, and the programmed state distribution is about 1.1 V wide. Threshold voltage distributions obtained from wafers without nanocrystals match the erased state V_t distribution width

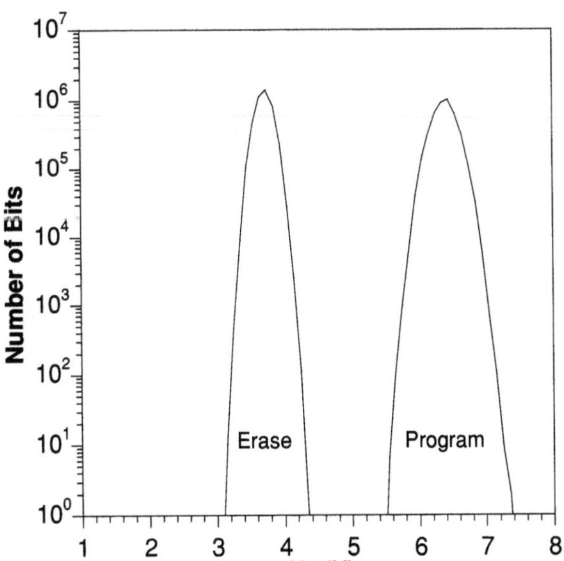

FIGURE 4.26. Threshold voltage distributions for both the erased and programmed state of a 4-Mb memory array produced using a 90-nm technology. The results indicate tight V_t distributions for both states and no extrinsic bits.

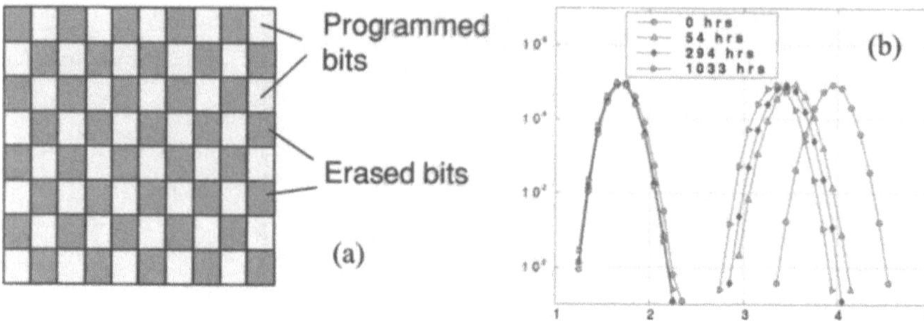

FIGURE 4.27. (a) Checkerboard pattern of programmed bits and (b) 200°C data retention bake after checkerboard programming of a 4-Mb memory array showing the evolution of V_t distribution of programmed and erased bits with time. No extrinsic bits are observed.

from wafers with nanocrystals, indicating that the nanocrystal process has minimal contribution to V_t variations in the array.

In order to test for lateral charge migration due to coalescence of nanocrystals, the individual arrays with different nanocrystal densities and sizes were programmed in a physical checkerboard pattern using a 6-V gate, 4-V drain, and −2.4-V well pulse for 10 μs, such that all adjacent bits were in opposite data states, as shown in Fig 4.27a. Nanocrystal coalescence is expected to enhance lateral charge spreading, thus degrading data retention performance due to hopping conduction. This should be manifested as a charge loss from the program state, as charge is redistributed over the trench isolation, reducing its impact on bitcell V_t. In the most severe case, lateral spreading could cause charge to cross into a neighboring bit in the erased state and cause charge gain and a corresponding increase in bitcell V_t. Since nanocrystal coalescence occurs randomly across the surface, not all bits in the array would be affected. One would expect affected bits to exhibit worse data retention behavior than the intrinsic population, thus forming extrinsic tails in the distribution that encroach on the usable V_t window for information storage. After the initial checkerboard programming, V_t distributions of both the programmed and erased bits were measured. No evidence of charge migration from the programmed to the erased bits was observed. The arrays were then subjected to a 200°C unbiased bake stress to accelerate charge movement. Figure 4.27b shows the evolution of V_t averaged across eight blocks (512 K bits) comprising one 4-MB array measured at several time intervals during the bake stress. The absence of extrinsic tails in the programmed and erased bits' V_t distributions indicates that there are no bits that display a faster charge loss compared to the intrinsic population. The shift of the programmed state V_t distribution to lower values is due to intrinsic charge loss from the memory at 200°C. This test reveals that even in large arrays, there is no leakage path between adjacent bitcells due to coalescence of nanocrystals. This is because the CVD process used for nanocrystal deposition results in partially self-organized nanocrystal formation, ensuring a minimum separation between adjacent nanocrystals [19].

Figure 4.28a shows the program–erase (P/E) endurance and data retention measurements on 0.25-μm 4-Mb arrays. The cycled (1000 P/E cycles) and

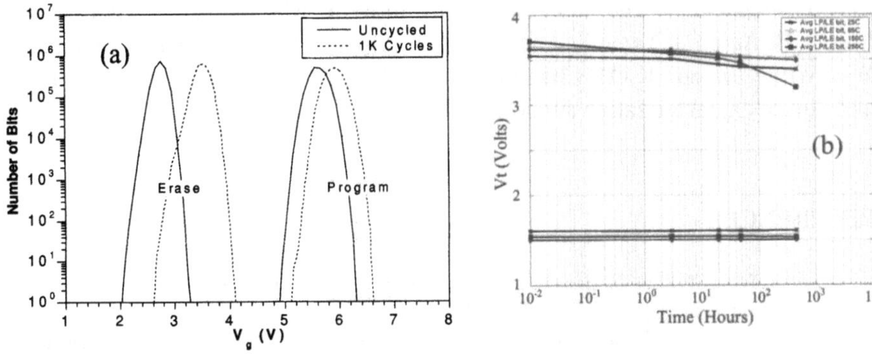

FIGURE 4.28. (a) Threshold voltage distributions for both the erased and programmed state of a 4-Mb memory array produced using a 0.25-um technology. The results indicate tight V_t distributions for both states and no extrinsic bits after cycling. (b) Data retention of the least programmed and least erased bits in an array at different temperatures.

uncycled erased V_t and programmed V_t distributions indicate tight distributions with no extrinsic bits upon cycling. V_t distributions for the erased and programmed state drift up in value with cycling due to charge trapping as observed in the single-bit data. Unbiased data retention measured on the least programmed and least erased bit (worst-case bits) is shown in Fig. 4.28b. The data show that there is minimal charge loss even at 150°C and that a 2-V window can be maintained.

4.6. SUMMARY

We have reviewed nanocrystal-based NVM devices. They provide an opportunity to scale conventional FG NOR Flash by mitigating the vulnerability to isolated tunnel oxide defects, which, in turn, enables one to scale oxide thicknesses and reduce operating voltages. Superior FN erase characteristics compared to SONOS-type memories enable one to erase with conventional FN at oxide thicknesses that are needed to mitigate READ disturb without resorting to hot-hole injection. Further, deeper energy trap levels permit local storage and 2-bit/cell operation for small gate lengths. A key aspect for NVM applications is possible charge trapping in the oxide stacks during HCI–FN operation that manifests as an increase in V_t of both memory states during repeated program/erase operations. This must be mitigated by development of high-quality oxides and/or design solutions. Key to this memory technology is the ability to form isolated silicon nanocrystals of the required size and densities and preserving them during subsequent processing. CVD methods have so far yielded the highest nanocrystal densities on oxide and also offer process opportunities for controlling size and density while ensuring partial self-ordering. Nitridation of silicon nanocrystals offers a way to passivate them during subsequent processing.

Acknowledgments. The authors wish to thank R. Muralidhar, B. Hradsky, S. Straub, B. Acred, J. Yater, S. Anderson, E.J. Prinz, G. Rinkenberger, T. Merchant, B.E. White, Jr., and K.-M. Chang for useful discussions, processing, integration, and device characterization support.

REFERENCES

1. S. Tiwari, F. Rana, H. Hanafi, A. Hartstein, E.F. Crabbe, and K. Chan, "A silicon nanocrystals based memory", *Appl. Phys. Lett.* **68**, 1377 (1996).
2. J. De Blauwe, "Nanocrystal nonvolatile memory devices", *IEEE Trans. Nanotech.* **1**, 72 (2002).
3. R. Muralidhar, R.F. Steimle, M. Sadd, R. Rao, C.T. Swift, E.J. Prinz, J. Yater, L. Grieve, K. Harber, B. Hradsky, S. Straub, B. Acred, W. Paulson, W. Chen, L. Parker, S.G.H. Anderson, M. Rossow, T. Merchant, M. Paransky, T. Huynh, D. Hadad, K.-M. Chang, and B.E. White Jr., "A 6 V embedded 90 nm silicon nanocrystal nonvolatile memory", *IEEE International Electron Devices Meeting IEDM 2003 Tech. Digest*, p. 26.2.1, (2003).
4. C.T. Swift, G.L. Chindalore, K. Harber, T.S. Harp, A. Hoefler, C.M. Hong, P.A. Ingersoll, C.B. Li, E.J. Prinz, and J.A. Yater, "An embedded 90 nm SONOS nonvolatile memory utilizing hot electron programming and uniform tunnel erase", *IEEE International Electron Devices Meeting IEDM 2002 Tech. Digest*, p. 927 (2002).
5. M.H. White, in *Nonvolatile Semiconductor Memory Technology*, edited by W.D. Brown and J.E. Brewer, IEEE, New York (1998).
6. B. Eitan, P. Pavan, I. Bloom, E. Aloni, A. Frommer, and D. Finzi, "NROM: A novel localized trapping, 2-bit nonvolatile memory cell", *IEEE Elec. Dev. Lett.* **21**, 543 (2000).
7. M. Sadd, J.A. Yater, J. Bu, C.M. Hong, W.M. Paulson, C.T. Swift, R. Singh, L. Parker, and M.G. Khazhinsky, "Validation of a Predictive SONOS Model", *Proceedings of the IEEE Non-volatile Semiconductor Memory Workshop*, (2003).
8. M. Sadd, J.A. Yater, B.E. White, C.T. Swift, S. Straub, R.F. Steimle, R. Rao, E.J. Prinz, R. Muralidhar, B. Hradsky, K. Harber, B. Acred, S. Bagchi, *Proceedings of the IEEE Non-volatile Semiconductor Memory Workshop*, (2004).
9. S.S. Ang, Y.J. Shi, and W.D. Brown, "Electrical characterization of rapid thermal nitrided and re-oxidized low-pressure chemical-vapor-deposited silicon dioxide metal-oxide-silicon structures", *J. Appl. Phys.* **79**, 1968 (1996).
10. L.D. Landau and E.M. Lifshitz, *Quantum Mechanics*. (Pergamon Press, Oxford, 1977), p. 559.
11. F. Rana, S. Tiwari, and J.J. Welser, "Kinetic modelling of electron tunneling processes in quantum dots coupled to field-effect transistors", Superlattices and Microstructures **23**, 757 (1998).
12. Y. Shi, K. Saito, H. Ishikuro, and T. Hiramoto, "Effects of traps on charge storage characteristics in metal-oxide-semiconductor memory structures based on silicon nanocrystals", *J. Appl. Phys.* **84**, 2358 (1998).
13. M. Sadd, R. Muralidhar, S. Madhukar, K. Scheer, D. Gentile, B. Hradsky, M. Rossow, R. Rao, M. Ramon, A. Konkar, J. Conner, S. Bagchi, B.E. White, A. Thean, J.P. Leburton, "Physical Modeling of the Reliability of a Nano-crystalline Memory Cell", *Proceedings of the IEEE Non-volatile Semiconductor Memory Workshop*, (2001).
14. M. Rosmeulen, E. Sleeckx, and K. De Meyer, "Silicon-rich-oxides as an alternative charge-trapping medium in Fowler-Nordheim and hot carrier type non-volatile-memory cells", *IEEE International Electron Devices Meeting IEDM 2002 Tech. Digest*, p. 189 (2002).
15. F. Mazen, T. Baron, G. Bremond, N. Buffet, N. Rochat, P. Mur and M.N. Semeria, "Influence of the chemical properties of the substrate on silicon quantum dot nucleation", *J. Electrochem. Soc.* **150**, G203, (2003).

16. R.A. Rao, R.F. Steimle, M. Sadd, C.T. Swift, B. Hradsky, S. Straub, T. Merchant, M. Stoker, S.G.H. Anderson, M. Rossow, J. Yater, B. Acred, K. Harber, E.J. Prinz, B.E. White Jr., and R. Muralidhar, "Silicon nanocrystal based memory devices for NVM and DRAM applications", *Solid-State Electronics* **48**, 1463 (2004).
17. T. Baron, A. Fernandes, J.F. Damlencourt, B. De Salvo, F. Martin, F. Mazen, and F.S. Haukka, "Growth of Si nanocrystals on alumina and integration in memory devices", *Appl. Phys. Lett.* **82**, 4151 (2003).
18. J.A. Venables, *Introduction to Surface and Thin Film Processes*, (Cambridge University, Cambridge, 2000).
19. S. Lombardo, R.A. Puglisi, I. Crupi, D. Corso, G. Nicotra, L. Perniola, B. DeSalvo, and C. Gerardi, "Distribution of the Threshold Voltage Window in Nanocrystal Memories with Si Dots formed by Chemical Vapor Deposition: Effect of Partial Self Ordering", *Proceedings of the IEEE Non-volatile Semiconductor Memory Workshop*, (2004).
20. M.W. Stoker, T.P. Merchant, R. Rao, R. Muralidhar, S. Straub, and B.E. White Jr., "A model of silicon nanocrystal nucleation and growth on SiO_2 by CVD", in *Materials and Processes for Nonvolatile Memories*, edited by A. Claverie, D. Tsoukalas, T-J. King, and J.M. Slaughter (Materials Research. Society. Symposium. Proceedings.Vol 830, Materials Research Society, Warrendale, PA, 2005, p. D5.7.
21. K.C. Scheer, R.A. Rao, R. Muralidhar, S. Bagchi, J. Conner, L. Lozano, C. Perez, M. Sadd, and B.E. White Jr.,. "Thermal oxidation of silicon nanocrystals in O_2 and NO ambient", J. Appl. Phys. **93**, 5637, (2003).
22. J.E. Hayes, US Patent #4,173,766 (1979).
23. R.F. Steimle, R. Rao, C.T. Swift, K. Harber, S. Straub R. Muralidhar, B. Hradsky, J.A. Yater, E.J. Prinz, W. Paulson, M. Sadd, C. Parikh, S.G.H. Anderson, T. Huynh, B. Acred, L. Grieve, M. Rossow, R. Mora, B. Darlington, K.-M. Chang, and B.E.White Jr., "Integration of Si nanocrystals into a 6V 4Mb Nonvolatile Memory Array", *Proceedings of the IEEE Non-volatile Semiconductor Memory Workshop*, (2004).

5

Novel Dielectric Materials for Future Transistor Generations

Gennadi Bersuker,[1] Byoung H. Lee,[1] Anatoli Korkin,[2] and Howard R. Huff[1]

5.1. INTRODUCTION

Responding to the market growth for computational power in various applications, semiconductor technology continues unabated in its drive toward higher transistor densities and faster transistors. The general direction of this trend is scaling down the critical dimensions of the integrated circuit (IC) components. Transistor scaling is governed by the duality of speed versus power: transistor speed, which depends on the drive current, should be increased while decreasing transistor power consumption. The static power consumption depends on the total transistor leakage current (it includes the subthreshold, p-n junction, and leakage currents), of which the gate leakage current may constitute the major component. According to the International Technology Roadmap for Semiconductors (ITRS) [1], the electrical gate oxide thickness (EOT) needed by 2007 for high-performance applications such as microprocessors (MPUs) should be less than 1 nm and the gate leakage current should be less than 10^3 A/cm^2 at 100°C. For low standby power applications, such as for cell phones, where the static power consumption is the major limiting factor, by the year 2006 the leakage current should be less than 1.5×10^{-2} A/cm^2 at 100°C, with the EOT $\leq$ 1.6 nm. The SiO_2 gate dielectric cannot support these requirements for the leakage current since the electron direct tunneling and the associated leakage current in the case of such thin dielectrics are too high. Although the exact limit of SiO_2 thickness for different electronic application is disputable, the dielectrics with the higher permittivity can offer a potential solution by providing the equivalent EOT for the larger thickness of the gate oxide [2]. This

[1] SEMATECH, Austin, TX 78741
[2] Nano & Giga Solutions, Gilbert, AZ 85296

has instigated a search for high-*k* gate dielectrics, which, however, has been complicated by both fundamental (intrinsic material properties) and manufacturing limitations [3, 4].

Indeed, replacing SiO_2 with a high-*k* material triggers multiple changes in the gate stack properties and transistor fabrication process, which are closely related to each other and to the intrinsic differences in the properties of the new dielectric materials from SiO_2. Shortly after an initial "excitement" about potential application of perovskite-type materials with a very higher dielectric constant, it was realized that many other factors besides high permittivity are important in order to obtain the required performance of the transistor. In case of barium strontium titanite (BST), the stopper was originated from the low-valence-band offset [5] and low-resistance degradation apparently caused by the high ion mobility under the electric bias [6]. The other materials, such as TiO_2 and Ta_2O_5 are thermodynamically unstable on silicon (e.g., by losing oxygen or by forming silicides) [7].

Another serious challenge for the introduction of the high-*k* materials into the IC manufacturing, which will be discussed below, is degradation of electron mobility in the transistor channel. The interaction (chemical coupling) of the high-*k* dielectric with Si can be expected to depend on the thermal budget of the given device fabrication process that may further complicate integration of new materials into the complementary metal–oxide–semiconductor (CMOS) process.

Consequently, introduction of novel gate dielectrics requires complex, "multidimensional" process optimization, which includes multiple electrical, materials, and process integration requirements. We will discuss some aspects of these challenges for the introduction of high-*k* gate dielectrics for semiconductor device applications.

5.2. WHY *k* VALUES OF HIGH-*k* MATERIALS ARE HIGH

The dielectric constant as a function of the material polarizability contains electronic and ionic contributions. The electronic polarizability is determined by the electron redistribution induced by the applied electric field. By shifting the equilibrium distribution of the electrons with respect to the nuclei, an external electric field generates an electric dipole moment in the system. This redistribution is caused by electron transitions, consistent with the allowed symmetry operations, to unoccupied electronic states with a different electron spatial configuration. In the second-order perturbation theory, the correction to the electron wavefunction associated with this transition is inversely proportional to the energy gap between the occupied and unoccupied states. Hence, the materials with smaller energy gaps between the occupied valence and unoccupied conduction bands may be expected, in general, to exhibit a greater admixture of the excited electronic states relative to the ground state by the applied electric field, resulting in increased polarizability and, accordingly, larger values of the dielectric constant *k*.

An additional increase of the dielectric constant may occur via ionic polarizability, caused by the dipole–active displacements of the chemically bonded ions.

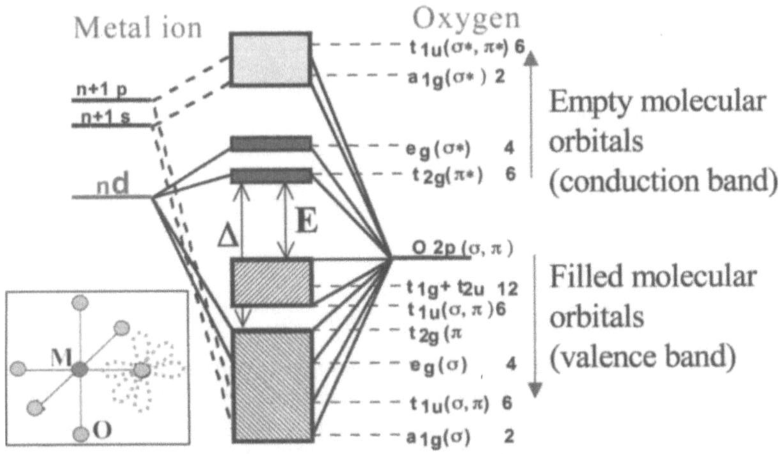

FIGURE 5.1. The molecular orbitals spectrum formed by the atomic electronic states of the octahedral metal–oxygen complex (special distribution of the oxygen p_x, p_y, and p_z states are shown in the inset). Δ represents the smallest energy gap between the bonding and antibonding orbitals responsible for the band gap E_g between the conduction and valence bands. Reprinted from *Materials Today*, Jan. 2004, p. 26, with permission from Elsevier.

The applied electric field induces larger ion displacements if a bond has a smaller force constant, indicative of a lower vibrational frequency ("softer" vibrations). Materials with transition metal ions can produce an increased ionic as well as electronic components of polarizability due to the presence of the open-shell d-electrons with highly anisotropic electronic spatial distributions.

To illustrate the role of d-electrons, we use the classical case of the Ti ion in the octahedral oxygen coordination [8]. One can see from Fig. 5.1, that the lowest conduction band consists of the antibonding orbitals formed by the metal d-states of t_{2g} symmetry and oxygen p-states. The threefold degenerate t_{2g} electronic term includes d_{xy}, d_{xz}, and d_{yz} states located in the xy, xz, and yz planes, respectively, as illustrated in Fig. 5.2(a). The electron distributions in these states do not have significant extensions toward the surrounding oxygen atoms, and, therefore, the overlap between these type of d wave functions of the transition ion and the p wavefunctions of the oxygen atoms is small forming weak π-type bonds. As a result, the splitting between the bonding and antibonding orbitals, Δ, determined by this overlap is relatively small. Accordingly, the lowest unoccupied orbitals (the lowest conduction band) lies rather close to the top of the valence band (formed by the nonbonding π-orbitals of the oxygen p-state). The band gap E_g for the d-electron materials is, therefore, small. On the other hand, the spatial distribution of the remaining two d-states, $d_{(z^2)}$ and $d_{(x^2-y^2)}$, favors greater overlap with the oxygen p-states resulting in higher energy for the second conduction band formed by these d-electrons.

The optical component of the dielectric constant (k_∞), which is associated with the electronic polarizability, was, indeed, observed to increase in the materials

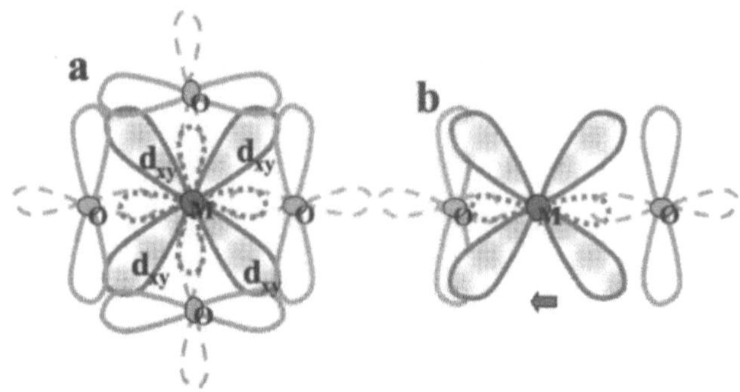

FIGURE 5.2. (a) Schematic of the metal d_{xy} (red) and oxygen p (green) electron distribution in the xy plane of the octahedral MO_6 complex. Broken lines correspond to the σ type *sp*-bonding, which provides major contribution to the force constant K_0 of the dipole-active T_{1u} vibrations. (b) The x component of the dipole-active T_{1u} vibrations softens due to an increase of the metal d-state and an oxygen p-state overlap. Reprinted from *Materials Today*, Jan. 2004, p. 26, with permission from Elsevier.

with smaller band gaps, as summarized in Table 5.1. This component, however, represents only a small portion of the dielectric constant in the frequency range of practical interest (up to 1000 GHz). The ionic contribution to the polarizability, associated with the electric dipole moment generated by ionic displacements, plays a more dominant role. It is worth noting that the magnitude and symmetry of these displacements are controlled by the same characteristic features of the d-states as the energy spectrum.

TABLE 5.1. Band gap, optical dielectric constant, and conduction band offset values for various dielectric materials.

	Gap (eV)	ϵ_∞	Conduction band offset (eV)
Si	1.1	12	
SiO_2	9	2.25	3.5
Si_3N_4	5.3	3.8	2.4
$SrTiO_3$	3.3	6.1	
$PbTiO_3$	3.4	6.25	
$BaZrO_3$	5.3	4	0.8
$PbZrO_3$	3.7	4.8	
Ta_2O_5	4.4	4.84	0.3
$SrBi_2Ta_2O_9$	4.1	5.3	0
TiO_2	3.05	7.8	0.05
ZrO_2	5.8	4.8	1.4
HfO_2	6	4	1.5
Al_2O_3	8.8	3.4	2.8
Y_2O_3	6	4.4	2.3
La_2O_3	6*	4	2.3
$ZrSiO_4$	6*	3.8	1.5
$HfSiO_4$	6*	3.8	

The vibrations of the metal–oxygen complex of T_{1u}-symmetry, responsible for the generation of an electric dipole moment, corresponds to the displacement of the metal ion with respect to the oxygen atoms, which increases the overlap of the corresponding metal d- and oxygen p-states; see Fig. 5.2(b). In the molecular orbital description, this overlap corresponds to the mixing of the t_{2g}-symmetry orbital (originated primarily from the metal d-states) with the t_{1u}-symmetry orbital of the oxygen p-states; see Fig. 5.1 (these orbitals contribute, respectively, to the T_{1u} and A_{1g} electronic terms of the metal–oxygen complex). This d–p overlap, which produces an additional electron charge transfer from the oxygen to the metal ion, has been shown to lead to a reduction of the force constant K_0 of the T_{1u} vibrations, which "facilitate" this overlap [9, 10]. This phenomenon [i.e., the softening (or even instability) of the high-symmetry configuration of molecular (or crystal) systems caused by the vibrations] is referred to as the pseudo-Jahn–Teller effect.

The softening effect is inversely proportional to the difference in energies, $E_{T\text{-}A}$, between the corresponding orbitals (in this case t_{1u} and t_{2g}, Fig. 5.1). The reduction of the force constant in the pseudo-Jahn–Teller effect is given by the following formula [10]:

$$K = K_0 - \frac{V_{T_{1u}}^2}{E_{T-A}}, \tag{5.1}$$

where $V_{T_{1u}}$ is the vibronic constant for the T_{1u} vibrations, which describes how responsive the nuclear configuration is to the electronic rearrangements, and K_0 is the force constant without the pseudo-Jahn–Teller effect determined by the main metal–oxygen σ bonds.

The reduction of the K value leads to the increase of the ionic polarizability χ_i:

$$\chi_i \propto \frac{Z_i^2}{K} \approx \frac{Z_i^2}{K_0 - V_{T_{1u}}^2/E_g}, \tag{5.2}$$

where Z_i is the ionic charge. Here we took into consideration that the E_{T-A} values are of the same order of magnitude as the dielectric band gap, E_g (see Fig. 5.1) because they are determined by the same d–p overlap mentioned above: $E_{T\text{-}A} \approx E_g$. Note the resonance character of this expression: The polarizability rises very fast when $V_{T_{1u}}^2/E_g$ approaches K_0.

Thus, the specific spatial distribution of the d-electrons, which leads to lower band gaps in the metal oxide compounds, is responsible for the higher dielectric constant in these materials; Fig. 5.3. Similar effects occur in the perovskite-type systems, which may exhibit ferroelectric behavior. In the case of ferroelectrics, $V_{T_{1u}}^2/E > K_0$, and $K < 0$. Therefore, the position of the metal ion in the center of the oxygen complex becomes unstable, and the metal ion shifts into one of the possible off-center energy minima forming a dipole moment. Interaction between the shifted ions may order them along one of the possible directions resulting in

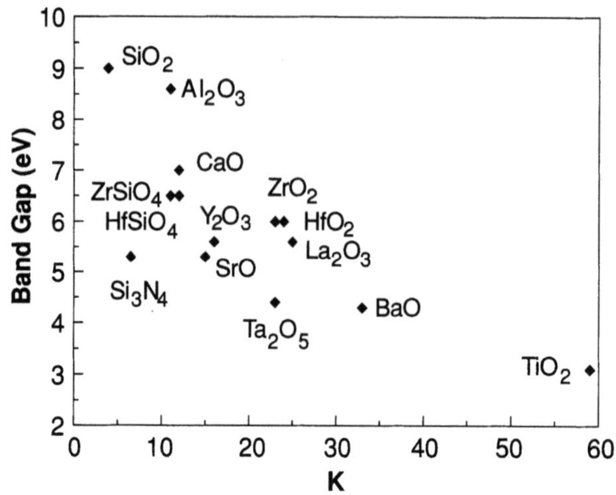

FIGURE 5.3. Variation of the dielectric constant with the band gap in binary oxides. (Courtesy of John Robertson) [11].

the formation of a polarized crystal (i.e., a permanent electrical dipole moment). This type of undesirable nonlinear electric behavior limits potential application of ferroelectric materials as gate dielectrics.

Although the above discussion of the octahedral metal configuration in general remains valid for the tetragonal, cubic, and, in certain cases, lower symmetry metal–oxygen complexes, the magnitude of the effects, as well as the symmetry of the atomic vibrations involved, may vary significantly. In the case of the octahedral complexes, the electric dipole-generating vibration (of the t_{1u} type) is directed toward the center of the octahedron faces. This arrangement allows the metal ion to remain much further away from the oxygen atoms located at the corners of the octahedron than in the case of the electric dipole-active vibrations in the MO_4 and MO_8 complexes, thereby reducing the metal–oxygen repulsion associated with the σ-type bonding. *Ab initio* calculations for ZrO_n, $n = 4, 6, 8$, systems show that, indeed, the force constant is smaller and the dielectric constant is larger in the ZrO_6 compound comparing to the others with $n = 4$ and 8 [12].

Although the relation between the dielectric constant and the band gap is fairly complex, in the first approximation both electronic and ionic components increase with decreasing band gap, which dramatically limits the choice of potential gate dielectrics materials since sufficiently large band gaps (<5 eV) and band offsets (<1 eV) are required for CMOS transistors. Another limiting factor related to the high dielectric constant is a high ion mobility, which has the same (similar) nature as ionic polarizability (e.g., decreased force constant and barrier for ion motion in the crystal). The primary candidates for SiO_2 replacement as the gate oxide, such as HfO_2 are well known and broadly used ion conductors [13], which raises concern in the reliability of high-*k*-based transistors.

We will discuss in the following sections how these fundamental material properties may affect selection of the high-*k* dielectrics and electrical characteristics of the transistors—in particular, threshold voltage and mobility.

5.3. CHOICE OF MATERIALS

Since increasing the permittivity of the gate dielectric has been the primary motivation for SiO_2 replacement, it is natural that the metal oxides with the largest values of dielectric constants were considered as potential candidates. Perovsikes, binary oxides with the chemical formula ABO_3, have a cubic crystal structure with coordination of the cations 6 and 12. The high dielectric constant of STO (200–300) is primary caused by a shallow potential (soft mode) of the 6-coordinated Ti cation. Replacement of some Sr cations by Ba, which has a larger ionic radius, increases further the dielectric constant. However, STO and $BaSrTiO_3$ (BST) have a small conductance band offset resulting in a low barrier for electrons [14, 15]. Nonlinear behavior of the dielectric properties or ferroelectric materials, such as spontaneous polarization and reduction of the dielectric constant with increased dielectric field, create additional problems for many potential candidates for gate dielectrics. It has been also established that thin-film dielectric materials have lower dielectric constants than bulk crystalline materials. This phenomenon is likely to have a nature similar to the reduction of the dielectric constant in the high electric fields, since the structural imperfectness such as grain boundaries and pores create the local electric field ad its gradients [16].

In order to be compatible with the CMOS technology, the gate dielectric must withstand process conditions such as rapid thermal anneal (e.g., 10 s at 1000°C) without degrading its electric properties caused by phase transition (e.g., crystallization) or/and diffusion. Since SiO_2 is thermodynamically stable compound ($\Delta H_f{}^o = -902.66$ kJ/mol) [17], most of the elements in the periodic table have lower stability than silicon oxide [e.g., have higher (less negative) heat of formation], which results in exothermicity of the following reaction:

$$MO_x + Si \rightarrow M + SiO_2.$$

The other reactions that can "contaminate" the oxide are silicide and silicate formation:

$$MO_x + Si \rightarrow MSi_y + O_2\uparrow,$$
$$MO_x + Si \rightarrow MSi_y + SiO\uparrow,$$
$$MO_x + Si \rightarrow M + MSi_vO_w,$$
$$MO_x + Si \rightarrow MSi_y + MSi_zO_w.$$

Silicide formation is particularly undesirable and it is likely a cause of instability of polysilicon on ZrO_2; which makes it less favorable than HfO_2 [18]. However, hafnium seems to form silicides as well at a low oxygen pressure [19].

Although the thermodynamic consideration based on the heats of formation of the bulk materials provides useful and relatively simple guidelines for selection of the possible candidates, which are likely to be stable on silicon, the complex process conditions, which include variation of temperature, pressure, and concentration of the precursors during the deposition, particularly at the atomic layer deposition (ALD) cycles, complicate the situation dramatically. With the tough requirements on the concentration of the defects, kinetically stable defects can remain in small but "damaging" concentrations in the deposited films and remain relatively stable during the subsequent thermal annealing. Thermodynamic equations based on the bulk materials also do not take into account the surface and interface effects which play important roles in the structure and energy of the thin films. For example, a single oxygen atom transfer from HfO_2 into Si is thermodynamically plausible due to the interface dipole formation [20], and according to the density functional calculations (DFT), the oxygen vacancies are more stable at the HfO_2 interface with silicon than in the bulk of the film [21].

5.4. EFFECTS OF ELECTRON TRAPPING

The presence of the *d*-electrons leads to specific structural features of the high-*k* materials, which determines their electrical properties. In particular, an important consequence of shifting from a covalent bonding in SiO_2 to an ionic bonding in transitional metal oxides is a potentially high density of as-grown structural defects in the high-*k* dielectrics.

Since high-*k* materials crystallize much easier than SiO_2, grain boundaries and dislocations potentially are the major sources of the charge traps. They also create channels for low-barrier ion diffusion and breakdown initiation points (via electric field concentration). Materials with a high dielectric constant also have lower formation energies of the charged defects, which modify the electric characteristics of transistors and facilitate the breakdown.

Some defects, which represent electron traps capable of capturing injected electrons, may contribute to the instability of the transistor electrical properties, in particular, the threshold voltage, as shown in Fig. 5.4 [22]. The electron trapping is reversible: By applying voltage of the opposite polarity, V_t can be returned to its initial prestress value. The detrapping kinetics is very similar to the trapping one (compare the slopes of the up and down V_t curves in Fig. 5.4b. The V_t value after detrapping is defined by the detrapping voltage. The V_t dependence on the stress time does not change with each subsequent stress cycle, and there are no noticeable degradations of the interface electrical properties (transconductance, G_m, and subthreshold slopes do not change). Therefore, one may conclude that the electron trapping occurs on the preexisting ("as-grown") defects, and a low-voltage electrical stress of practical importance does not generate new traps. This reversible behavior may be expected when the trapped electron resides mostly on

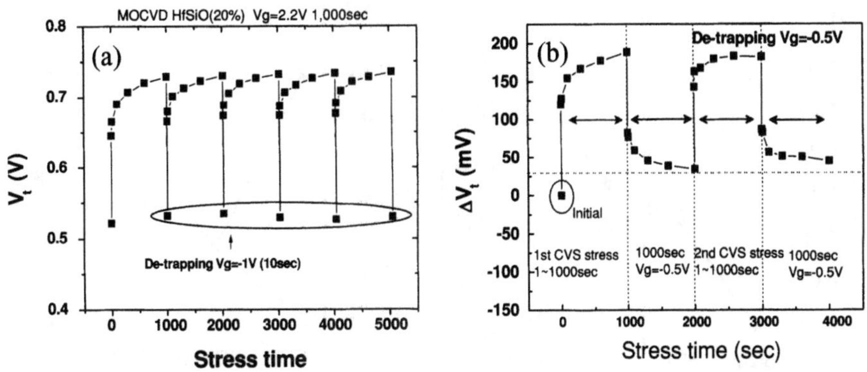

FIGURE 5.4. Variation of the negative metal oxide semiconductor (NMOS) transistor threshold voltage during stress cycles, which include 1000-s substrate injection stress followed by (a) 10-s and (b) 1000-s stress of the opposite bias under the specified voltage conditions.

the *d*-orbital, which is delocalized over the metal ion and its nearest neighbors: In this case, no significant structural relaxation should be expected, and the electron trapping and detrapping may not lead to structural damage.

As can be seen in Figs. 5a and 5b, initial fast increase of V_t during the first second of the constant voltage stress is followed by a slower V_t growth. To investigate the fast transient electron trapping, a pulse I_d–V_g technique in the microsecond time range was employed [23]. An example of these measurements is shown in Fig. 5.5a, where the drain current decrease, ΔI_d, is attributed to the effective increase of the device threshold voltage $\Delta V_t = Q/C$ due to the charge, Q, trapped in the dielectric (the trap locations with respect to the electrode define the capacitance C) during the gate voltage pulse: $\Delta I_d = (W/L)C_0\mu\Delta V_t$, where W and L are the transistor width and length, respectively, C_0 is the gate capacitance, and μ is the carrier mobility. This explains the higher pulse current compared to the conventional DC current measurements (Fig. 5.5b): The latter longer time (up to

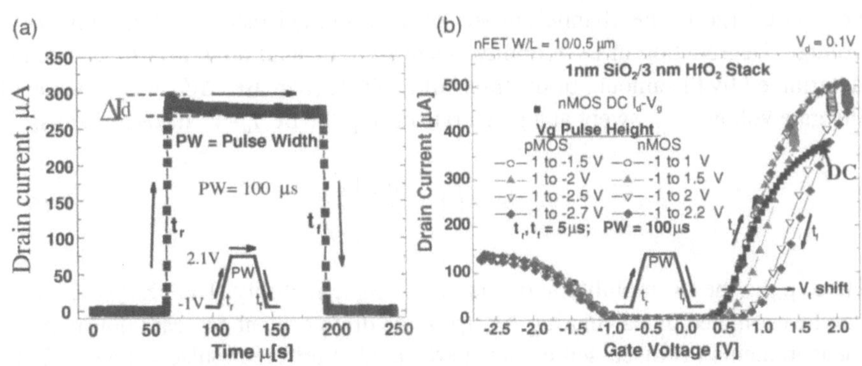

FIGURE 5.5. (a) Example of the drain current change during the pulse of 100 μs width. (b) Variation of the drain current with the pulse voltage during the pulse rise, fall, and width times. The drain current was recorded during 100-μs gate voltage sweep from − 1V to the peak values and back to the starting value. Conventional NMOS DC I_d–V_g curve is shown for comparison.

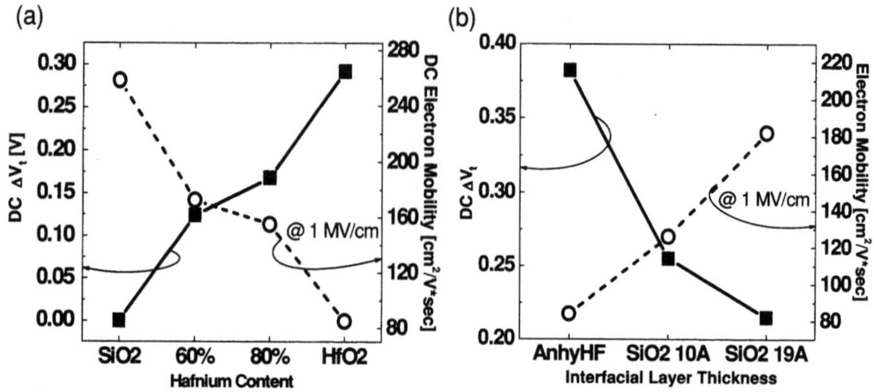

FIGURE 5.6. Dependence of the DC drain current degradation with respect to the pulse measurement (in the units of ΔV_t) and effective electron mobility (at the effective field of 1 MV/cm) versus (a) HfO_2 dielectric composition (% SiO_2) and (b) starting (pre-HfO_2 deposition) thickness of the bottom interfacial layer. The higher Hf content increases the effective trap density in the dielectric, whereas the thinner interfacial layer presents a thinner barrier for the electron injection from the substrate leading to higher gate current density.

a second) results in significant electron trapping. Measured ΔI_d values can be employed for calculating the parameter ΔV_t, which can be considered a figure of merit for the fast transient charging effect [24]:

$$\Delta V_t = (\Delta I_d / I_d)(V_g - V_t), \tag{5.3}$$

where V_g is the gate voltage.

Fast electron trapping during DC measurements {as measured by ΔV_t [Eq. (5.3)]} is partially responsible for low DC drain current in high-k transistors (as compared to the SiO_2 devices of equivalent EOT), which leads to lower mobility values extracted from the conventional DC I_d–V_g measurements; Fig. 5.6. This apparent mobility degradation is associated with underestimation of the inversion charge in the channel due to an unaccounted increase of the threshold voltage. Degradation of DC mobility is shown to depend on $\partial \Delta V_t / \partial V_g$, which is determined by the amount of the threshold voltage increase, ΔV_t, at each value of the gate voltage, V_g, swept at a given rate during the DC I_d–V_g measurements:

$$\mu_a \approx \mu \left(1 - \frac{\partial \Delta V_t}{\partial V_g}\right). \tag{5.4}$$

Here, μ_a is the DC mobility and μ is the intrinsic mobility. The $\partial \Delta V_t / \partial V_g$ value can be estimated by calculating ΔV_t from the drain current decrease during pulse measurements (as discussed earlier) performed at different pulse voltages V_g. In each of the samples from the variety of the studied gate stacks, it was found that $\partial \Delta V_t / \partial V_g$ = constant, which can be explained by fact that the ΔV_t variation with the V_g values is primarily determined by the position of the trapped charge with respect to the electrode. In the samples with significant fast transient electron trapping, the apparent mobility reduction may constitute up to 30% of the intrinsic mobility values.

To study fast transient charging, we adopt a simplified model where trapping of each electron is independent from others and does not affect the injection current density. Under these conditions, the kinetics of fast transient electron trapping is described by the following equation:

$$\frac{\partial n}{\partial t} = \frac{1}{\tau}(N_0 - n). \tag{5.5}$$

Here, n is the density of the filled traps, N_0 is the total trap density, and τ is the characteristic trapping time, which is inversely proportional to the probability of a single electron trapping event: $1/\tau \equiv p = \sigma J/q$, where σ is the defect capture cross section and J is the injection current density. By fitting the solution of this equation,

$$n = N_0(1 - e^{-t/\tau}), \tag{5.6}$$

to the measured time dependence of the drain current during the pulse, ΔI_d (Fig. 5.5a), which reflects the time dependence of the electron trapping and accompanying increase of ΔV_t [see Eq. (5.3)], one can obtain an estimate for the characteristic trapping time, gτsgnd, subsequently, σ[25]. In the example in Fig. 5.5a, these values are $\tau \approx 20$ μs and $\sigma \approx 10^{-12}$–10^{-13} cm^2.

In the time scale of the fast charging process, the direct electron tunneling to the traps, which may only weakly depend on temperature, controls the trapping; Fig. 5.7. On the other hand, the electron detrapping occurs via a temperature activated process proportional to $\exp(-E_t/kT)$. From the slope of the temperature dependence of ΔV_t, which was calculated, as described earlier, based on the pulse measurements data in Fig. 5.8, we have estimated the energy of the traps, E_t, to be on the order of 0.20 to 0.30 eV with respect to the bottom of the conduction band. Thus, the defects participating in the fast transient charging represent shallow electron traps with extremely large capture cross section.

Similar analysis [see Eq. (5.6)] performed for the slow charging process in Fig. 5.4a yields the value for the characteristic electron trapping time on the order of 100 s and effective capture cross section of 10^{-18}–10^{-19} cm^2 if one assumes that the electrons available for trapping were supplied by the same gate current of the density J. Such significant difference of the σ values of the slow and fast charging processes could suggest that the former process is assisted by a different type of defect. However, the above σ values are too small to represent any meaningful physical defect indicating that the electron capture from the dielectric conduction band is apparently ineffective. In addition, similar kinetics of the trapping and detrapping processes (Fig. 5.4b) points to a very limited lattice relaxation caused by the electron capture by the "slow" traps, similar to the case of the "fast" traps. The "slow" traps can be detrapped as effectively as "fast" traps under the same applied gate bias, indicating that both types of trap may have very similar energies. These considerations suggest that different electron trapping mechanisms rather

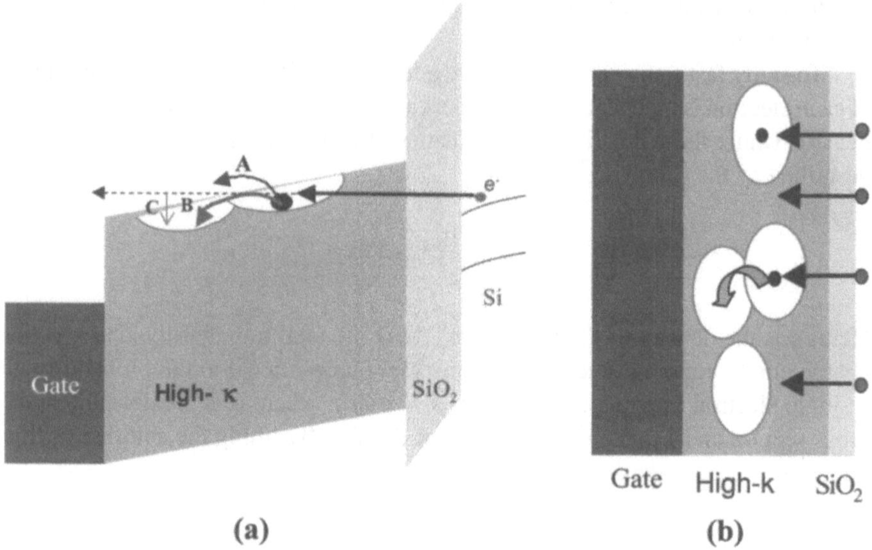

FIGURE 5.7. Schematic of the electron trapping/detrapping processes: (a) Energy band diagram with (A) detrapping of the electron captured in the direct tunneling process, (B) electron hopping between the traps, and (C) electron capture from the dielectric conduction band. (b) Top view on the gate stack depicting electron capture by the traps accessible in the direct tunneling process and migration of the captured electron between the traps.

than different nature of traps may be responsible for the fast and slow charging processes.

The observed slow charging, then, may reflect capture of the "secondary" electrons; that is, the electrons supplied by the traps charged in the fast direct tunneling

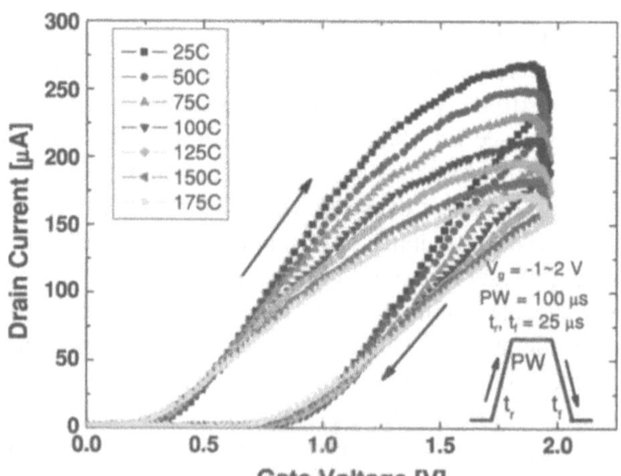

FIGURE 5.8. Pulse I_d–V_g curves measured at different temperatures.

process; Fig. 5.7. Due to their large capture cross section, these shallow traps may overlap, which would facilitate electron migration directly between the traps (as opposed to the apparently ineffective process of capture of the electrons excited from the trap to the conduction band, process C in Fig. 5.7). Indeed, the density of the current of the "secondary" electrons, J_s, required to sustain the observed slow charging rate $\tau \approx 100$ s (in the case presented in Fig. 5.5) for the traps with the capture cross section σ on the order of $\sim 10^{-12}$–10^{-13} cm^2, can be estimated from Eq. (5.5) to be $J_s \approx 10^{10}$–10^{11} electrons/s. Taking into account that the electron trap density, n, is usually on the order of 10^{12} cm^{-2}, the above magnitude of J_s results in the probability of the electron migration between the traps $R_s = J_s/n \approx 10^{-1}$–$10^{-2}$ electrons/s. A strong increase of the V_t shift with temperature [26] indicates that these electrons are supplied by the temperature-activated process. Assuming, for the purpose of a simple evaluation, the Boltzmann temperature dependence for this process, $R_s \propto \omega \exp\{-(E_t)/kT\}$, where ω is the vibrational frequency of the shallow trap of the energy E_t (traps may have slightly different energies, in particular, due to their proximity to the dielectric/electrode interface), one can show that the above estimate for R_s can be satisfied even by very low ω values. Therefore, migration of the electrons captured during the fast charging process to other available (unoccupied) shallow traps may represent the major process responsible for the threshold voltage change during the constant-voltage stress.

Comparison of the constant-voltage stress data for NMOS and PMOS devices points to the existence of another type of defect, deep electron trap, whose energy levels lie near the valence band. Discharging of these traps (which are occupied with electrons under the zero gate bias condition) and electron migration between them under certain gate voltages may also contribute to V_t instability.

The atomic structure of the shallow defects still remains unclear and is a subject of intense theoretical studies. Due to relatively deep energy levels and large relaxation energy, [27] (see Fig. 5.9)], neutral and positively charged oxygen vacancies are unlikely candidates for the shallow traps. The U-type centers represented by the negatively charge oxygen vacancies were suggested as a candidate for the shallow traps [28], although it is not clear whether this type of defect can exhibit a large effective capture cross-section value extracted from the electrical measurements. Localization of injected electrons in the d-states split off from the conduction band due to the Jahn–Teller effect and distorted at the grain boundaries was considered as a possible mechanism of electron trapping [29]. We suggest that the similar split off of the d-state may also occur in the interface region, where chemical environment of the metal cations is different from the bulk film. The electron traps may be also associated with impurities—in particular, Zr atoms usually present in hafnia within a few atomic percent. The Zr substitution in HfO_2 was calculated using plane-wave density functional theory with the spin polarized generalized gradient approximation [30]. The calculations have shown (Fig. 5.10) that Zr defects represented shallow traps with the energy of ≤ 0.2 eV, and an electron trapped at this defect was distributed over a large area with a radius of

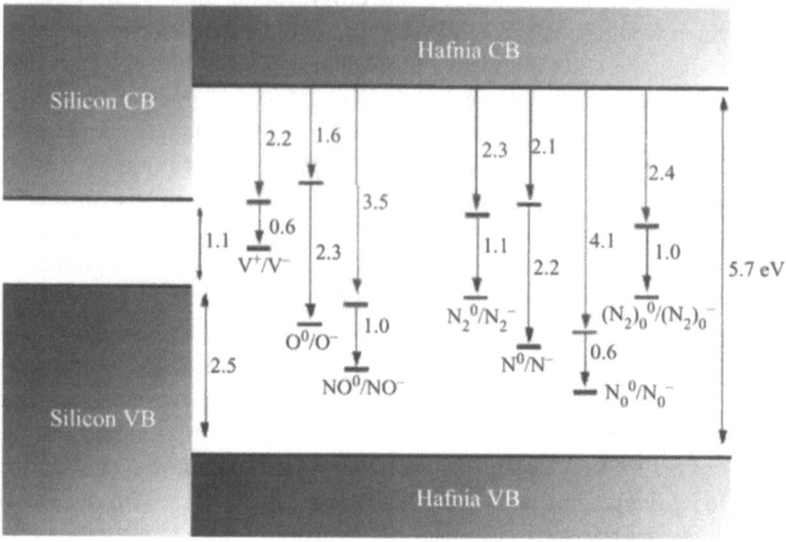

FIGURE 5.9. Electron affinities for the oxygen vacancies (V), interstitials (O), and N-related defects in different charge states in monoclinic HfO_2 given in both vertical and relaxed (deeper) forms (all values are in electron volts).

more than 0.5 nm, which is consistent with the above electrical estimates for the electrical characteristics of the traps active in fast transient charging.

Whatever the actual physical structure of the shallow traps is, their critical feature of interest for this study is a very large effective dimension, in excess of 1 nm and possibly up to a few millimeters. The ability of the dielectric film to capture injected electrons is expected to diminish when the film thickness becomes

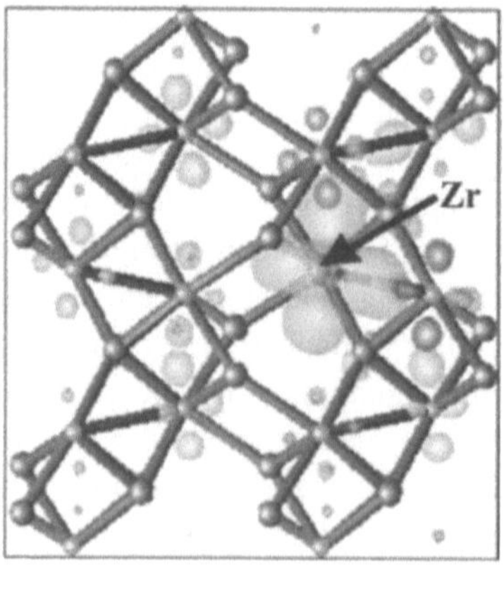

FIGURE 5.10. Spin density isosurface of the extra electron trapped at Zr atom. The view is along the *c* axis. Small "blobs" around Hf atoms indicate electron density delocalization over the large area. The electron density on the cations has distinct *d* character.

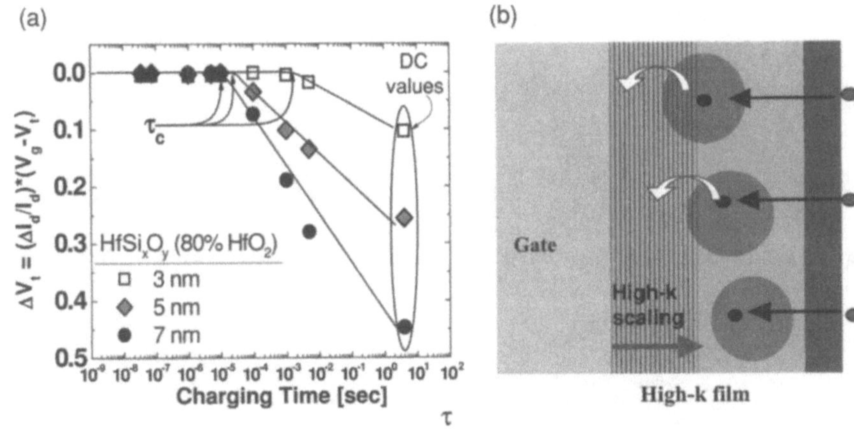

FIGURE 5.11. (a) Electron trapping versus pulse time for the HfSiO gate stacks of different high-*k* film thickness; (b) Schematic of the electron detrapping phenomenon for thinner high-*k* films.

comparable to the effective size of the trap (from the standpoint of localization of the trapped electron). This effect is clearly seen in the pulse measurement data performed in a set of Hf-silicate films of different thicknesses; Fig. 5.11a. In thinner films, more traps become unable to localize the captured electrons; which may "leak" to the gate electrode; Fig. 5.11b. Therefore, aggressive scaling of the gate dielectric physical thickness may provide trap-free device electrical characteristics.

5.5. STRUCTURAL PROPERTIES OF GATE STACK AND MOBILITY DEGRADATION

Transistors with high-*k* gate oxide show lower carrier mobility than SiO_2 FETs. The presence of the high-*k* material near the Si substrate may affect the intrinsic mobility of the carriers in the channel via different scattering mechanisms associated, in particular, with the nonuniformity of the surface potential along the transistor channel, fixed charges in the dielectric, quality of the amorphous interfacial layer in between the high-*k* dielectric and substrate, roughness of the gate electrode/high-*k* interface, soft optical phonons in the high-*k* dielectric, and so forth. We briefly discuss each of the above-mentioned potential contributions to mobility degradation, although other factors, not discussed here, might happen to be of no less importance.

The *d*-electrons involved in the bonding are delocalized to a great degree within the first coordination sphere of the metal ion (Fig. 5.2) and, therefore, the latter is sensitive to the positions and chemical bonding of all the nearest-neighbor oxygen atoms, and via them, to the next shell of metal ions. The resulting structure may exhibit a tendency to create a long-range order (crystallization). A polycrystal

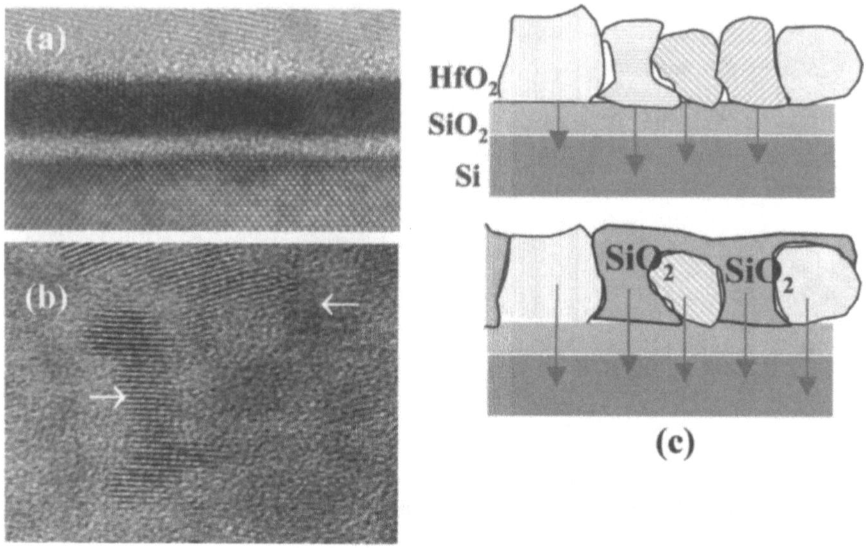

FIGURE 5.12. (a) TEM image of the cross section in the gate area of a fully processed transistor fabricated with a 3-nm HfO_2 gate dielectric and polysilicon gate electrode. (b) Top-view TEM of the Hf silicate (60% SiO_2) annealed at 1000°C for 10 s in N_2 ambient. Phase separation resulted in the formation of the grains of crystallized HfO_2 (as indicated by arrows). (c) Schematic of the effect of random grain orientation and phase separation on the vertical effective electric filed along the transistor channel.

structure, however, is, in general, an undesirable feature since it introduces, among other negative features, nonuniformity in the range of the grain size in the gate stack; see Fig. 5.12a. In order to mitigate crystallization, the metal oxide system may be diluted, in particular, with the addition of SiO_2 or Al_2O_3 (thereby forming metal silicates or aluminates) [31]. However, silicate structures, which have lower *k* values than the corresponding metal oxides, usually exhibit phase separation resulting in the formation of crystallized high-*k* and amorphous SiO_2 regions; Fig. 5.12b [32]. As in the case of the polycrystal structure of the undiluted high-*k* films, this creates a nonuniform dielectric film composition under the gate area (see Fig. 5.12c), which leads to nonuniformity of the surface potential along the transistor channel and, hence, may suppress carrier mobility.

Deposition of the high-*k* film facilitates growth of the amorphous interfacial SiO_2 layer in between the Si substrate and high-*k* dielectric (since an excess of oxygen is utilized to minimize the formation of the oxygen vacancies in the case of metal-organic chemical vapor deposition (MOCVD) process and the H_2O or O_3 cycles involved in the atomic layer deposition (ALD) process). It has been shown that oxygen can be also transferred from HfO_2 into the silicon, resulting in formation of the oxygen vacancy in the high-*k* dielectric and the interface dipole [27]. The interfacial SiO_2 layer, usually in the physical thickness range of 0.6–1.1 nm or higher, depending on the processing conditions, was considered to limit the gate stack scaling in the EOT range below 1 nm. It was found, however, that the *k* value

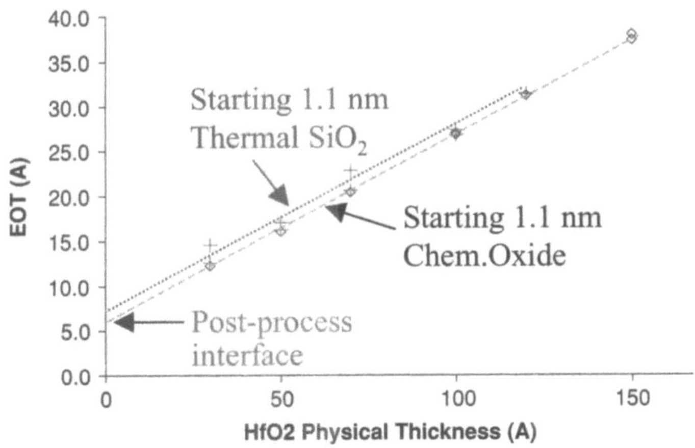

FIGURE 5.13. EOT of the gate stack versus physical thickness of the ALD HfO_2 for the gate stacks with thermal and chemical oxide starting interfaces.

of the interfacial layer was significantly higher than stoichiometric SiO_2, which was attributed to the oxygen deficiency of this layer, caused by the high-*k* deposition process [33]. Meanwhile, an oxygen-deficient SiO_x layer can be also a source of the hole and electron traps equivalent to the oxygen vacancies in SiO_2 dielectric.

In order to study the effect of high-*k* materials on the properties of the underlying thin SiO_2 film, the ALD HfO_2 films in the thickness range of 3–15 nm were deposited on thermal and chemical oxide interfacial layers of a similar physical thickness of 1.1 nm. The chemical oxide layer was formed by the pre-high-*k* deposition O_3 treatment of the Si substrate, and the thermal SiO_2 layer was fabricated by the etch-back process of the 2.0-nm ISSG (in situ Steam generated oxide, 950°C) film. The interfacial layers' *k* value was found to increase by a factor of 1.6 with respect to $k = 3.9$ value of the stoichiometric oxide; Fig. 5.13. This increase of *k* value is accompanied by distortion of the O K-edge spectra of the interfacial layer: growth of the low-energy shoulder in their spectra (sharper edge onset intensity as a function of energy) (Fig. 5.14) reflects higher density of the unoccupied localized states below the bottom of the oxide conduction band, which could be associated with oxygen vacancies [34]. Band-gap reduction leads to higher electronic contribution to the material polarizability.

Oxygen removal from the interfacial oxide to the high-*k* film was shown to be thermodynamically advantageous when hafnia oxidation proceeded in an oxygen-deficient environment [35], which is also present during the neural gas purge in the ALD process. It was demonstrated for ZrO_2, that the oxidation state of the zirconium appears to decrease with the surface proximity, suggesting that oxides such as ZrO, Zr_2O_3, are possibly present in the surface layers, thus creating an oxygen concentration gradient through the film oxide [36].

DFT calculations performed using local density approximation (LDA) approach demonstrate that oxygen removal from the immediate vicinity of the

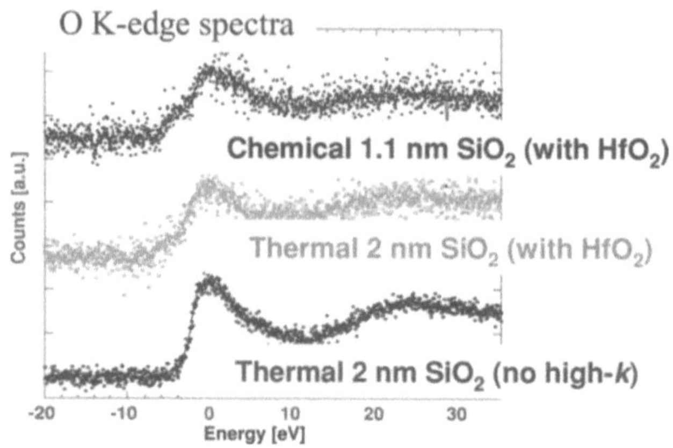

FIGURE 5.14. O K edge spectra taken on the oxide films with and without overlaying HfO_2 layer. (The spectra were collected by S. Stemmer and M. Agustin of UCSB.)

underlying silicon substrate (position 1 in Fig. 5.15) is by 0.5–1.0 eV more favorable than from the "bulk" of the interfacial SiO_2 layer (position 2 in Fig. 5.15). This is consistent with earlier conclusions based on TEM and electrical results [33] that the portion of the SiO_2 interfacial film adjusted to the Si substrate may have a significantly higher k value (approaching the k value of Si, ≈ 11) than the bulk of the interfacial SiO_2 film.

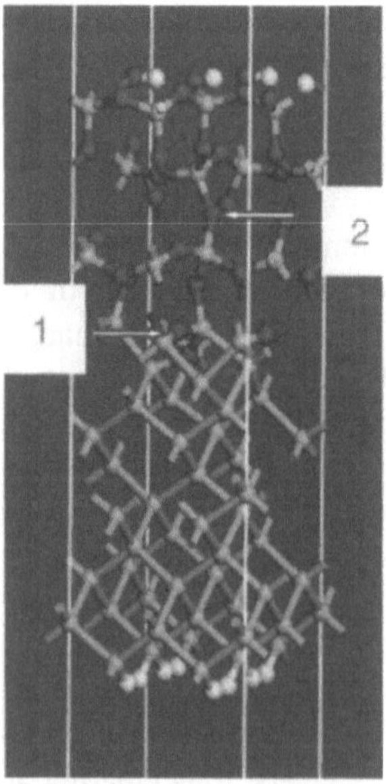

FIGURE 5.15. Calculated fragment of the gate stack, which includes the SiO_2 layer in the proximity to the Si substrate. Positions 1 and 2 represent oxygen atoms near the interface with Si and in the "bulk" of the SiO_2 layer, respectively.

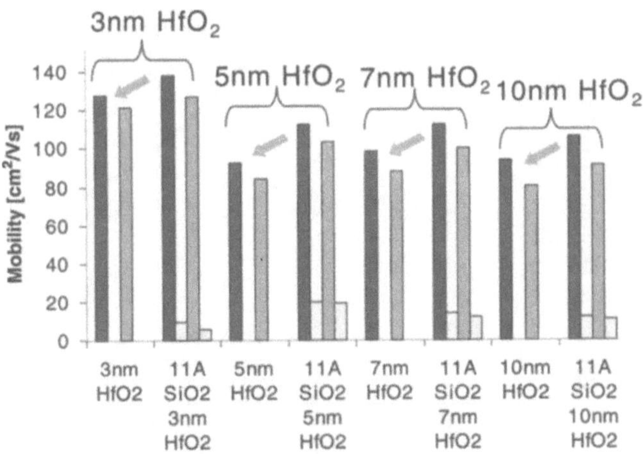

FIGURE 5.16. Peak and high-field mobility values for the gate stacks of various HfO_2 thicknesses with chemical oxide and thermal SiO_2 (labeled as 11 A SiO_2) interfaces.

The fixed bulk and interface charges in the gate stack can be estimated by fitting the V_{fb} versus EOT functional dependence to the measured values. In the case of the samples discussed above, the gate stack with the chemical oxide interfacial layer exhibits a much higher interface charge density than the thermal oxide stack (3.9×10^{11} cm^2 versus 2.2×10^{10} cm^2, respectively), whereas the bulk charge in both types of gate stack were about the same, $\approx 10^{19}$ cm^{-3}. This may reflect lower stability of a less dense, low-temperature chemical oxide with respect to oxygen removal. The higher density of the interface charge correlates to lower both peak and high-field mobility values in the transistors with the chemical oxide gate stacks; Fig. 5.16. The interface state density in both types of stack was similar, about (2–4) $\times 10^{10}$ cm^{-2}, as measured by the charge pumping method.

The interaction between the poly-Si gate electrode and high-k gate dielectric may result in the formation of an SiO_2-like interfacial layer (Fig. 5.12a), which increases the total EOT of the gate stack and may introduce additional electrically active defects in the high-k film. Due to these and other process-related issues with the poly-Si gate (in particular, poly depletion and boron penetration problems), metal gate electrodes with band-edge work function values are required for the advanced gate stacks. Although metal gate stacks exhibit lower EOT (by 0.5–0.6 nm) and much less electron trapping compared to the poly-Si gate stacks of similar thicknesses, the intrinsic mobility of the former was demonstrated in some instances to not be better and even inferior; Fig. 5.17. Transmission electron microscope (TEM) images and electron energy loss spectroscopy (EELS) and energy dispersive X-ray (EDX) analysis data [37] show that the composition and structure of the gate dielectric in both types of stacks are very similar, with the exception of its interfaces with the electrodes (Fig. 5.18b). The metal gate seems to exhibit significant interaction with the dielectric and the poly-Si/dielectric interface looks smooth due to formation of the SiO_2 intermediate layer (Fig. 5.18b). The edge

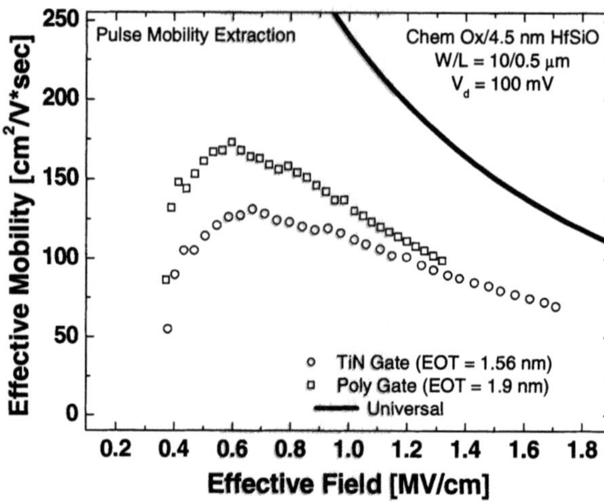

FIGURE 5.17. Peak and high field mobility values for the gate stacks of various HfO_2 thicknesses with chemical oxide and thermal SiO_2 interfaces.

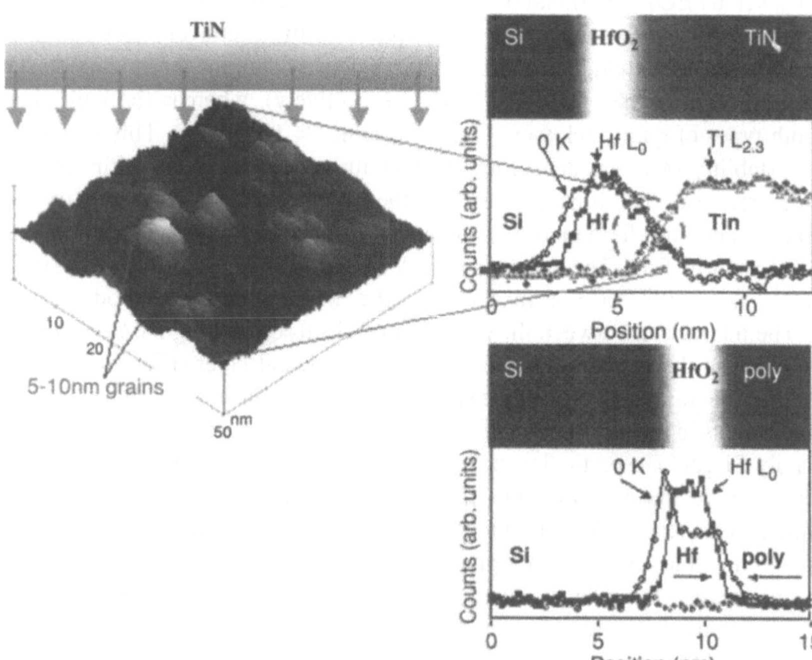

FIGURE 5.18. AFM HfO_2 surface roughness and HAADF-STEM images and EELS data collected across the gate stacks with metal and poly-Si gate electrodes.

onset and splitting of the Ti $L_{2,3}$ signal point to the TiN material penetrating deep into the HfO_2 film rather than Ti atoms bonding to the high-*k* matrix, which would result in a TiO_2 signal. This conclusion is supported by the atomic force microscopy (AFM)-based observation of a rough surface of the ALD HfO_2 film prior to electrode deposition, with the peak-to-peak microroughness of up to 50% of the film thickness. This analysis suggests that the roughness of the electrode/dielectric interface may contribute to mobility degradation in the metal/high-*k* gate stacks, for instance, via the remote surface scattering mechanism (e.g., Coulomb scattering of the channel carriers caused by the roughness-related variation of the position of the charge in the electrode) [38]. Consequently, the effect of the electrode on transistor characteristics is expected to be strongly influenced by the specifics of the electrode fabrication process—in particular, the electrode material and deposition method (e.g., ALD vs. chemical vapor deposition), poly-Si gate deposition conditions, thermo budget, and so forth. As a result, relative performances of the metal and poly-Si gate devices could be different from the example discussed above [39]. In addition, gate stack structural factors, metal and poly-Si gates differ in their effectiveness to screen the soft optical phonon coupling to the channel electrons [40].

Soft optical vibrations responsible for the higher *k* values in the metal oxides may effectively couple with the charge carriers in the transistor channel contributing to the degradation of their mobility [41]. However, the presence of the interfacial layer significantly suppresses negative effect of the soft phonons on the device performance since the carriers coupling to the phonons exponentially decreases with the interface thickness. In general, the phonon-limited mobility is expected to exhibit a strong temperature dependence, which could be used for the evaluation of the soft phonons' contribution in the high-*k* gate stacks. After correction of the total effective mobility μ_{eff} for the transient charging and subtraction of the contributions from the Coulomb and standard interface roughness scattering, the remaining mobility limiting term, $1/\mu_{remaining}$, is usually attributed to the soft optical phonons:

$$\frac{1}{\mu_{\text{remaining}}} = \frac{1}{\mu_{\text{eff}}} - \frac{1}{\mu_{\text{coul}}} - \frac{1}{\mu_r} = \frac{1}{\mu_{\text{ph}}} \tag{5.7}$$

Contrary to the phonon-limited mobility in the SiO_2 stack, in the case of the high-*k* dielectric, the μ_{ph} term in Eq. (5.7) is shown to have a very weak temperature dependence (Fig. 5.19), which is explained by a particular relation between the values of the soft phonon energy and measurement temperatures. Small values of μ_{ph} in Fig. 5.19 may be alternatively explained if one considers that the $\mu_{remaining}$ term could be dominated by temperature-independent factors related to dielectric nonuniformity due to its polycrystalline structure, interfacial layer quality, and gate electrode interface roughness, as discussed earlier in this section. This still leaves open the question of the significance of the soft optical phonon's contribution to mobility degradation relative to other factors.

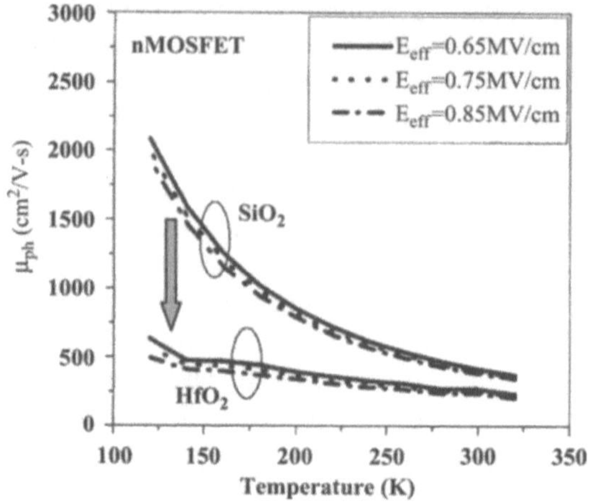

FIGURE 5.19. Temperature dependence of the phonon-limited mobility component in the SiO_2 and HfO_2 gate dielectrics. (After [33]).

5.6. CONCLUSION

The very same electronic feature of the high-*k* dielectrics that facilitates their high dielectric constant values (i.e., the presence of the *d*-electrons characterized by anisotropic spatial distribution) is also responsible for the intrinsic limitations of these materials. Structural defects causing fixed charge and electron trap centers and the dielectrics' tendency for crystallization and phase separation leading to nonuniformity of its electrical properties along the transistor channel present serious challenges for the implementation of the high-*k* materials as the gate dielectric. The quality of the process-grown SiO_2-like layer at the Si–substrate interface, as well as the dielectric's interaction with the gate electrode complicate integration of these materials into the transistor manufacturing process.

REFERENCES

1. International Technology Roadmap for Semiconductors (ITRS), 2001 Edition, 2003, Semiconductor Industry Association, available at; http///www.itrs.net/.
2. A.I. Kingon, J.-P. Maria, and S.K. Streiffer, *Nature*, **406**, 1032 (2000).
3. H.R. Huff, A. Hou, C. Lim, Y. Kim, J. Barnett, G. Bersuker, G.A. Brown, C.D. Young, P.M. Zeitzoff, J. Gutt, P. Lysaght, M.I. Gardner, and R.W. Murto, *Microelectr. Eng.* **69**, 152 (2003).
4. B.H. Lee, C.D. Young, R. Choi, J.H. Sim, G. Bersuker, C.Y. Kang, R. Harris, G.A. Brown, K. Matthews, S.C. Song, N. Moumen, J. Barnett, P. Lysaght, K.S. Choi, H.C. Wen, C. Huffman, H. Alshareef, P. Majhi, S. Gopalan, J. Peterson, P. Kirsh, H. Li, J. Gutt, M. Gardner, H.R. Huff, P. Zeitzoff, R.W. Murto, L. Larson, and C. Ramiller, *IEDM Tech. Dig.* 859 (2004).
5. J. Robertson, *J. Vac. Sci. Technol. B* **18**, 1785 (2000).
6. M. Grossmann, S. Hoffmann, S. Gusowski, R. Waser, S.K. Streiffer, C. Basceri, C.B. Parker, S.E. Lash, and A.I. Kingon, *Integr. Ferroelec.* **22**, 83 (1998).

7. S. Stemmer and D.G. Schlom, in *Nano and Giga Challenges in Microelectronics*, edited by J. Greer, A. Korkin, and J. Labanowski, Elsevier, Amsterdam, 2003, p. 129.
8. S. Sugano and Y. Tanabe, *Multiplets of Transition Metal Ions in Crystals*, Academic Press, New York 1970.
9. R. Englman, *The Jahn–Teller Effect in Molecules and Crystals,* Wiley–Interscience, New York, 1972.
10. I.B. Bersuker *Electronic Structure and Properties of Transition Metal Compounds*, Wiley–Interscience, New York, 1996.
11. J. Robertson, *Appl. Surface Sci.* **190**, 2 (2002).
12. V. Fiorentini and G. Gulleri, *Phys. Rev. Lett.* **89**, 266,101 (2002).
13. V.V. Kharton, A.A. Yaremchenko, E.N. Naumovich, and F.M.B. Marques, *J. Solid Electrochem.* **4**, 243 (2000).
14. X. Zhang, A.A. Demkov, H. Li, X. Hu, and J. Kulik, *Phys. Rev. B* **68**, 125,323 (2003).
15. J. Robertson, *Solid-State Electr.* **49**, 283 (2005).
16. T. Ostapchuk, J. Petzelt, V. Železný, A. Pashkin, J. Pokorný, I. Drbohlav, R. Kužel, D. Rafaja, B.P. Gorshunov, M. Dressel, Ch. Ohly, S. Hoffmann-Eifert, and R. Waser, *Phys. Rev. B.* **66**, 235,406 (2002).
17. M.W. Chase, Jr., NIST-JANAF Themochemical Tables, Fourth Edition, J. Phys. Chem. Ref. Data, Monograph 9, 1998.
18. M. Gutowski, J.E. Jaffe, C-L. Liu, M. Stoker, R.I. Hegde, R.S. Rai, and P.J. Tobin, *Appl. Phys. Lett.* **80**, 1897 (2002).
19. S.J. Wang, P.C. Lim, A.C.H. Huan, C.L. Liu, J.W. Chai, S.Y. Chow, J.S. Pan, Q. Li, and C.K. Ong, *Appl. Phys. Lett.* **82**, 2047 (2003).
20. H. Takeuchi, H.Y. Wong, D. Ha, and T-J King, IEEE Int. Elect. Dev. Meeting, 829 (2004).
21. J.L. Gavartin, L. Fonseca, G. Bersuker, and A.L. Shluger, unpublished results.
22. J.H. Sim, R. Choi, B.H. Lee, C. Young, P. Zeitzoff, G. Bersuker, *Solid State Dev. Mater.* 214 (2004).
23. A. Kerber, E. Cartier, L.Å. Ragnarsson, M. Rosmeulen, L. Pantisano, R. Degraeve, Y. Kim, and G. Groeseneken, VLSI Technology Symposium, Kyoto, 2003, p. 159.
24. G. Bersuker, J.H. Sim, C.D. Young, R. Choi, P.M. Zeitzoff, G.A. Brown, B.H. Lee, and W. Murto, *Microelectr. Reliab.* **44**, 1509 (2004).
25. G. Bersuker, J.H. Sim, C.D. Young, R. Choi, B.H. Lee, P. Lysaght, G.A. Brown, P.M. Zeitzoff, M. Gardner, R.W. Murto, and H.R. Huff, *Mater. Res. Soc. Symp. Proc.* **811**, 31 (2004).
26. S. Zafar, A. Callegari, E. Gusev, and M.V. Fischetti, *J. Appl. Phys.* **93**, 9298 (2003).
27. J.L. Gavartin, A.L. Shluger, A.S. Foster, and G.I. Bersuker, *J. Appl. Phys.***97**, 53,704 (2005).
28. C. Shen, M.F. Li, X.P. Wang, H.Y. Yu, Y.P. Feng, A.T.-L. Lim, Y.C. Yeo, D.S.H. Chan, and D.L. Kwong, *IEDM Conf. Proc.* 733 (2004).
29. G. Lucovsky, C.C. Fulton, Y. Zhang, Y. Zou, J.Luning, L.F. Edge, J.L. Whitten, R.J. Nemanich, H. Ade, D.G. Schlom, V.V. Afanase'v, A. Stesmans, S. Zollner, D. Triyoso, and B.R. Rogers, *IEEE Trans. Mater Dev. Reliab.* **5**, 65 (2005).
30. G. Bersuker, B.H. Lee, H.R. Huff, J. Gavartin and A. Shluger, in "Defects in High-k Gate Dielectric Stacks" ed. E. Gusev, NATO Science Series, v. 220, p. 227 (2006).
31. G.D. Wilk, R.M. Wallace, and J.M. Anthony, *J. Appl. Phys.* **87**, 484 (2000); L. Manchanda, M.D. Morris, M.L. Green, R.B. van Dover, F. Klemens, T.W. Sorch, P.J. Silverman, G. Wilk, B. Bush, and S. Aravamudhan, *Microelectr. Eng.* **59**, 351 (2001); G. Lucovsky, Y. Zhang, G.B. Rayner, Jr., G. Appel, H. Ade, and J.H. Whitten, *J. Vac. Sci. Technol. B* **20**, 1739 (2002).
32. S. Stemmer, Z.Q. Chen, C.G. Levi, P.S. Lysaght, B. Foran, J.A. Gisby, and J.R. Taylor, *Jpn. J. Appl. Phys.* **42**, 3593 (2003); P. Lysaght, B. Foran, S. Stemmer, G. Bersuker, J. Bennett, R. Tichy, L. Larson, and H.R. Huff, *Microelectr. Eng.* **69**, 182 (2003).
33. G. Bersuker, J. Barnett, N. Moumen, B. Foran, C.D. Young, P. Lysaght, J. Peterson, B.H. Lee, P.M. Zeitzoff, and H.R. Huff, *Jpn. J. Appl. Phys.* **43**, 7899 (2004).

34. J.B. Neaton, D.A. Muller, and N.W. Ashcroft, *Phys. Rev. Lett.* **85**, 1298 (2000).
35. W.L. Scopel, Antônio J.R. da Silva, W. Orellana, and A. Fazzio, *Appl. Phys. Lett.* **84**, 1492 (2004).
36. G.R. Corallo, D.A. Asbury, R.E. Gilbert, and G.B. Hoflund, *Phys. Rev. B* **35**, 9451 (1987).
37. P.S. Lysaght, , J.J. Peterson, B. Foran, C.D. Young, G. Bersuker, and H.R. Huff, *Mater. Sci. Semicond. Process.* **7**, 259 (2004).
38. S.-I. Saito, K. Torii, Y. Shimamoto, S. Tsujikawa, H. Hamamura, O. Tonomura, T. Mine, D. Hisamoto, T. Onai, J. Yugami, M. Hiratani, and S. Kimura, *Appl. Phys. Lett.* **84**, 1395 (2004); F. Gamiz and J. B. Roldan, *J. Appl. Phys.* **94**, 392 (2003).
39. S. Datta, G. Dewey, M. Doczy, B.S. Doyle, B. Jin, J. Kavalieros, R. Kotlyar, M. Metz, N. Zelick, and R. Chau, *IEDM Conf. Proc.* 653 (2003).
40. R. Kotlyar, M.D. Giles, P. Matagne, B. Obradovic, L. Shifren, M. Stettler, E. Wang, *IEDM Tech. Dig.* 391 (2004).
41. M.V. Fischetti, D.A. Neumayer, and E.A. Cartier, *J. Appl. Phys.* **90**, 4587 (2001).

6

Scanning Force Microscopies for Imaging and Characterization of Nanostructured Materials

Bartosz Such, Franciszek Krok, and Marek Szymonski

6.1. INTRODUCTION

The growing tendency to miniaturize devices used in every-day life (e.g., integrated circuits or technologies for optical information transfer), as well as exciting prospects of new emerging technologies (quantum electronics, biochips, etc.) are prompting a huge interest in nanotechnology and the science of nanostructured materials. In particular, there is a need to develop efficient technologies for the preparation of surfaces with desired structure and electronic properties. In light of those advances, designing analytical and nondestructive tools for the characterization of surface electronic and chemical properties with nanometer resolution is of utmost importance. Such tools are important for both nanostructure characterization and nanomanipulation of small-size assemblies of atoms and/or molecules on crystal surfaces in order to construct nanodevices. Furthermore, local determination of surface morphology and mechanical as well as electronic properties might be important for practical functioning of various nanosensors and nanodevices in biological objects.

Centre for Nanometer-Scale Science and Advanced Materials (NANOSAM), Institute of Physics, Jagiellonian University, ul. Reymonta 4, 30-059 Krakow, Poland such@if.uj.edu.pl; krok@if.uj.edu.pl; ufszymon@cyf-kr.edu.pl

6.2. SCANNING PROBE MICROSCOPY

Scanning probe microscopy (SPM) is commonly used for imaging and characterizing surface structure in the range between hundreds of micrometers and a few nanometers with atomic structure details. In some way, it represents the development of the concept of a profilometer.

The heart of every SPM is a sharp tip acting as a probe (see Fig. 6.1). The detailed design of its support as well as the interaction that is probed by the tip are dependent on the specific type of SPM. A tip is scanned over a sample (or a sample is scanned against a tip) by a piezoelectric scanner, usually in the form of a tube. The tube is divided into four sectors along its axis, which bend according to the voltage applied across them. Since the tube is long (up to several centimeters) and typical scan ranges are small (below 100 μm) the curvature of the tip path is negligible. The piezotube is also used for varying the tip–surface distance. Modern designs give an accuracy on the order of picometers in both vertical and horizontal directions. However, for some applications (such as biological studies), scan ranges exceeding 100 μm are required. For that purpose, SPM designs in which the scanning motion is accomplished by two perpendicular piezoelements are used. In order to control the tip position with an accuracy of picometers, the entire instrument must be rigorously separated from external vibrations. Various systems of spring suspension and eddy current damping are used.

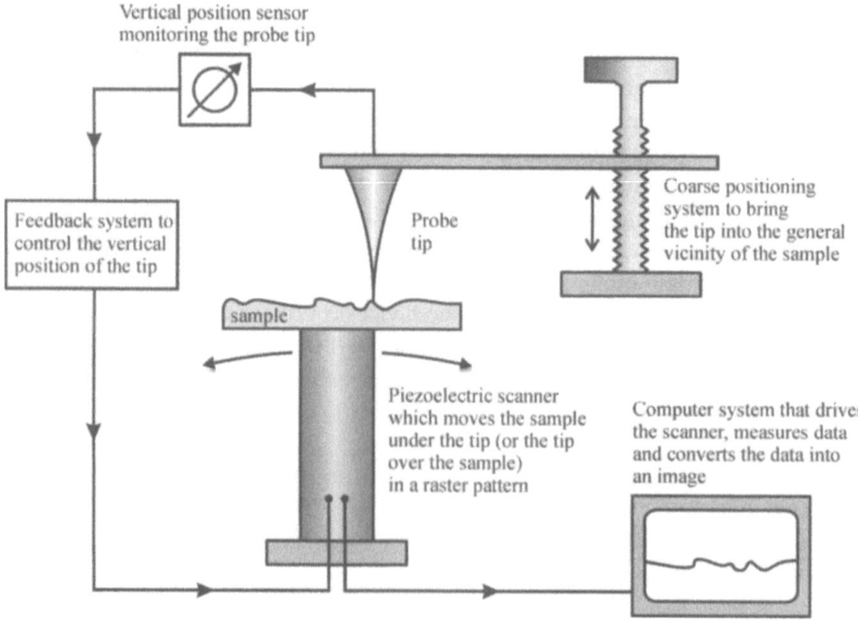

FIGURE 6.1. A schematic diagram illustrating the main components of SPM.

In principle, various fundamental interactions between the probe (atoms of the tip) and the sample atoms could be applied for practical operation of the SPM. Depending on the particular choice of such interactions we can distinguish between different types of scanning probe microscopy:

- Scanning tunneling microscopy (STM) [1, 2], chronologically it is the first technique of scanning probe microscopy, based on measurements of a tunneling current flowing between the tip and the conducting sample.
- Spin-polarized scanning tunneling microscopy (SP-STM) allowing for studying of conducting magnetic structures [3, 4].
- Atomic force microscopy (AFM) [5], known also as contact-mode atomic force microscopy, based on the analysis of the interaction between the tip and the surface being in direct contact.
- Dynamic force microscopy (DFM) [6], also known as a noncontact atomic force microscopy (nc-AFM).
- Friction force microscopy (FFM) or so-called lateral force microscopy (LFM) [7, 8] sensing friction forces between atoms of the surface and the tip on a nanometer-size area.
- Kelvin probe force microscopy (KPFM) [9] allowing for determination of the contact potential difference between the tip and the investigated sample.
- Magnetic force microscopy (MFM) allowing for high-resolution imaging of magnetic materials [10, 11].
- Scanning near-field optical microscopy (SNOM) and other variations not listed here.

In order to operate the scanning probe microscope the following conditions must be fulfilled [12]:

1. **The very end of the tip apex must be atomically sharp**. The quality of the tip is of utmost importance for any type of the SPM. This was already recognized in the very first, classic work on STM by Binnig and co-workers [1], who estimated that even if the tip radius is on the order of 100 Å, features of size below 100 Å should be possible to resolve. However, even for a tip formed by cutting a wire that has much larger radius of curvature than required, there are likely to be atomic-sized protrusions extending in the form of nanotips (see Fig. 6.2). They are sufficient for obtaining atomic resolution in the case of STM. Furthermore, it was also found that such nanotips could be produced simply by direct tip collision with the investigated sample. Unfortunately, it is impossible to control such a process. Thus, the most common tip preparation techniques are chemical etching [13] and field-emission processes [14–16] in the case of STM and ion sputtering, chemical etching, and annealing [17] in the case of AFM. It should be stressed that those procedures are still far from being perfect, and the tip quality control, in particular during scanning,

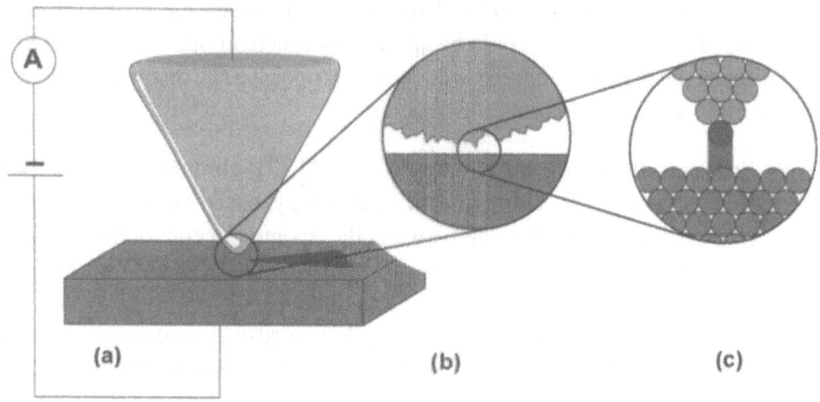

FIGURE 6.2. (a) A tip for STM is brought close to a surface and biased; (b) despite of the fact that it has a large radius of curvature, there are atomic-scale "nanotips" on its apex; (c) the front atom at tip apex is responsible for majority of interaction.

has yet to be developed. It will be shown in Section 6.5.8) how the tip's chemical composition and its geometry could have a principal effect on the quality of the images, as well as on their interpretation.

2. **The tip–sample interaction used for sensing must rapidly decrease with increasing separation distance.** This condition is necessary for obtaining a high resolution in the "z" direction, perpendicular to the surface. In the case of tunneling microscopy, the current depends on the distance exponentially. So with typical values of the work function, the resulting current changes by up to three orders of magnitude for a distance change corresponding to typical single atomic step [1]. In the case of AFM in a contact mode, short-range repulsive interactions are sampled, which result from the overlap of electron clouds of the tip atoms and the atoms on the sample surface. In contrast to contact-mode AFM, DFM uses attractive interactions; however, some major components of the tip–surface force slowly vary with distance. Consequently, in order to approach the tip close enough to the sample and to obtain the required sensitivity, the cantilever is set into oscillatory motion. As a result, the tip can temporarily approach the surface very closely at its maximum deflection, and then the frequency and amplitude of the cantilever oscillations become precise measures of the tip–surface forces.
3. **The interaction must change by an amount easily measurable during scanning of the tip above or on the sample surface.** This condition refers to the magnitude of the interaction changes in the course of tip scanning over the sample surface. Large variations in the interaction force magnitude are essential for obtaining a high lateral resolution (i.e., the resolution in x-y plane parallel to the surface). In principle, both the tunneling current and the forces measured by the DFM could be used for obtaining a true atomic resolution, although it is not the case for an

AFM operating in a contact mode. The tip–surface contact area in AFM is much larger than the area of an elementary lattice cell. For example, for a rather hard system of a tungsten carbide tip in contact with a diamond surface, and a load force in the range of 0 nN (only an adhesion force is operating) to 12 nN, the contact area is between 3.9 nm^2 and 7.4 nm^2 [18]. It simply means that a single atomic vacancy at the surface could not be seen due to a force averaging over that relatively large area. Therefore, in the best case, only the surface structure atomic periodicity could be imaged with contact AFM. Other types of SPM are usually unable to provide atomic resolution imaging. For example, MFM typically offers a modest resolution on the order of 100 nm [19].

4. **For efficient performance of the feedback loop, the dependence of the interaction on the distance should be monotonic, at least within a certain range.** This condition implies that the dependence of the tip–surface interaction on the distance must be monotonic, enabling an univocal interpretation of the measured images. More importantly, however, it is necessary for driving a feedback loop system controlling the tip–surface distance. If the interaction versus distance dependence is monotonic, at least within a certain range, the loop electronics could easily ignore small instabilities that are inevitable during the scanning process. As such, those instabilities do not significantly disturb the control system.

6.3. MODES OF SPM OPERATION

There are two principal modes of SPM operation: the constant interaction mode (current for STM, force for AFM, detuning for DFM, etc.) and the constant-distance mode. A schematic view of the principles of both modes is shown in Fig. 6.3. In the constant-distance mode, the tip is scanned over the sample surface without changing its z-coordinate (perpendicular to the surface). The instrument feedback

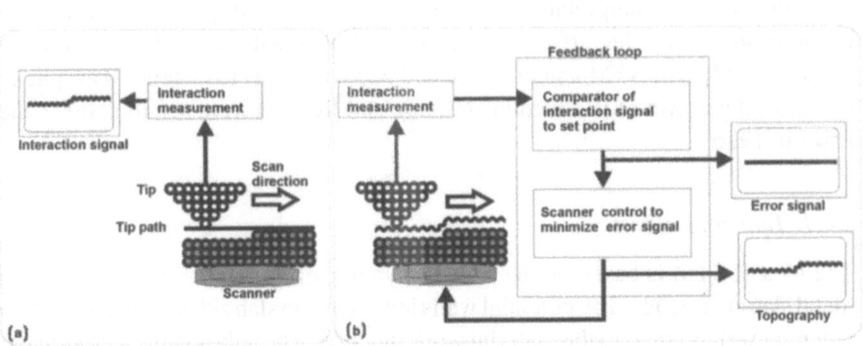

FIGURE 6.3. SPM principle of operation for constant distance (a) and constant interaction (b) modes.

loop is not active and the result of the measurement—the two-dimensional map of the interaction magnitude—is not affected by noise and/or artifacts that would otherwise be generated by the loop. This mode of operation could only be applied to very flat surface planes and for relatively small scanning areas. Scanning across structures higher than the tip–surface distance (on the order of several angstroms) could result in a collision between the tip and the surface, thus causing irreversible tip damage and disturbing the scanning process.

Usually, the constant interaction mode is more common than the former one. The tip movement is driven by a feedback loop system, which keeps the interaction constant. An advantage of such a mode is that it could be applied for scanning over surfaces covered with relatively large structures without danger of tip and/or sample damage. The image is obtained by recording variations of the feedback system signal driving a piezoelement controlling the z coordinate. Such an image is often called the "surface topography," but, in reality, it may not reflect a "true topography," particularly for atomic-scale images.

The feedback system by itself can distort the signal and, therefore, a so-called "mixed mode" is applied for atomic structure imaging. In that mode, the feedback system is active, but its gain is very small and it is used only to keep the average tip–surface distance constant. The varying magnitude of the interaction is used as the measured signal. This mixed method has significant advantages over the previous ones, since the active feedback loop protects the tip against crashes with topography features on the investigated surface but it does not react on subtle atomic structure irregularities, which could be imaged as in the case of the constant-distance mode [20].

6.4. SCANNING TUNNELING MICROSCOPY

A STM was chronologically the first successful scanning probe method. The first STM data were presented in 1982 by Binnig et al. in their groundbreaking work [1]. The new technique quickly gained attention from surface scientists, resulting in a quick development of microscope designs, growth in the amount of data collected with that technique, and an increase in understanding of contrast formation mechanisms. The interpretation of STM images is usually performed within the frame proposed by Tersoff and Hamann [2, 21], based on Bardeen's theory of tunneling [22]. The STM's importance and impact on surface science was quickly recognized and two of its inventors (Binnig and Rohrer) were awarded the Nobel Prize in 1986.

6.4.1. Tunneling Effect

The idea of STM is based on the well-known tunneling effect. The electrons in a metal slab are restricted by potential walls located at the slab sides. The height of the potential wall is equal to the material work function. The solution of Schrödinger's equation inside the slab delivers the electron wavefunction in the form of a plane

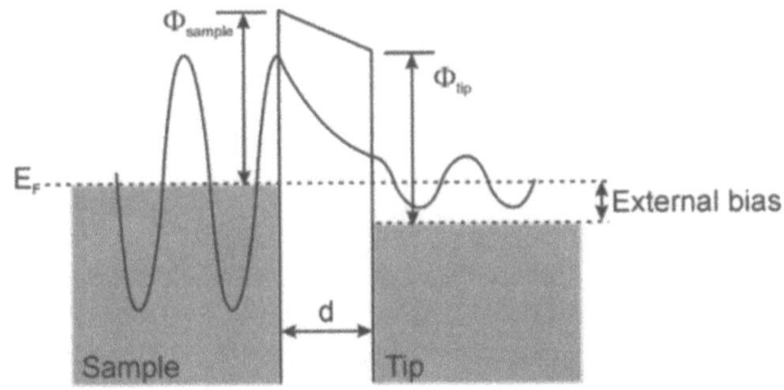

FIGURE 6.4. An electron wavefunction is extending out of an electrode. If another electrode is brought close and there are unoccupied states at given energy, an electron can tunnel.

wave. Moreover, at the boundary, the wavefunction does not disappear abruptly, but decays exponentially into the vacuum region. That means that there is a certain probability of finding electrons outside of the metal slab. If two metal pieces (say, a sample and a tip) are placed very close to each other, the wavefunctions can overlap. Moreover, if there is a small external bias V between them, the electron wavefunction located close to the Fermi level of one electrode can extend to the other, where empty states just over Fermi level are located (see Fig. 6.4). Therefore, an electron from the first electrode can move, or tunnel, to the other, producing a current that can be measured. The probability of finding the electron on the other side of the finite height barrier is nonzero.

If the barrier height is assumed to be

$$\Phi = \frac{1}{2}\left(\Phi_{\text{sample}} + \Phi_{\text{tip}}\right), \tag{6.1}$$

where Φ_{sample} and Φ_{tip} are the sample and tip work functions, respectively, quantum mechanics will tell us that the current must be proportional to

$$I \propto \sum_{E=E_F-U}^{E_F} |\Psi_n(0)|^2 e^{-2kd}, \tag{6.2}$$

where the sum goes through all of the initial states between energy equal to the difference between Fermi energy E_F and external bias U to the Fermi energy, $\Psi_n(0)$ is the electron wavefunction of the source electrode at its border, and d is the width of the tunneling barrier. If the external bias U is much smaller than the work function (i.e., $U \ll \Phi$), then k can be written as

$$k \approx \frac{\sqrt{2m\Phi}}{\hbar}, \tag{6.3}$$

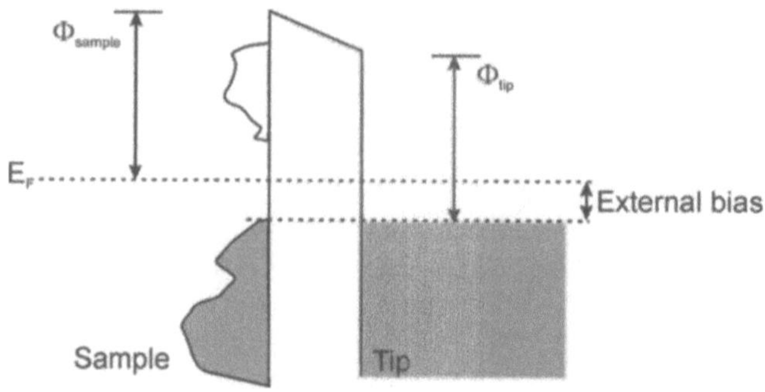

FIGURE 6.5. If there is a complicated density of states in either the tip or at the sample, a tunneling current will reflect that. For the situation in this image, the tunneling current would not flow until a certain value of external bias is exceeded.

where m is the free-electron mass and $\hbar$ is the Planck constant divided by 2π. Then the tunneling current can be estimated [1] as

$$I \propto \sum_{E=E_F-U}^{E_F} |\Psi_n(0)|^2 e^{-1.025\sqrt{\Phi}d} \tag{6.4}$$

if the distance is measured in nanometers and the energy in electron volts.

The tunneling current is usually in the nanoamper range and it is exponentially dependent on the distance between the sample and the tip. For typical values of Φ ($\sim$5 eV), the tunneling current changes by an order of magnitude with a distance change of 0.1 nm. Therefore, that effect gives an exceptional measure of the distance between the two electrodes.

In reality, the situation is more complicated. The current is dependent on the work functions of both tip and sample materials and, as presented by Tersoff and Hamann [2, 21], on the tip and on the sample local density of states (LDOS) (see Fig. 6.5). Therefore, in general, the STM image does not represent the "true topography" of the sample surface. Rather, it is a map of its LDOS and, particularly for images taken with subnanometer resolution, the protrusions in the STM images (regions where a higher tunneling current is recorded) can be completely different from geometric positions of the surface atoms. The well-known example of such a phenomenon has been reported for the surface of Highly Ordered Pyrolitic Graphite (HOPG) [23], where only half of the surface atoms were imaged with STM. Also, many conflicting reports on the structure of reconstructed III–V semiconductor (001) surfaces (GaAs, InSb, etc.) were caused by false interpretations of the STM data (for a discussion, see references [24, 25]).

6.4.2. Examples of STM Imaging

As an example of STM measurements, the images of a c(8 × 2) reconstructed InSb(001) surface are presented below. The same surface will also be discussed in Section 6.5.7 concerning a DFM mode of scanning probe operation. The large-scale

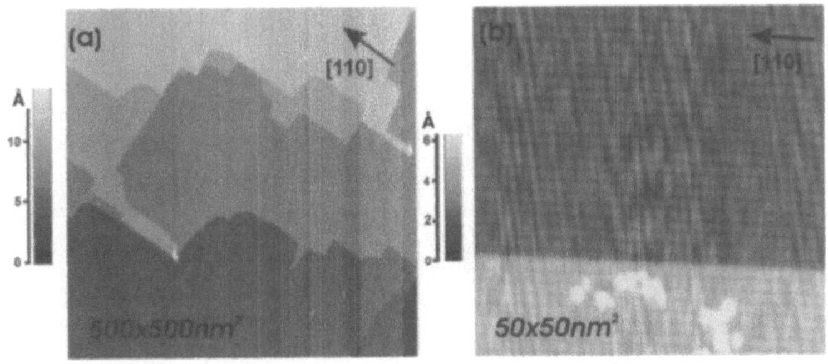

FIGURE 6.6. STM images of c(8 × 2)InSb(001) surface: (a) large-scale, unoccupied states ($U = 1\ V, I = 0.76$ nA); (b) 50 × 50 nm^2, occupied states ($U = -1\ V, I = 0.355$ nA). Reprinted from *Surface Science*, Vol. 506, Numbers 1-2, Kolodziej J. J., Such B., Czuba P., Krok F., Piatkowski P., Szymonski M., Scanning-tunneling/atomic-force microscopy study of the growth of KBr films on InSb(001), pp. 12–22. Copyright 2002, with permission from Elsevier.

image (Fig. 6.6a) shows perfectly flat terraces separated by sharp edges. Only some local contaminations are present.

It is an inevitable conclusion that this image represents the real topography of the sample. In Fig. 6.6b, a smaller area of the same surface is shown. A terrace edge is visible in the lower part of the image and the entire image area is covered by stripes running along the ⟨110⟩ direction. The stripes are connected with a strongly anisotropic reconstruction of the surface. Since that image was taken under conditions in which the occupied surface states are imaged as protrusions (at sample bias: −1 V), only the orbitals connected with antimony atoms are depicted. Therefore, that pattern definitely does not represent the topography of the surface.

6.5. ATOMIC FORCE MICROSCOPY

Despite all of the advantages, STM has a very serious drawback: it can only be applied to conductive samples. Since the very first days of scanning probe microscopy, it has been obvious that the extension of the method to analyze insulating samples was necessary. To achieve this goal, a new physical quantity had to be measured instead of the tunneling current. The tip–surface interaction force was the natural choice. In the early days of AFM development, this force was thought to be the van der Waals interaction force exclusively. In general, those interaction forces have a short-range repulsive core and a long-range attractive part (see Fig. 6.7). Somewhere at a distance of a few angstroms from the surface there is a minimum in the interaction curve. There are two regions where the force can be used as the feedback signal: in the repulsive core area when the tip is in contact with the surface (although, recently, realizations have been presented in which the tip in dynamic mode is vibrating in the repulsive force region [26, 27]), and in the attractive region, preferably at a tip–surface distance larger than the distance corresponding to a maximum in the attractive force. However, in the attractive

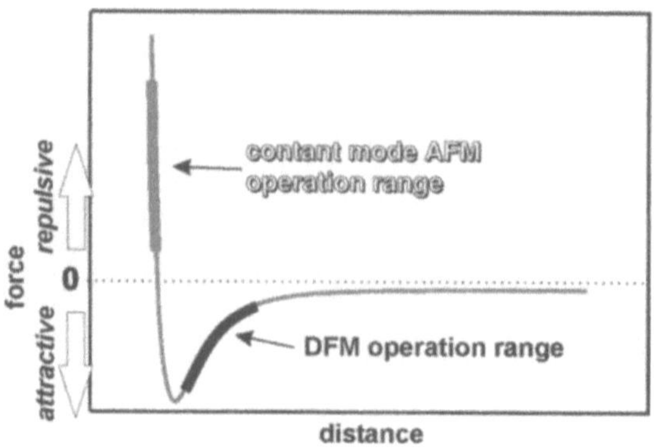

FIGURE 6.7. Dependence of tip–surface force on distance simulated according to the Lenard–Jones potential with the regions in which AFM and DFM operate.

regime, the dependence of the force on the distance is relatively weak, and in order to achieve the required sensitivity (and in order to avoid a so-called "snap to contact" phenomenon), the static force measurement was replaced by dynamic methods.

In 1986, Binnig and co-workers [5] proposed the method of investigating tip–surface interaction forces. The tip was mounted at the end of a flexible cantilever and brought into contact with the surface. While the cantilever scanned the surface, the topographic features caused tip deflection, which can be measured accurately. There are a few methods to detect cantilever deflection. Originally, Binnig and co-workers used the STM tip to detect the cantilever deflection. Later, optical interferometry was employed, in which a laser beam reflected from the top of the cantilever was used for detecting very small changes of the cantilever position. The other method uses the piezoresistive effect [28]. The cantilever has an integrated conductive path that is connected as a part of a Wheatstone bridge. The cantilever deflection changes the resistance of the path, which can be measured.

Nevertheless, the most popular way is the optical detection mode in which a laser beam is reflected from the upper, mirrorlike, part of the cantilever into a position-sensitive photodetector (see Fig. 6.8).

The influence of the tip–surface interaction force on the tip deflection characteristics enables us to retrieve a signal, which could be called "topographic."

6.5.1. *Contact AFM*

In the contact-mode of AFM (c-AFM) [5], the tip is in close proximity to the sample surface subject to a strongly repulsive interaction (see Fig. 6.7). This characteristic of the interaction results from the overlap of the tip electron clouds and the substrate atoms. Following the Pauli principle, the electrons must be promoted to higher-energy states; thus, the resulting force is short range and strongly repulsive. Due to averaging of the interaction over an area larger than the individual atom size, this mode of operation is not able to provide atomic resolution.

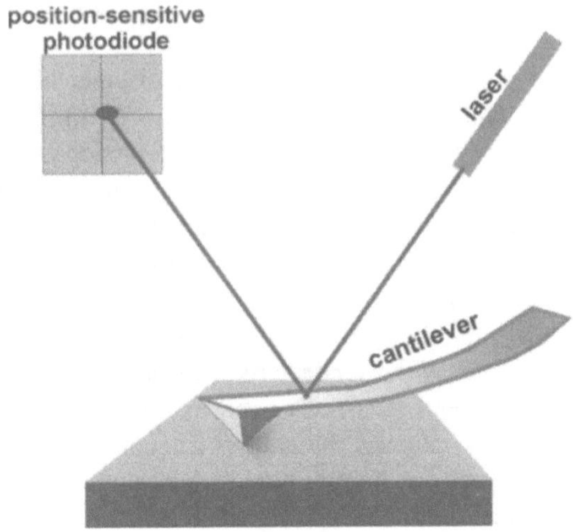

FIGURE 6.8. Schematic view of the optical detection system for monitoring the AFM cantilever deflection.

Most of currently used AFMs also allow for measurements of friction forces between the tip and the substrate, so-called lateral force microscopy (LFM) or friction force microscopy, (FFM) [7, 8]. The measurement of the friction force is performed in the contact mode of AFM. Sliding of the tip over the substrate results in torsional deflection of the cantilever, which could be measured by an optical system similar to the one used for monitoring the vertical deflection. The magnitude of the torsional deflection is proportional to the local friction force sensed by the tip. For scanning over a flat surface, this directly determines the local value of the friction coefficient. For more complicated surface morphologies, however, the interpretation of this measurement is much more complicated. Alternate scans along the same line but in opposite directions could help to distinguish between deflections caused by complicated substrate morphology from the local variations of the frictional properties.

6.5.2. *Dynamic Force Microscopy*

The DFM method (often called noncontact atomic force microscopy, nc-AFM) has been developed as an extension to the contact mode of AFM. It enables imaging surfaces with better resolution than in the contact mode (atomic resolution was obtained for the first time in 1995 [29, 30]) and reduces unwanted modifications of the imaged surface while preserving the most important advantage of AFM: the possibility of imaging nonconducting samples. Over the last few years, DFM succeeded in atomic resolution imaging of different materials such as ionic crystal surfaces (NaCl [30], KBr [31] and CaF_2 [20]), sapphire [33], Si (111) 7×7 reconstructed surfaces [29, 34], [110] and [100] surfaces of III–V semiconductors

[30], or adsorbed molecular overlayers [35, 36]. The rapid development of this technique can be traced in recent reviews [37–39].

In DFM, the tip is moved over the surface at a distance corresponding to the attractive part of the interaction potential (see Fig. 6.7). In this range, the force could not be used as a measured signal since the interaction is only slowly varies with distance (condition 2 from Section 6.2). In order to solve this problem, the tip is mounted on a cantilever vibrating at a certain distance from the surface.

Typically, the tip is vibrating with an amplitude of several tens of nanometers temporarily approaching the sample as close as 0.4 nm. This is much closer than could be possible in a static mode even with the use of very rigid cantilevers.

The cantilever vibrates with a characteristic resonant frequency due to stimulation by a piezoelectric element. Usually a self-exciting system is used for which the stimulation frequency matches the natural frequency of the cantilever. The vibrations of the cantilever can be monitored by an optical detection system as described earlier (see Fig. 6.8) or by a piezoresistive material. The influence of the tip–surface interaction force on the tip vibration characteristics enables the detection of a signal, which is called "topographic." The real meaning of that signal will be discussed later.

In the so-called "ac-detection mode" [40], a change in the vibration amplitude is measured. Alternatively, the "FM-detection mode" [6] is used in which the variation of the cantilever vibration frequency is measured during scanning while the amplitude is kept constant with an additional feedback system. The former mode is commonly used in instruments for ambient environments, whereas the latter is used in microscopes designed for vacuum operation. This is due to the fact that the quality factor, Q, for the cantilevers operating in vacuum ($>10{,}000$) is much higher than the ones operating in air (>10). For ac detection, the increase of the Q factor results in a narrowing of the bandwidth. Thus, in vacuum applications, the system could not respond sufficiently fast for changes in tip–surface interaction. In the case of the FM-detection scheme, the band is determined by the demodulator parameters that could be suitably chosen.

A block scheme of the DFM is presented in Fig. 6.9. In this particular example, the sample is mounted in a fixed position, whereas the cantilever is moved with the scanner. In several commercial instruments, the reverse situation is used.

6.5.3. Detection of the Cantilever Vibrations

Fast cantilever frequency change measurements are essential for successful operation of the DFM. Since the basic information on the tip–surface interaction is contained in a fast varying (60–250 kHz) voltage signal, which must be probed very often (say, every 1 ms). The system must be sensitive for frequency changes as small as a few millihertz. A great improvement in quality of such measurements could be achieved with the help of a so-called "phase-lock loop" (PLL) device [41, 42]. A schematic diagram of the PLL is shown in Fig. 6.10. The concept of the PLL is the same as the one often used in radioelectronics (e.g., for the detection of FM radio signals).

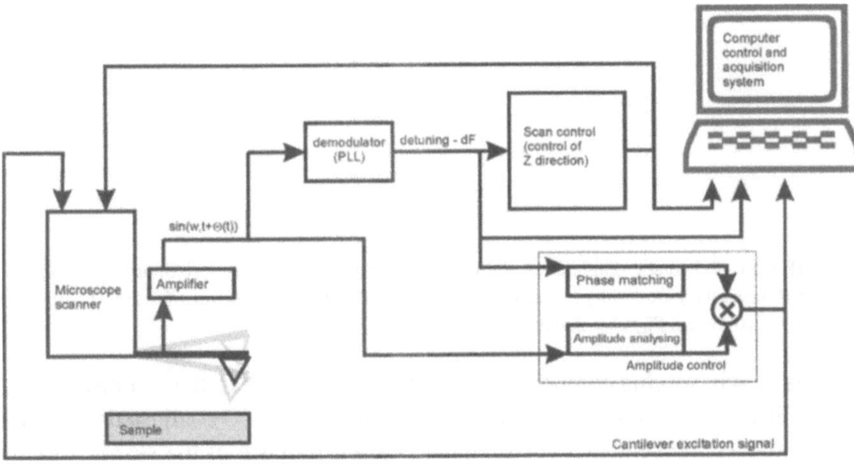

FIGURE 6.9. Block scheme of the DFM.

Initially, the system must be synchronized with the resonant frequency of the free cantilever (no interaction; the tip is retracted away from the sample). This is used as the reference signal. Next, the system can measure the frequency of the cantilever vibrations modified by the interaction with the sample during scanning. The cantilever signal is then represented by a function: $\sim\sin[\omega_c t + \Theta(t)]$, where ω_c is the original cantilever frequency and $\Theta(t)$ represents the time-dependent phase modulation due to the tip–surface interaction. This signal is delivered into a phase comparator. The comparator multiplies that signal with the signal from a voltage-controlled oscillator (VCO), alternating with frequency ω_r. As a result, a output signal is composed of two components: the fast alternating one, with a frequency equal to the sum of the incoming frequencies ($\sim\sin[(\omega_c + \omega_r)t + \Theta(t)]$),

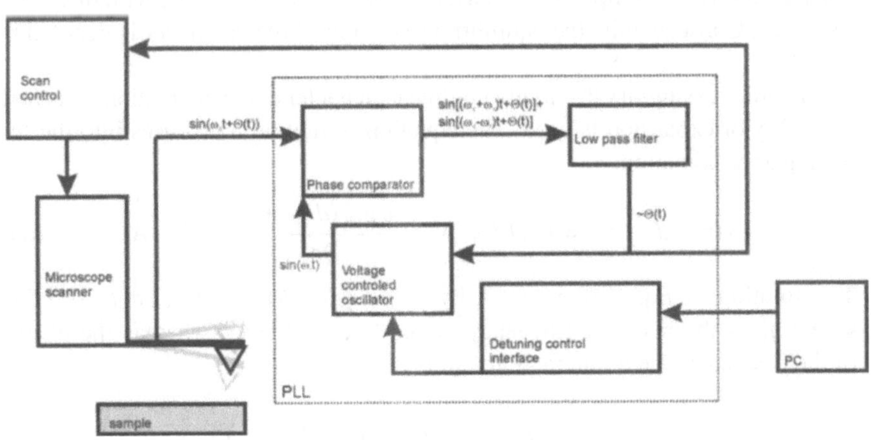

FIGURE 6.10. Schematic diagram of the PLL as used in the commercial Nanosurf easyPLL.

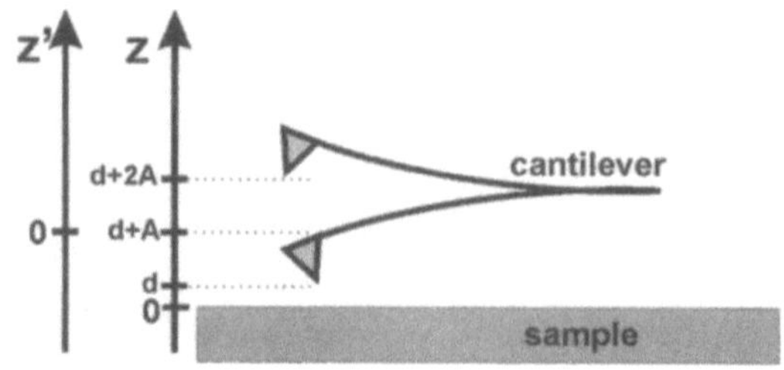

FIGURE 6.11. Schematic diagram for the cantilever vibrating in a close proximity of the sample surface. Reprinted Figure 1 with permission from, Giessibl F.J., *Phys. Rev. B*, Vol. 56, p. 1610. Copyright 1997 by the American Physical Society.

and the slowly varying component proportional to the sine of the incoming signal phase difference ($\sim\sin[(\omega_c - \omega_r)t + \Theta(t)]$). Since the VCO is adjusted to balance the cantilever frequency ($\omega_c = \omega_r$), the phase difference of the slowly varying component is equal to the modulation $\Theta(t)$. Subsequently, the output signal passes through the low-pass filter, transmitting the slow component only. For small phase modulations, the signal which is $\sim\sin(\Theta(t))$ can be approximated by $\Theta(t)$ and used for controlling the tip movement along the z direction of the microscope.

6.5.4. *Tip–Surface Interaction of a Vibrating Cantilever*

In early work on DFM [6] it was proposed that the change of the cantilever vibration frequency is proportional to the gradient of the tip–surface interaction forces. It follows from the following consideration. The cantilever motion could be described by the harmonic oscillator equation [43]

$$m^*\ddot{z} + k(z - d - A) = F_{\text{int}}(z), \tag{6.5}$$

where m* is the effective cantilever mass, k is its spring constant, $F_{\text{int}}(z)$ is the force of interaction between the tip and the sample surface, d denotes the closest approach distance between the tip and the surface, and A is the amplitude as indicated in Fig. 6.11. Consequently, the equilibrium position of the cantilever is denoted by $d + A$.

In order to simplify the highly nonlinear character of the interaction, a second-order Taylor expansion is used. The equation of motion transforms into the following approximation:

$$m^*\ddot{z} + k(z - d - A) = F_{\text{int}}(d + A) + \frac{\partial F_{\text{int}}(d + A)}{\partial z}(z - d - A). \tag{6.6}$$

The solution of Eq. (6.6) is given by $z(t) = (d + A) + A\cos(2\pi(f_0 + \Delta f)t)$, where f_0 is the resonant frequency of the free cantilever, whereas the change of the frequency due to interaction, Δf, is given by

$$\Delta f = f_0\left(\sqrt{1 - \frac{1}{k}\frac{\partial F_{\text{int}}(d + A)}{\partial z}} - 1\right). \tag{6.7}$$

A second approximation ($\sqrt{1-x} \approx 1 - x/2$) reduces the above equation to

$$\Delta f = \frac{f_0}{2k} \frac{\partial F_{\text{int}}(d+A)}{\partial z}. \tag{6.8}$$

In the above considerations, it is also assumed that the equilibrium point of the free cantilever is the same as the one for the cantilever close to the surface.

Equation (6.8), although frequently cited, is not correct—in particular, for high-resolution imaging of atomic structures where short-range chemical forces have to be considered. Only in the case of extremely small amplitudes (in comparison to the interaction range) could Eq. (6.8) hold. A more accurate solution has been proposed by Giessibl [44] using the perturbation theory. It was shown that the cantilever frequency change, Δf, due to a small perturbation induced by the tip–surface interaction force, $F_{\text{int}}(z)$, is well described by

$$\Delta f = -\frac{f_0}{kA^2} \left\langle F_{\text{int}}(z') \right\rangle, \tag{6.9}$$

where $z' = z - (d - A)$ as denoted in Fig. 6.11. Giessibl [44] used a power form of the interaction force:

$$F_{\text{int}}(z) = -Cz^{-n}, \tag{6.10}$$

where C is a constant and $n \geq 1$. It is known that most of the analytical potentials used for modeling of the tip–surface interaction (van der Waals, Morse, Lenard–Jones) are well approximated by the power potential. Since the average interaction from Eq. (6.9) must be calculated over a full period of vibration, Eq. (6.9) becomes

$$\Delta f(d, k, A, f_0, n) = \frac{1}{2\pi} \frac{f_0}{kA} \frac{C}{d^n} \int_0^{2\pi} \frac{\cos x}{\left[1 + (A/d)(\cos x + 1)\right]^n} \, dx. \tag{6.11}$$

For very small amplitudes ($A \ll d$), Eq. (6.11) can be reduced to Eq. (6.8). At the other extreme ($A \gg d$), which corresponds to the most realistic situation, Eq. (6.11) could be transformed to

$$\Delta f(d, k, A, f_0, n) = \frac{1}{\sqrt{2}\pi} \frac{f_0}{kA^{3/2}} \frac{C}{d^{n-1/2}} \int_{-\infty}^{+\infty} \frac{1}{\left[1 + y^2\right]^n} \, dy. \tag{6.12}$$

Therefore, the cantilever detuning, Δf, could be calculated for a given form of the interaction potential.

Equation (6.12) can be tested experimentally, as presented in Fig. 6.12. The dependence of cantilever frequency detuning, Δf, on the distance between the tip's lower turning point and the probed surface plane has been measured for a

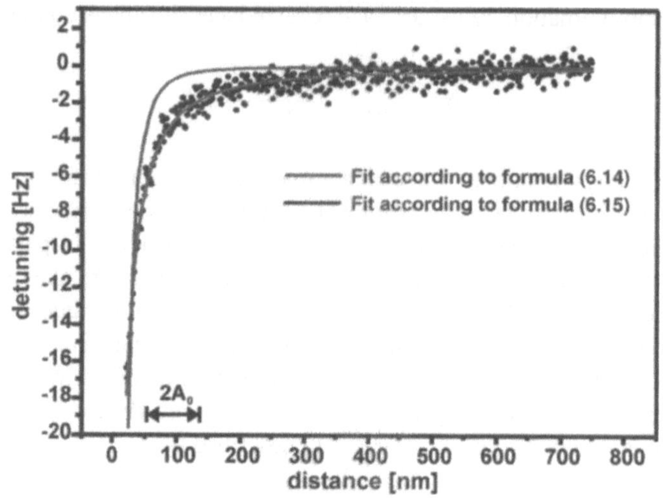

FIGURE 6.12. Force distance dependence for a blunt tip on CaF_2. Curves representing the best fit of Eq. (6.14) (red line) and Eq. (6.15) (blue line) are displayed together with the experimental points. The oscillation amplitude is marked as "$2A_0$."

CaF_2 crystal. The measurement was performed with a deliberately blunt Si tip (van der Waals force noticeable even at the distance of 200 nm). If the tip shape is similar to the one presented in Fig. 6.13, it can be assumed that only the spherical part is involved in the van der Waals interaction.

The van der Waals interaction between a sphere and a flat surface can be described as [45, 46]

$$F_{vdW} = -\frac{H}{6}\frac{R}{d^2}, \tag{6.13}$$

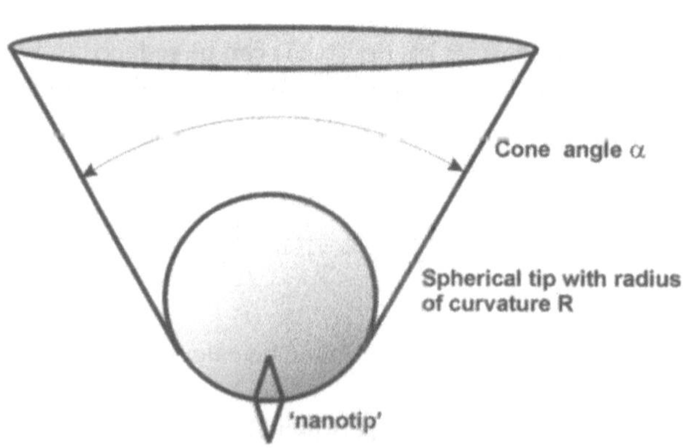

FIGURE 6.13. Model of a tip. Based on the graphics from Ref. 45.

where H is the Hamaker constant and R is the radius of curvature of the tip. The Hamaker constant for a system Si–CaF_2 is equal to $6,7 \times 10^{-20}$ J [47]. Since the force dependence on the tip–sample distance is known, Δf can be calculated. According to Eq. (6.12), detuning is equal to

$$\Delta f(d) = \frac{1}{12}\frac{H}{\sqrt{2}}\frac{f_0}{kA_0^{3/2}}\frac{1}{d^{3/2}}, \tag{6.14}$$

where f_0 is the resonant frequency of the tip, A_0 is the oscillation amplitude, and k is the cantilever spring constant. Similarly, according to the simpler formula [Eq. (6.8)], the detuning is given by

$$\Delta f(d) = \frac{f_0 H R}{6k}\frac{1}{d^3}. \tag{6.15}$$

In Fig. 6.12, both approaches are presented: curves representing the best fit of Eq. (6.14) (red line) and Eq. (6.15) (blue line) are displayed together with the experimental points. It is quite clear that the fit according to Eq. (6.14) is significantly better than the Eq. (6.15) fit. In spite of the fact that the tip apex is large, the van der Waals forces are detectable even at a very long distance from the surface. When the experimental parameters are taken as $H = 6.7 \times 10^{-20}$ J, $f_0 = 243$ kHz, $k = 8$ N/m, and $A_0 = 40$ nm, fitting gives an estimation of tip radius near 1.8 μm, which is consistent with the expectation of a very blunt tip. On the other hand, Eq. (6.8) gives a useful approximation that is valid if the oscillation amplitude is smaller than the interaction range. For instance, this approximation is widely used for describing the cantilever vibrations in kelvin probe force microscopy (see Section 6.6).

Giessibl has also shown that there are two conditions concerning interaction forces for stable performance of DFM microscopes [44]: (1) The gradient of the interaction force must be larger than the cantilever spring constant and (2) the resulting force acting on the cantilever at the position of closest approach must be directed away from the sample.

Condition 1

$$\frac{\partial \mathrm{F_{int}}}{\partial z} = \frac{\partial^2 \mathrm{V_{int}}}{\partial z^2} > \mathrm{k} = \frac{\partial^2 \mathrm{V_{harmonic}}}{\partial z^2},$$

results from the fact that the tip is moving in response to a sum of two interaction potentials: the first one determined by the spring constant of the cantilever and the second one, which is the tip–surface interaction potential. Therefore, there are two minima on the plot of the resulting total interaction potential. As the cantilever tip is moved closer to the surface, the tip–surface potential becomes increasingly important, breaking Condition 1 and causing a rapid transition to direct contact of the tip with the surface: "snap to contact" [48].

Condition 2, $-kz' + F_{\text{int}}(-z') < 0$ (where $-z'$ is the closest approach coordinate) could be secured by an appropriate choice of the vibration amplitude.

6.5.5. *Tip–Surface Interaction Forces*

Initially, only van der Waals interactions were taken into account for a theoretical description of DFM. It is clear, however, that such an approach is quite unsatisfactory for the interpretation of DFM images. More recently, three contributions to tip–surface interactions have been considered: long-range electrostatic forces, van der Waals forces, and short-range "chemical" forces. Guggisberg and co-workers [45] have presented an example of their relative ranges and strengths. It is obvious that the exact form of all the interactions is dependent on the real shape of the tip. The knowledge of the tip shape constitutes one of the most important problems in understanding DFM contrast formation. It is usually guessed by a trial-and-error procedure in which simulated images (or preferably frequency–distance curves) are compared to experimental ones. There are different approaches to model the tip [44, 45, 49, 50], but many researchers assume that the tip is constructed as a cone, with a spherical cap at the end with a certain radius and an atomic-scale "nanotip" that is embedded in the sphere. The "nanotip" is usually treated atomistically since it has the key role in atomic contrast formation. If an experiment is aimed at achieving atomic resolution, long-range electrostatic and van der Waals forces should be minimized.

Electrostatic forces are always attractive and therefore enlarge the detuning Δf. They stem from different work functions of the tip and sample materials but are also due to charges present on the surface. Minimizing these forces, and therefore detuning as well, is achieved by biasing the system during scanning. This works well when the surface under investigation is electrically homogeneous. When this is not the case, a setup similar to the one used in Kelvin probe force microscopy (will be reviewed briefly in Section 6.6) has to be used. If the surface is composed of two distinct materials with different work functions (like an insulating islands on a semiconductor: KBr/InSb), it is possible to find a bias that gives the same detuning (but not minimal) over both materials and, as a result, does not introduce an additional height difference between the materials [57].

Van der Waals interactions can be minimized by using sharp (macroscopically) tips. However, van der Waals forces are always present and can have a positive effect on the stability of the scanning process. Since the tip can locally experience repulsive chemical interactions, a large positive van der Waals background prevents the system from sudden excursions in the course of scanning.

However, the most important interaction for atomic resolution is the chemical force acting between the atom placed on the tip apex and the atom on the surface just underneath. That force (ionic, covalent, etc.) is dependent on the chemical composition of the "nanotip" and the surface and acts on distances on the order of a few angstroms—a distance similar to interatomic distances in typical crystals. Due to its short range, in typical DFM instruments in which the cantilever amplitude is several nanometers, chemical forces act in the lower turning point only. That leads to a nonlinearity of the potential in which the tip is moving. Some examples of the importance of chemical interactions are presented in following subsections.

Another vividly discussed issue in the field of DFM is the understanding of damping of cantilever oscillation and energy dissipation. The topographic signal is obtained from the analysis of the frequency of the cantilever oscillating with constant amplitude. However, in order to keep the amplitude constant, the cantilever has to be constantly excited. Therefore, the excitation amplitude is a measure of the dissipated energy. The dissipation can be quite large—even 10 eV/cycle [52] was reported, but is typically in the order of meV/cycle.

A damping signal can be recorded simultaneously to a topographic signal, but it is completely independent. Loppacher and co-workers have shown [53] that even if topographic images are virtually identical, damping images can look completely different. This even affects atomic-scale images. Since the damping signal is monotonic (e.g., is growing upon approaching the surface), it was proposed to be used as a feedback signal [54].

The exact physical nature of damping is not fully understood. Some authors point to the fact that oscillations can be damped due to mathematical properties of the equations governing the motion of the cantilever. However, it seems that real energy dissipation takes place. It is suggested that the approaching tip induces an atomic-scale instability, a "soft mode," in which the potential of an unstable atom can have two minima that change with tip–surface distance. The oscillation of the cantilever causes an atom to flip between them, which does not happen at the same time during approach and retraction of the tip. This leads to histeresis which results in energy dissipation. The amount of dissipated energy predicted by this model is similar to ones recorded in experiments [55].

6.5.6. *Dynamic Force Microscopy for Ionic Insulator Surfaces*

Alkali halides are often used as model materials in SFM. Their structure is well known, their (001) surfaces are easy to prepare by cleavage, and they are very stable, especially under ultrahigh-vacuum conditions. The first atomically resolved images of alkali halides were published in 1997 [31]. Features in large-scale DFM images (where achieving atomic resolution is impossible due to the fact that every pixel is averaged from more than a surface atom) can usually be interpreted as topographic. Atomic-scale images are more complicated because of notable imaging artefacts.

In Fig. 6.14, an $18.1 \times 18.1\text{-nm}^2$ surface area of a (001) KBr crystal is imaged with an atomic resolution. The image in Fig. 6.14a is presented "as measured" except for a plane-background subtraction for better display. Figure 6.14b was subsequently filtered in order to enhance the atomic pattern. The surface was irradiated by an electron beam before imaging. The electron irradiation led to desorption of crystal material and, as a consequence, created pits of monolayer depth in the topmost layer of the crystal [32, 56, 57]. Pit edges are oriented along main crystallographic directions on the (001) surface: $\langle 100 \rangle$ and $\langle 010 \rangle$. The fast scan direction was vertical (y direction). The fact that the corners do not have

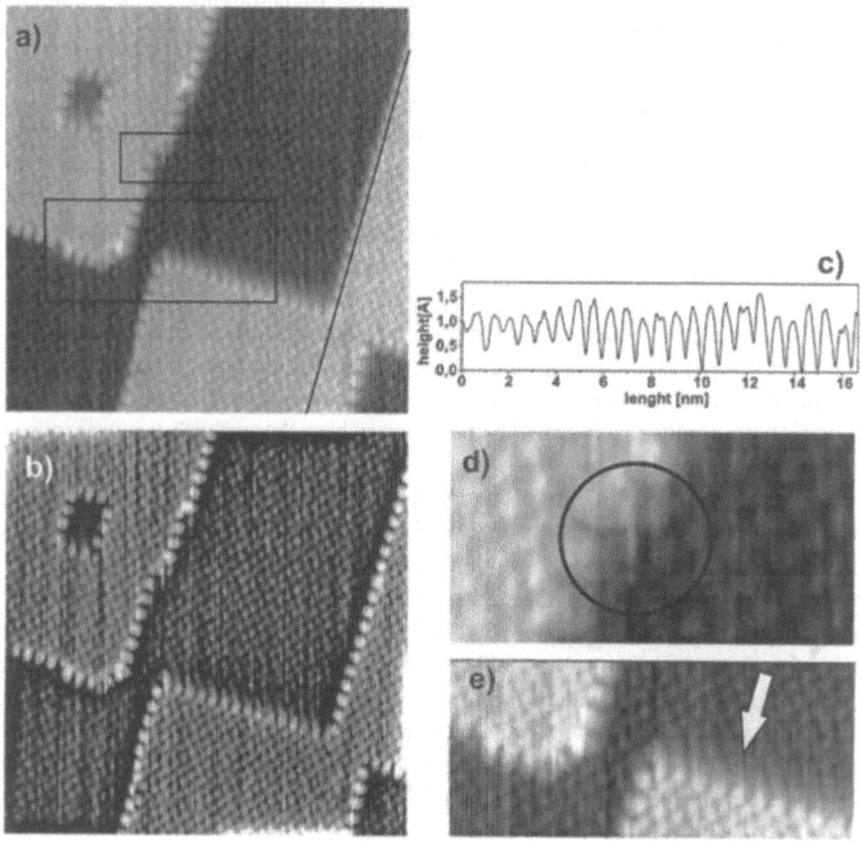

FIGURE 6.14. DFM images of a KBr(001) surface: (a) 18.1 × 18.1 nm^2 surface topography imaged with an atomic resolution; (b) the same as (a), but the image is filtered in order to enhance the atomic pattern; (c) a cross section along the black line indicated on the right side of image (a); (d) a closeup area indicated by the upper black box in (a)—the feature seen in the center of the ring is caused by tip-induced mobility of the step edge atom; (e) a closeup area from the lower box in (a)—an arrow indicates a "shadow" caused by the tip asymmetry.

perfectly right angles is a result of thermal drift and/or piezotube creep, which are virtually impossible to avoid at room temperature. Atomic rows in the image, however, are oriented along the ⟨101⟩ and ⟨011⟩ directions, indicating the surface arrangement of the same type of ions (either alkali or halogen ion rows). This is due to the fact that DFM does not measure the "true" topography, only the surface of constant frequency shift (i.e., the surface of the constant interaction force). The main interaction responsible for creating the atomic pattern is the short-range ionic force between ions on the surface lattice and the ones inevitably contaminating the tip apex. Thus, independent of the ionic radius of the ions, only one kind is visible at the time (alkali ions in the case of halogen contamination of the tip apex and halogen ions for the opposite situation). Since the alkali halide surface lattice is perfectly symmetric, it is impossible to say which kind of ions are imaged at a given time. If the front ion at the tip apex suddenly changes, the image will shift by half the lattice constant [58].

CaF_2 is an example of an ionic crystal for which the sublattice identification was achieved for the first time [20]. That was possible due to the nonequivalent positions of both ions in the lattice and, consequently, images of calcium and fluorine sublattices looked different. The authors of that work demonstrated two interesting features of DFM microscopy of ionic crystal surfaces: (1) Since the ion charge determines the character of the ionic force responsible for the formation of the image contrast, one could find the tip apex conditions for which calcium ions from the topmost layer were imaged as holes, whereas fluorine ions located deeper in the lattice were imaged as protrusions; (2) using a perfectly clean (or oxidized) Si tip, the apex could not interact with the ionic surface by the short-range ionic force; thus, high-resolution imaging required the tip to be contaminated by the sample material (likely to occur in the case of accidental tip–surface contact) [59, 60]. More recently, quite similar observations were made for AIII–BV semiconductor surfaces, as described in the next subsection.

In Fig. 6.14, a few characteristic artifacts of DFM imaging could be spotted. First, ions at edges are imaged as much brighter than the ions on flat terraces. Figure 6.14c presents a cross section along a black line marked in Fig. 6.14a; atomic corrugations are up to 0,15 nm on the edge, but only half of that in a terrace. That is the indication of surface atom relaxation. Under the tip approach, which in the time scale of atomic movement is very slow, both atoms on the tip and the surface relax in order to adapt to a new potential landscape. Ions that are imaged as protrusions are attracted toward the tip. Certainly, edge atoms have a lower coordination number and, thus, are more weakly bound to the surface. Therefore, their relaxation is much stronger than the relaxation of atoms in a flat terrace. As a result, they are much closer to the tip and appear as much brighter (i.e., higher in the z direction).

Figure 6.14 demonstrates two other typical imaging artifacts. A closeup of the upper rectangle drawn in Fig. 6.14a is shown in Fig 6.14d. A sharp line section along the fast scanning direction (vertical) is noticeable in the marked circle. This shows that there was a strong interaction there, but only during one scan line. Such a feature is a sign of tip-induced mobility of an edge atom, which was detached from the edge and presumably transported with the tip. Similarly, the next effect is presented in Fig. 6.14c: The asymmetry in the shape of the tip caused the edge of the terrace to be imaged differently for each scanning direction.

6.5.7. *Dynamic Force Microscopy of AIII–BV Semiconductor Surfaces*

Interesting examples of using DFM as a complementary tool for STM are experiments concerning structures of reconstructed (001) surfaces of AIII–BV semiconductors. Those compounds have a zinc-blende crystal structure. The (001) face is not their natural cleavage face and exhibits many different structures depending on the preparation and stoichiomentry of the surface [61]. Generally, the surfaces terminated with a BV element can be obtained by Molecular Beam Epitaxy (MBE) methods, whereas AIII terminated structures can be also prepared by ion sputtering and annealing. The most typical sputter-prepared surfaces are reconstructed with c(8 × 2) symmetry, which is common for many (001) AIII–BV compounds, such as GaAs, InAs, and InSb. Many models have been proposed for explaining that reconstruction, mainly based on STM data. However, in contrast to a success story of explaining Si(111) 7 × 7 reconstruction by STM, BIII-element-rich (001) surfaces of AIII–BV semiconductors were found to be much more difficult [62–65]. Their large and complex unit cells make an unambiguous interpretation of the experimental data rather doubtful. More recently, Lee and co-workers [86] based on theoretical calculations and Kumpf and co-workers [24, 25], based on surface X-ray diffraction measurements, proposed a so-called "ζ model," which was confirmed by real space imaging with DFM [66, 67].

Those DFM experiments were focused on InSb(001) and GaAs(001) surfaces. The epi-ready wafers were placed into the vacuum system and degassed by annealing for several hours at 700 K and 850 K for InSb and GaAs, respectively. Subsequently, cycles of 700-eV Ar^+ ion sputtering and annealing were repeated until the required structure was achieved as checked by low-energy electron diffraction. The ion beam was rastered over the sample, giving an average current density of 0.5 $\mu A/cm^2$. During scanning, samples were kept at temperatures of 750 K and 870 K for InSb and GaAs, respectively. After the cleaning procedure, the samples were left to cool for up to 20 h at a pressure of 2×10^{-10} mbar. Then cooled samples were transferred under an ultrahigh vacuum to a STM/DFM microscope chamber for imaging. The microscope used was Park Scientific Instruments VP2 equipped with a NanoSurf easyPLL demodulator for improving performance in frequency modulation mode of noncontact microscopy. Commercially available piezoresistive silicon cantilevers were used as probes.

Sputter-cleaned c(8 × 2)InSb(001) surfaces have large atomically flat terraces with predominantly monolayer step edges usually oriented along the ⟨110⟩ and ⟨1–10⟩ directions, as was already shown in the STM image of Fig. 6.7. A closeup of the DFM image (see Fig. 6.15) reveals the atomic structure of the surface, including vacancies. The structure is strongly anisotropic and appears as rows of atoms oriented in the ⟨110⟩ direction. However, atomically resolved DFM images of c(8 × 2)InSb(001) do not show the same pattern every time; instead, three distinct patterns are observed [66]. Moreover, the patterns change abruptly during scanning in an uncontrolled way. This phenomenon could only be explained after analyzing DFM image formation more deeply, as discussed in Section 6.5.8.

6.5.8. Chemical Sensing with DFM

As was described earlier, in DFM a tip is mounted to a flexible cantilever oscillating at its resonant frequency over the surface. The oscillation amplitude used is in the order of a dozen to a few dozen nanometers, whereas at its closest approach, the tip is as close as a few angstroms to the sample. In that range of distances, various forces act on the tip that alter the oscillation frequency. Van der Waals and electrostatic forces should be avoided due to their long ranges and weak dependence

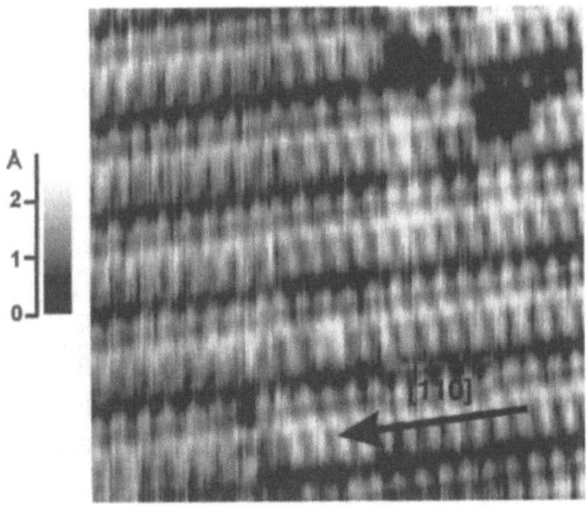

FIGURE 6.15. A $10 \times 10\text{-nm}^2$ DFM image of a c(8 × 2) InSb(001) surface. Reprinted from *Progress in Surface Science*, Vol. 74, Number 1, Szymonski, M., Krok F., Struski P., Kolodziej J., Such B., Ion-beam-induced surface modification and nanostructuring of $A_{III}B_V$ semiconductors, pp. 331–341. Copyright 2003, with permission from Elsevier.

on distance. They often can be assumed constant within the atomic-size image. In contrast, chemical forces are extremely short ranged and vanish within a few nanometers from the surface [45]. Therefore, atomically resolved DFM images taken in the topographic mode show the surface of constant interaction, which stems from the spatial distribution of chemical forces and geometric position of surface atoms. Certainly, for a surface of two-component materials, the chemical interaction changes substantially for each constituent. As a result, an image of such a surface is usually quite different from the topographic relief, as in the previously described case of alkali halide surfaces where ions of only one kind are imaged as protrusions [20, 32]. However, for AIII–BV compounds, which are predominantly covalent (but partially ionic), early results showed that a similar mechanism can also take place.

Among the first surfaces observed successfully with atomic resolution were (110) surfaces of AIII–BV semiconductors [30]. That is the natural cleavage face of those crystals and atom positions on that surface differ only slightly from the perfect bulk crystal positions. The DFM images show, however, only one kind of atom, which was attributed to group V elements. The understanding of contrast formation came with theoretical works by Ke and co-workers and Tobik and co-workers [68, 69]. Ke and co-workers investigated tip morphology effects on the contrast formation on GaAs(110) surfaces. Previously, it was believed that the interaction is the strongest where the highest charge density is located. This would result in imaging a BV element as a protrusion on AIII–BV surfaces. In contrast, the above-cited authors proposed that in some circumstances the interaction is stronger at the points where a high density of states above the Fermi level is present (i.e., leading to imaging the AIII element as a protrusion).

The chemical composition of the tip apex is the key to understanding image formation. It is virtually impossible to avoid contaminating the tip with the sample material. In the case of GaAs, it is either Ga or As atom. As a result, an As sublattice is imaged as a protrusion with a Ga-terminated tip, whereas a Ga sublattice is imaged as protrusion for an As-terminated tip. The latter fact is notable as Ga atoms actually lie 0.7 Å *lower* than As atoms.

It is crucial to understand that protrusions do not necessarily mark the geometric positions of atoms. They show the area of tip–surface attraction, which can actually be between the atoms (see also the HOPG case from Ref. 70). A very similar effect has been seen in the simulation images of InP(110) [69] and is expected to be found for other III–V compounds such as InSb and InAs. Tip–surface interactions for (001) surfaces of III–V compounds are not particularly different from the ones on their (110) faces. Therefore, in principle, various sublattices could be imaged depending on tip composition, similarly to the previously mentioned case of CaF_2 [20].

As discussed previously, three distinct patterns were found on c(8 × 2) InSb(001) surfaces as shown in Fig. 6.16. Since a well-established sample preparation procedure was repeated and diffraction experiments showed no difference, it was reasonable to assume that the structure of the surface was the same for all measurements. Moreover, changes of contrast patterns happened abruptly. It was demonstrated [66, 67] that in such a change there were only a few atoms involved and the pattern change was due to different chemical interactions.

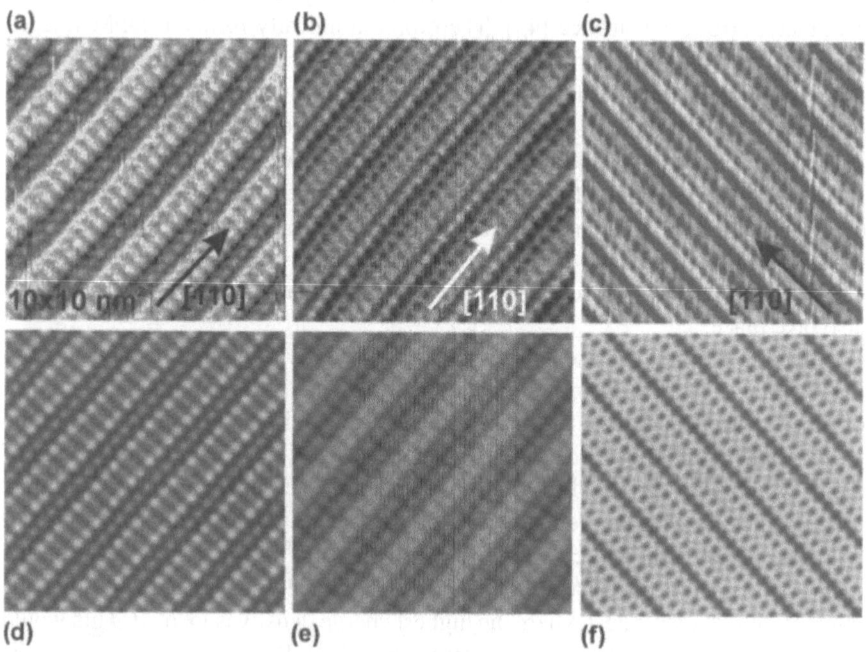

FIGURE 6.16. Three patterns visible in DFM atomically resolved images of c(8 × 2)InSb(001) surface: (a)–(c) recorded images of patterns 1, 2, and 3; (d)–(f) simulated geometric images of indium, antimony, and both sublattices. Reprinted Figure 1 with permission from, Kolodziej J.J., Such B., Szymonski M., Krok F., *Phys. Rev. Lett.*, Vol. 90, p. 226101. Copyright 2003 by the American Physical Society.

Since all three patterns seem to carry different chemical information, they should be compared to existing models of c(8 × 2)InSb(001) and especially with the latest ζ model [24, 25]. The detailed analysis shows that all patterns can be explained by the ζ model. Since images of different sublattices are expected, the DFM patterns were compared to the geometrical structure of sublattices in the model. There is a striking correspondence between the three patterns and the surface geometry of the indium sublattice, antimony sublattice, and the sum of both sublattices. It is seen that the tip apex composition-dependent DFM possesses the ability to resolve various chemical species present on the surface. This may lead to the development of scanning force spectroscopy (analogous to the widely used and successful scanning tunneling spectroscopy), which could give additional insight into the nature of atom–atom interaction.

Specific force distance curves are necessary in order to acquire full information on tip–surface interaction. Such curves would allow one to fully verify the assumed tip model and lead to an unambiguous interpretation of data. However, due to stability requirements, they are only possible at low temperatures, and only few have been reported so far [71, 72]. Detailed knowledge about the tip structure is extremely important since contrast patterns can change even on Si(111) 7 × 7, depending on whether the tip is oxidized or reactive [73, 74]. Moreover, the tip–sample bias applied to reduce long-range electrostatic forces can have a profound effect on contrast formation. Arai and Tomitori have recently showed that at a very small tip–surface distance, the application of a certain bias can tune the tip and surface electronic states, resulting in a site-specific appearance of the attractive force [75].

6.6. KELVIN PROBE FORCE MICROSCOPY

Kelvin probe force microscopy (KPFM) is an extension of the DFM technique, offering surface chemical sensitivity at the nanometer scale by means of surface contact potential difference measurements [9, 76]. The KPFM technique enables one to separate the electrostatic term from the total tip–sample interaction and therefore allows both topography and contact potential images of the sample to be acquired simultaneously. The method of contact potential measurements, which is crucial for KFPM, is based on the original idea of Lord Kelvin [77], who in 1898 demonstrated an experimental determination of the surface potential (the work function) using a vibrating-plate capacitor. In that experiment, the work function difference between two materials (metals or semiconductors) forming the two sides of a parallel-plate capacitor is measured. If two plates with different work functions are held in close proximity to each other, a force acts between them due to the potential difference V_{CPD}. The contact potential difference (CPD) between two materials (e.g., between the tip and the sample) is defined as

$$V_{\mathrm{CPD}} = \frac{\Phi_{\mathrm{tip}} - \Phi_{\mathrm{sample}}}{-e}, \tag{6.16}$$

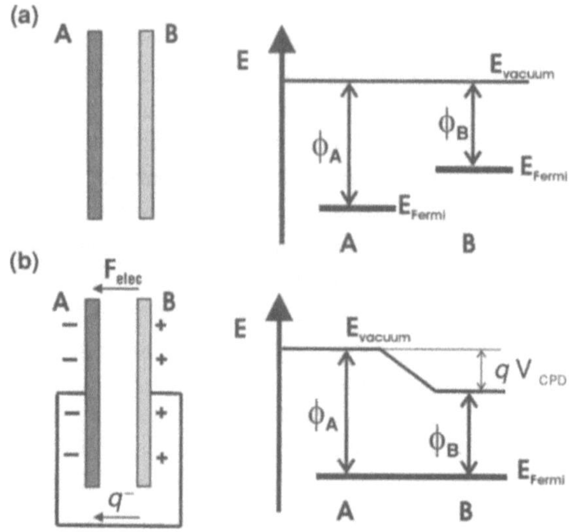

FIGURE 6.17. Energy diagram explaining the development of the electrostatic force between two plates of metal forming a plate capacitor.

where Φ_{tip} and Φ_{sample} are the work functions of the tip and the sample, respectively, and e is the elementary charge.

When two different materials are not connected, their local vacuum levels are aligned, but there is a difference in their Fermi levels (Fig. 6.17a). Upon electrical connection, electrons will flow from the plate with the smaller work function to the plate with the higher one until the Fermi levels are equal (Fig. 6.17b). The two plates are then charged and there is a difference in their local vacuum levels. Hence, the charging of the electrodes causes an electrostatic force to develop. The force can be nullified by applying an external bias between the plates equal to the contact potential difference.

Assuming that the vibrating cantilever and sample surface play the role of the plates of the capacitor, the method of surface potential measurements can be implemented in the well-developed DFM instrument. For a better understanding of the concept of KPFM, let us first consider the situation in which the cantilever vibrates over the surface without scanning (a point spectroscopy measurement).

Based on the vibrating capacitor model, the electrostatic long-range force between the tip and the sample can be written as [78]

$$F_{elec} = \frac{1}{2}(V - V_{CPD})^2 \frac{\partial C}{\partial z}, \tag{6.17}$$

where V, V_{CPD}, and $\partial C/\partial z$ are the applied voltage, the contact potential difference, and the capacitance gradient between the tip and the sample, respectively. According to the above formula, the tip–sample V_{CPD} could be determined by minimizing $F_{elec}(U)$ while keeping a constant tip–sample distance (constant $\partial C/\partial z$). In Fig. 6.18, the Kelvin probe force spectroscopy (KPFS) performed on the InSb(001)

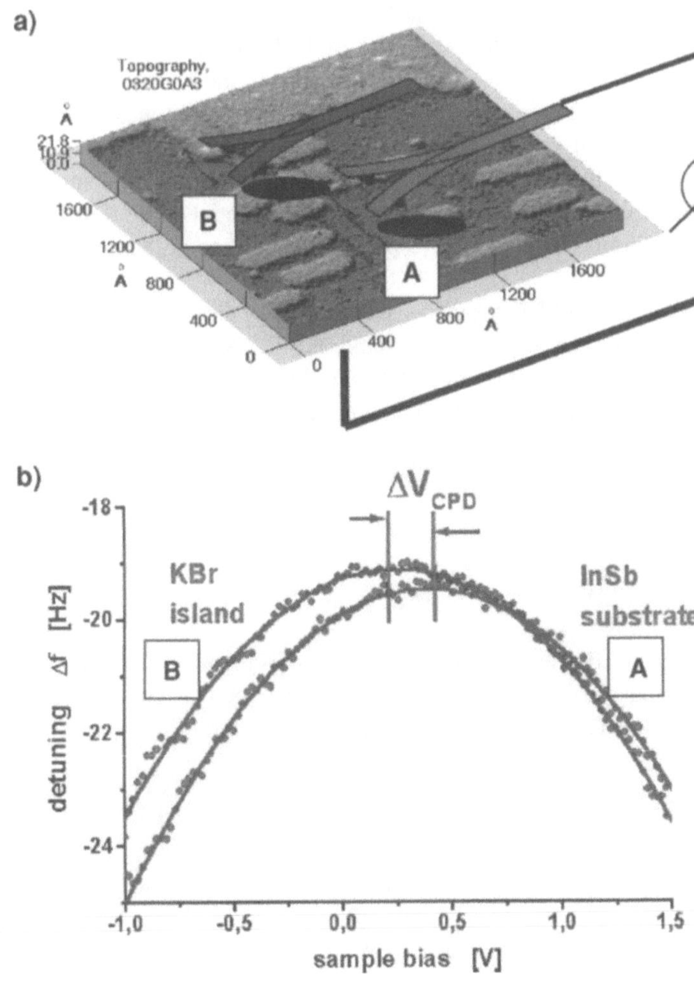

FIGURE 6.18. The principle of the KPFS performed on the KBr/InSb system. In (a), the topography of the KBr/InSb surface with the two points marked with A (substrate) and B (KBr island) where the dependence of detuning versus sample bias voltage were taken and are presented in (b).

surface covered with a submonolayer of KBr film is presented. At first, the surface is imaged with a DFM. The sample surface consists of atomically flat terraces of InSb substrate with protruded features (islands) corresponding to the grown KBr film. Subsequently, the scan range was set to zero and the probe was moved to the desired location (over the substrate surface in A) for the KPFS. Then, the feedback loop was disabled and the sample bias was scanned over a certain range. Simultaneously, the error signal (detuning) from the PLL demodulator was collected. Since the tip–sample distance was kept constant during the KPFS measurements (disabled feedback loop), the detuning signal versus bias voltage curve was characteristic of a pure electrostatic force in the tip–sample system. Subsequently, the measurement was repeated over the KBr island (point B). In both cases, the parabolic dependence of the detuning signal on the bias voltage was obtained according to Eq. (6.17) with the maxima corresponding to the sample bias for which the V_{CPD} between the chosen areas on the surface and the tip was nullified. Although the absolute values of the tip work function were unknown, the difference in the surface potentials of the two different locations on the sample surface was measured. This experiment showed that the surface potential of the KBr/InSb system is lower by about 210 mV with respect to the surface potential of the bare InSb substrate.

As demonstrated, KPFS measures the local surface potential at a single point on a sample surface. On the other hand, KPFM conveniently acquires the whole map of the sample surface potential simultaneously with the topographic image. A typical KPFM measurement is conducted in the following way. The sample is biased with sinusoidal voltage $V_{\mathrm{sample}} = V_{\mathrm{dc}} + V_{\mathrm{ac}}\sin(\omega t)$. As shown already, the total tip–sample interaction, F_{total}, is composed of the following contributions:

$$F_{\mathrm{total}} = F_{\mathrm{vdW}} + F_{\mathrm{short}} + F_{\mathrm{elec}} = F_{\mathrm{topo}} + \frac{1}{2}(V_{\mathrm{sample}} - V_{\mathrm{CPD}})^2 \frac{\partial C}{\partial z}, \qquad (6.18)$$

where F_{vdW} is the van der Waals force and F_{short} is the chemical, short-range force. Because of the sinusoidal voltage, V_{sample}, applied between the conductive tip and the surface, the force (F_{total}) acting on the tip becomes

$$\begin{aligned} F_{\mathrm{total}} &= F_{\mathrm{topo}} + \left(\frac{1}{4}\frac{\partial C}{\partial z} V_{\mathrm{ac}} + \frac{1}{2}\frac{\partial C}{\partial z}(V_{\mathrm{dc}} - V_{\mathrm{CPD}})^2\right) \\ &\quad + \frac{\partial C}{\partial z} V_{\mathrm{ac}}(V_{\mathrm{dc}} - V_{\mathrm{CPD}})\sin\omega t - \frac{1}{4}\frac{\partial C}{\partial z} V_{\mathrm{ac}}^2 \cos 2\omega t \\ &= F_{\mathrm{topo}} + F_{\mathrm{cap}} + F_{\omega} + F_{2\omega}. \end{aligned} \qquad (6.19)$$

The DC components of F_{total} are composed of the van der Waals and short-range interactions, a so-called "topographic" force, F_{topo}, and the electrostatic interaction, F_{cap}, due to capacitance (capacitive force). The above time-independent components are used to regulate the height of the tip in topography imaging, but the term F_{topo} alone contains the true height information. The terms F_{ω} and $F_{2\omega}$ are components that vary with the angular frequency ω and 2ω, respectively. F_{ω} originates from the electrostatic interaction due to the CPD. $F_{2\omega}$ originates from the electrostatic interaction due to the charges induced by the applied ac field.

In the KPFM, the separation of the electrostatic term from the total tip–sample interaction is obtained by nulling the F_{ω} component (the compensation of the CPD). However, the method of the electrostatic interaction compensation has some influence on the topographic image. Even for $V_{\mathrm{dc}} = V_{\mathrm{CPD}}$, the F_{cap} does not vanish, as it is proportional to the ac component of the sample bias ($=\frac{1}{4}(\partial C/\partial z)V_{\mathrm{ac}}^2$). Fortunately, its contribution to the topographic signal can be easily estimated by the collection of the spatial distribution of the $F_{2\omega}$ signal, which has the same amplitude

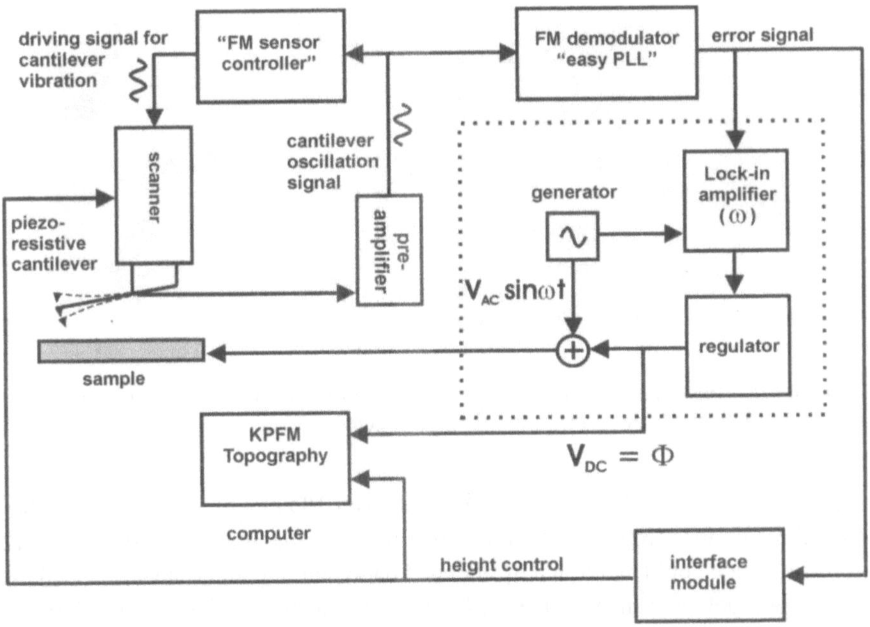

FIGURE 6.19. Block diagram of modified VP2 Park Scientific Instruments device used for the KPFM measurements. The part in the dashed envelop is a homemade setup for the CPD compensation.

as F_{cap}. In Fig. 6.19, the modified AFM system used for KPFM measurements is shown. It is a modified VP2 AFM/STM Park Scientific Instruments device, using a piezoresistive cantilever as a probe.

Within the AFM part of the setup, the sample topography is acquired using noncontact frequency modulation (FM) DFM as described earlier. Within the KPFM part, a sample is biased with an AC voltage with frequency ω (of few hundreds hertz and of about 1 V amplitude) and a DC component. The sample bias induces an oscillatory electrostatic force between the tip and the sample (F_{elec}) The time-dependent interaction with an angular frequency ω between the tip and the sample induces a variation in the FM demodulator output. The ω component of the error signal is detected by a lock-in amplifier, and a feedback loop is used to add a DC voltage (V_{dc}) to the sample to maintain the F_{elec} at its minimum. The obtained V_{dc} (= V_{CPD}) represents the local sample surface potential.

There is still another mode (also based on a lock-in technique) employed for the CPD compensation: the amplitude modulation (AM) mode [79]. In this mode, the V_{dc} map, representing the local sample surface potential, is obtained by measuring the amplitude changes of the cantilever oscillation at ω. Since the amplitude is proportional to the force, the electrostatic force itself is measured in the AM mode. This is in contrast to the FM mode in which the measurement of the frequency shift implies that the gradient of the electrostatic interaction is measured. The choice between those two detection modes influences both spatial resolution and accuracy of the CPD signal measurements. As shown by Glatzel

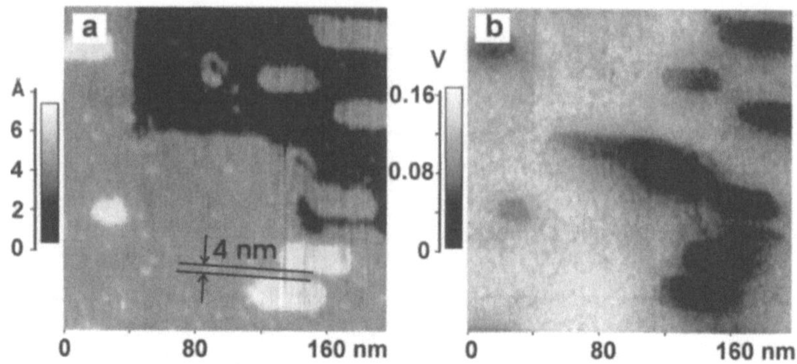

FIGURE 6.20. Topographic (a) and CPD (b) images of KBr film epitaxially grown on InSb(001) surface. The KBr islands are visible in (a) as brighter features.

and co-workers [79], the use of the FM mode for CPD imaging results in a higher spatial resolution. However, the FM mode requires the application of a high AC voltage, which could result in a tip-induced band bending at the surface and skew the surface potential measurements. On the other hand, the AM mode offers the possibility to use low AC voltages with relatively high-energy resolution. However, the long-range nature of the electrostatic interaction results in a strong averaging effect of the cantilever, itself decreasing the spatial resolution. Figure 6.20 presents the topographic and the CPD images of the KBr/InSb(001) system, acquired with KPFM in FM mode [80]. Two KBr islands that are separated by 4 nm from each other can easily be distinguished in the CPD image.

The long-range character of the electrostatic interaction sets conditions on the geometry of the tip used for the CPD imaging. It has been shown that by choosing an appropriate cantilever–tip geometry, the resolution can be improved significantly [81, 82]. The highest spatial resolution in CPD images are acquired with long, sharp tips. The prospect of using carbon nanotubes as tips for KPFM is worth mentioning, since it would have a profound effect on improving KPFM as a unique tool to study chemical composition of a sample surface at nanometer and subnanometer scales.

By compensating for the electrostatic interaction, KPFM can image the true topography of heterogeneous structures contrary to what is measured with the conventional DFM technique. For DFM, the surface topography is acquired by keeping the total interaction between the vibrating cantilever and the sample constant. In the case of imaging heterogeneous structures, the visible "topography" is highly influenced by the difference of the electrostatic interaction between the probe and the imaged surface areas of different composition. For the KBr/InSb system presented in Fig. 6.21, the "height" of KBr islands measured on the DFM images (the sample bias kept constant) is about 1.95 Å, whereas KPFM reads a "height" of about 2.42 Å. The last value is in good agreement with the height of monatomic terrace step on KBr(001) crystal surface (~2.52 Å) as measured with DFM alone.

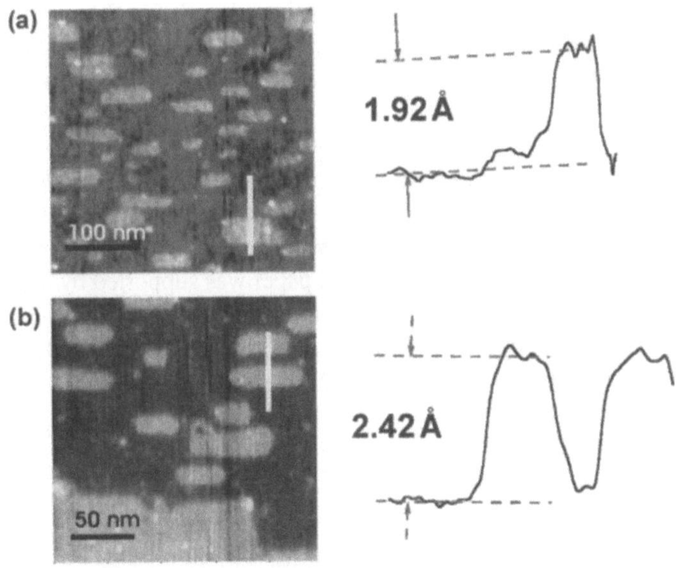

FIGURE 6.21. The topographic images of KBr islands grown on InSb substrate taken with standard DFM technique: (a) sample bias constant and (b) with KPFM. The cross sections of the island are taken along the corresponding white lines.

Although atomic resolution has been reported for KPFM surface imaging [83–85], the understanding of the contrast on an atomic scale in the CPD signal is far from being clear. Moreover, recent experimental results by Arai and Tomitori [75] revealed that in the total interaction between the tip and the sample surface, in addition to the electrostatic term, the short-range interaction is bias dependent. In practice, when performing KPFS measurements with the tip close to the surface (on the order of few angstroms), the detuning versus bias voltage deviates from its parabolic curve, exhibiting a few additional minima. The hopping between the local minima while imaging with the KPFM can provide a plausible explanation of the contrast in the CPD signal on an atomic scale. However, the obtained CPD contrast cannot be directly related to the contrast of the surface contact potential.

6.7. CONCLUSIONS

Looking at the rapid development of scanning probe methods in recent years, it is likely that they will be a group of most important instruments for everyone who is interested in the field of nanoscience and nanotechnology. Growing interests in the bottom-up approach will push new groups into the scanning probe community. Moreover, during the years, we have witnessed a remarkable technical progress in SPM. This development included not only achieving ultimate resolution limits but, more importantly, introducing new modes of operation for SPM and making them more accessible to a wider audience. In this chapter, only the most basic modes, such as STM and DFM, have been introduced, but there are dozens of modifications of SPM allowing for probing various interactions. This makes the SPM an extremely versatile tool, which will soon be essential for every material scientist.

Acknowledgment. The authors would like to thank dr. Jacek J. Kolodziej of the Jagiellonian University for many fruitful discussions and Mr. Alex Labuda of McGill University, Montreal, Canada, for his comments and careful reading of the manuscript. The preparation of this tutorial paper was supported by the European Commission, under contract No. MTKD-CT-2004-003132, 6th FP—Marie Curie Host Fellowships for Transfer of Knowledge: "Nano-engineering for Expertise and Development–NEED".

REFERENCES

1. G. Binnig, H. Rohrer, Ch. Gerber, and E. Weibel, *Phys. Rev. Lett.* **49**, 57 (1982).
2. J. Tersoff and D.R. Hamann, *Phys. Rev. B* **31**, 805 (1985).
3. M. Bode, M. Getzlaff, and R. Wiesendanger, *Phys. Rev. Lett.* **81**, 4256 (1998).
4. S. Heinze, M. Bode, A. Kubetzka, O. Pietzsch, X. Nie, S. Blügel, and R. Wiesendanger, *Science* **288**, 1805 (2000).
5. G. Binnig, H. Rohrer, Ch. Gerber, and E. Weibel, *Phys. Rev. Lett.* **49**, 57 (1982).
6. T.R. Albrecht, P. Grütter, D. Horne, and D. Rugar, *J. Appl. Phys.* **69**, 668 (1991).
7. C C.M. Mate, G.M. McClelland, R. Erlandsson, and S. Chiang, *Phys. Rev. Lett.* **59**, 1942 (1987).
8. E. Gnecco, R. Bennewitz, T. Gaylog, and E. Meyer, *J. Phys.: Condens. Matter* **13**, R619 (2001).
9. M. Nonnenmacher, M.P. O'Boyle, and H.K. Wickramasinghe, *Appl. Phys. Lett.* **58**, 2921 (1991).
10. Y. Martin and H.K. Wickramasinghe, *Appl. Phys. Lett.* **50**, 1455 (1987).
11. J.J. Sáenz, N. Garcia, P. Grütter, E. Meyer, H. Hinzelmann, R. Wiesendanger, L. Rosenthaler, H.R. Hibder, and H.-J. Güntherodt, *J. Appl. Phys.* **62**, 4293 (1987).
12. U.D. Schwarz, H. Hölscher, and R. Wiesendanger, *Phys. Rev. B* **62**, 13089 (2000).
13. A.J. Melmed, *J. Vac. Sci. Technol. B* **9**, 601 (1990).
14. M. Nagai, M. Tomitori, and O. Nihsikawa, *Jpn. J. Appl. Phys.* **36**, 3844 (1997).
15. D. Huang, F. Yamaguchi, and Y. Yamamoto, *Jpn. J. Appl. Phys.* **37**, 3824 (1998).
16. B. Barwiński and S. Sendecki, *Appl. Surf. Sci.* **119**, 111 (1997).
17. M. Tomitori and T. Arai, *Appl. Surf. Sci.* **140**, 432 (1999).
18. M. Echanescu, R.J.A. van den Oetelaar, R.W. Caprick, D.F. Ogletree, C.F.J. Flipse, and M. Salmeron, *Phys. Rev. Lett.* **81**, 1877 (1998).
19. M. Bode, M. Dreyer, H. Getzlaff, M. Kleiber, A. Wadas, and R. Wiesendanger, *J. Phys.: Condens. Matter* **11**, 9387 (1999).
20. A.S. Foster, C. Barth, A.L. Shluger, and M. Reichling, *Phys. Rev. Lett.* **86**, 2373 (2001).
21. J. Tersoff and D.R. Hamann, *Phys. Rev. Lett.* **50**, 1998 (1983).
22. J. Bardeen, *Phys. Rev. Lett.* **6**, 57 (1961).
23. H.A. Mizes, Sang-il Park, and W.A. Harrison, *Phys. Rev. B* **36**, 4491 (1987).
24. C. Kumpf, L.D. Marks, D. Ellis, D. Smilgies, E. Landemark, M. Nielsen, R. Feidenhans'l, J. Zegenhagen, O. Bunk, J.H. Zeysing, Y. Su, and R.L. Johnson, *Phys. Rev. Lett.* **86**, 3586 (2001).

25. C. Kumpf, C. Kumpf, L.D. Marks, D. Ellis, D. Smilgies, E. Landemark, M. Nielsen, R. Feidenhans'l, J. Zegenhagen, O. Bunk, J.H. Zeysing, Y. Su, and R.L. Johnson, *Phys. Rev. B* **64**, 075307 (2001).
26. S. Hembacher, F. J. Giessibl, J. Mannhart, and C. F. Quate, *Proc. Natl. Acad. Sci.* **100**, 12539 (2003).
27. S. Hembacher, F. J. Giessibl, and J. Mannhart, *Science* **305**, 380 (2004).
28. F.J. Giessibl and B.M. Trafas, *Rev. Sci. Instrum.* **65**, 1923 (1994).
29. F.J. Giessibl, *Science* **260**, 67 (1995).
30. Y. Sugawara, M. Ohta, H. Ueyama, and S. Morita, *Science* **270**, 1648 (1995).
31. M. Bammerlin, R. Lüthi, E. Meyer, A. Baratoff, J. Lü, M. Guggisberg, Ch. Gerber, L. Howald, and H.-J. Güntherodt, *Probe Microsc.* **1**, 3 (1997).
32. R. Bennewitz, S. Schär, V. Barwich, O. Pfeiffer, E. Meyer, F. Krok, B. Such, J. Kolodziej, and M. Szymonski, *Surf. Sci. Lett.* **474**, L197 (2001).
33. C. Barth and M. Reichling, *Nature* **414**, 54 (2001).
34. S. Kitamura and M. Iwatsuki, *Jpn. J. Appl. Phys.* **35**, L668 (1996).
35. A. Sasahara, H. Uetsuka, and H. Onishi, *Phys. Rev. B* **64**, 121406 (2001).
36. S.A. Burke, J.M. Mativetsky, R. Hoffmann, and P. Grutter, *Phys. Rev. Lett.* **94**, 096102 (2005).
37. R. Garcia and R. Perez, *Surf. Sci. Rep.* **47**, 197 (2002).
38. W.A. Hofer, A.S. Foster, and A.L. Shluger, *Rev. Mod. Phys.* **75**, 1287 (2003).
39. F.J. Giessibl, *Rev. Mod. Phys.* **75**, 949 (2003).
40. Y. Martin, C.C. Williams, and H.K. Wickramasinghe, *J. Appl. Phys.* **61**, 4723 (1987).
41. Ch. Loppacher, M. Bammerlin, F. Battiston, M. Guggisberg, D. Müller, H.R. Hibder, R. Lüthi, E. Meyer, and H.-J. Güntherodt, *Appl. Phys. A* **66**, S215 (1998).
42. U. Dürich, O. Züger, and A. Stadler, *J. Appl. Phys.* **72**, 1778 (1992).
43. H. Hölscher, U.D. Schwarz, and R. Wiesendanger, *Appl. Surf. Sci.* **140**, 344 (1999).
44. F.J. Giessibl, *Phys. Rev. B* **56**, 16010 (1997).
45. M. Guggisberg, M. Bammerlin, Ch. Loppacher, O. Pfeiffer, A. Abdurixit, V. Barwich, R. Bennewitz, A. Baratoff, E. Meyer, and H.-J. Güntherodt, *Phys. Rev. B* **61**, 11151 (2000).
46. C. Argento and R.H. French, *J. Appl. Phys.* **80**, 6081 (1996).
47. L. Bergström, *Adv. Colloid Interf. Sci.* **70**, 125 (1997).
48. N.A. Burnham and R.J. Colton, *J. Vac. Sci. Technol. A* **7**, 2906 (1989).
49. S. Ciriaci, A. Baratoff, and I.P. Batra, *Phys. Rev. B* **42**, 7618 (1990).
50. A.I. Livshits, A.L. Shluger, A.L. Rohl, and A.S. Foster, *Phys. Rev. B* **59**, 2436 (1999).
51. R. Bennewitz, M. Bammerlin, M. Guggisberg, Ch. Loppacher, A. Baratoff, E. Meyer, and H.-J. Güntherodt, *Surf. Interf. Anal.* **27**, 462 (1999).
52. B. Gotsmann, C. Seidel, B. Anczykowski, and H. Fuchs, *Phys. Rev. B* **60**, 11051 (1999)
53. Ch. Loppacher, R. Bennewitz, O. Pfeiffer, M. Guggisberg, M. Bammerlin, S. Schär, V. Barwich, A. Baratoff, and E. Meyer, *Phys. Rev. B* **62**, 13674 (2000).
54. M. Gauthier and M. Tsukada, *Phys. Rev. B* **60**, 11716 (1999).
55. L.N. Kantorovich and T. Trevethan, *Phys. Rev. Lett.* **93**, 236102 (2004).
56. B. Such, P. Czuba, P. Piątkowski, and M. Szymoński, *Surf. Sci.* **451**, 203 (2000).
57. B. Such, J. Kolodziej, P. Czuba, P. Piatkowski, P. Struski, F. Krok, and M. Szymonski, *Phys. Rev. Lett.* **85**, 2621 (2000).
58. R. Bennewitz, A.S. Foster, L.A. Kantorovich, M. Bammerlin, Ch. Loppacher, S. Schär, M. Guggisberg , E. Meyer, and A.L. Shluger, *Phys. Rev. B* **62**, 2074 (2000).
59. C. Barth, A.S. Foster, M. Reichling, and A.L. Shluger, *J. Phys.: Condens. Matter* **13**, 2061 (2001).
60. A.L. Shluger, L.A. Kantorovich, A.I. Livshits, and M.J. Gillan, *Phys. Rev. B* **56**, 15332 (1997).
61. P. Drathen, W. Ranke, and K. Jacobi, *Surf. Sci.* **77**, L162 (1978).
62. Q-K Xue, T. Hashizume, and T. Sakurai, *Prog. Surf. Sci.* **56**, 1 (1997).
63. P. John, T. Miller, and T.-C. Chiang, *Phys. Rev. B* **39**, 1730 (1989).

64. M.O. Schweitzer, F.M. Leibsle, T.S. Jones, C.F. McConville, and N.V. Richardson, *Surf. Sci.* **280**, 63 (1993).
65. N. Jones, C. Norris, C.L. Nicklin, P. Steadman, S.H. Baker, A.D. Johnson, and S.L. Bennett, *Surf. Sci.* **409**, 27 (1998).
66. J.J. Kolodziej, B. Such, F. Krok, and M. Szymonski, *Phys. Rev. Lett.* **90**, 226101 (2003).
67. B. Such, J.J. Kolodziej, P. Czuba, F. Krok, P. Piatkowski, P. Struski, and M. Szymonski, *Surf. Sci.* **530**, 149 (2003).
68. S.H. Ke T. Uda, I. Stich, and K. Terakura, *Phys. Rev. B* **63**, 245323 (2001).
69. J. Tobik I. Stich, and K. Terakura, *Phys. Rev. B* **63**, 245342 (2001).
70. H. Holscher, W. Allers, U.D. Schwarz, A. Schwarz, and R. Wiesendanger, *Phys. Rev. B* **62**, 6967 (2000).
71. R. Hoffmann, L.N. Kantorovich, A. Baratoff, H.J. Hug, and H.-J. Guntherodt, *Phys. Rev. Lett.* **92**, 146103 (2004).
72. M.A. Lantz, H.J. Hug, R. Hoffmann, P.J.A. van Schendel, P. Kappenberger, S. Martin, A. Baratoff, and H.-J. Guntherodt, *Science* **291**, 2580 (2001).
73. M.A. Lantz, H.J. Hug, P.J.A. van Schendel, R. Hoffmann, S. Martin, A. Baratoff, A. Abdurixit, H.-J. Guntherod, and Ch. Gerber, *Phys. Rev. Lett.* **84**, 2642 (2000).
74. M.A. Lantz, H.J. Hug, R. Hoffmann, S. Martin, A. Baratoff, and H.-J. Guntherodt, *Phys. Rev. B* **68**, 035324 (2003).
75. T. Arai, and M. Tomitori. *Phys. Rev. Lett.* **93**, 256101 (2004).
76. J.M.R Weaver, and D.W. Abraham, *J. Vac. Sci. Technol. B* **9**, 1559 (1991).
77. W.T. Kelvin, *Phil. Mag.* **46**, 82 (1898).
78. J. Lu, M. Guggisberg, R. Luthi, M. Kubon, L. Scandella, Ch. Gerber, E. Meyer, and H.-J. Guntherodt, *Appl. Phys. A* **66**, S273 (1998).
79. Th. Glatzel, S. Sadewasser, and M.Ch. Lux-Steiner, *Appl. Surf. Sci.* **210**, 84 (2003).
80. S. Sadewasser and M.Ch. Lux-Steiner. *Phys. Rev. Lett.* **91**, 266101 (2003).
81. F. Krok, J.J. Kolodziej, B. Such, P. Czuba, P. Struski, P. Piatkowski, and M. Szymonski, *Surf. Sci.* **566–568**, 63 (2004).
82. A. Kitamura, S. Hosaka, and R. Imura, *Appl. Phys. Lett.* **66**, 3510 (1995).
83. S. Kitamura, K. Suzuki, and M. Iwatsuki, *Appl. Surf. Sci.* **140**, 265 (1999).
84. K. Okamoto, Y. Sugawara, and S. Morita, *Appl. Surf. Sci.* **188**, 381 (2002).
85. T. Shiota and K. Nakayama, *Jpn. J. Appl. Phys.* (Pt. 2) **41**, L1178 (2002).
86. S.H. Lee, W. Moritz, M. Scheffler, *Phys. Rev. Lett.* **85**, 3890 (2000).

7

Simulation of Nano-CMOS Devices: From Atoms to Architecture

A. Asenov, A.R. Brown, B. Cheng, J.R. Watling, G. Roy and C. Alexander

7.1. INTRODUCTION

The progressive scaling of transistors in complementary metal–oxide–semiconductor (CMOS) technology to achieve faster devices and higher device density and to reduce the cost per function has fueled the phenomenal growth and success of the semiconductor industry—captured over the past 40 years by Moore's famous law. The International Technology Roadmap for Semiconductors (ITRS) predicts, as illustrated in Table 7.1, that 7-nm physical-gate-length CMOS transistors will be in mass production in 2018. The Roadmap of the leading integrated circuit (IC) manufacturer, IBM, goes further (see Table 7.2), predicting that the physical length of the transistors will reach 3 nm by 2025. Indeed, transistors with a 45 nm channel length are in mass production now in the 90 nm technology node and functioning transistors with a 4-nm channel length have been demonstrated already by NEC at IEDM 2003. Although it is clear that the scaling of the CMOS transistors will continue in the next two decades, it is widely recognized that intrinsic parameter fluctuations introduced by the discreteness of charge and matter will be a major factor limiting the integration of such devices with molecular dimensions in giga-transistor count chips.

Figure 7.1 shows that MOS field-effect transistors (MOSFETs) are becoming truly atomistic devices. The conventional way of describing, designing, modeling, and understanding semiconductor devices, illustrated in Fig. 7.1a assuming

Device Modelling Group, Dept. Electronics & Electrical Engineering, University of Glasgow, Glasgow, G12 8LT, Scotland

TABLE 7.1. Extract from the International Technology Roadmap for Semiconductors 2003.

	Near-term technology node characteristics Year of production						
	2003	2004	2005	2006	2007	2008	2009
Technology node		90			65		
DRAM 1/2 Pitch (nm)	100	90	80	70	65	57	50
Physical gate length (nm)	45	37	32	28	25	22	20
Oxide thickness	1.3	1.2	1.1	1	0.9	0.8	0.8

	Long-term technology node characteristics Year of production					
	2010	2012	2013	2015	2016	2018
Technology node	45		32		22	
DRAM 1/2 Pitch (nm)	45	35	32	25	22	18
Physical gate length (nm)	18	14	13	10	9	7
Oxide thickness	0.7	0.7	0.6	0.6	0.5	0.5

continuous ionized dopant charge and smooth boundaries and interfaces, is no longer valid. The granularity of the electric charge and the atomicity of matter, as illustrated in Fig. 7.1b, begin to introduce substantial variation in individual device characteristics. The variation in number and position of dopant atoms in the active region of nano-MOSFETs makes each transistor microscopically different and already introduces significant variations from device to device. In addition, the gate oxide thickness becomes equivalent to several atomic layers, with a typical interface roughness of the order of two atomic layers. This will introduce a variation in the oxide thickness within an individual transistor of more than 50%, resulting in each transistor having a microscopically different oxide thickness pattern. The unique oxide roughness pattern in each decanano-MOSFET will affect the device electrostatics, the surface-roughness-limited mobility, and the gate tunneling from device to device. The granularity of the gate material and the photoresist, together with other factors, will introduce unavoidable line-edge roughness in the gate pattern definition and statistical variations in geometry between devices.

TABLE 7.2. IBM Roadmap, Dec. 2003.

	Year of production		
	2019	2022	2025
Technology node	15	10	7
Printed gate length	9	6	4
Physical gate length (nm)	6	4	3

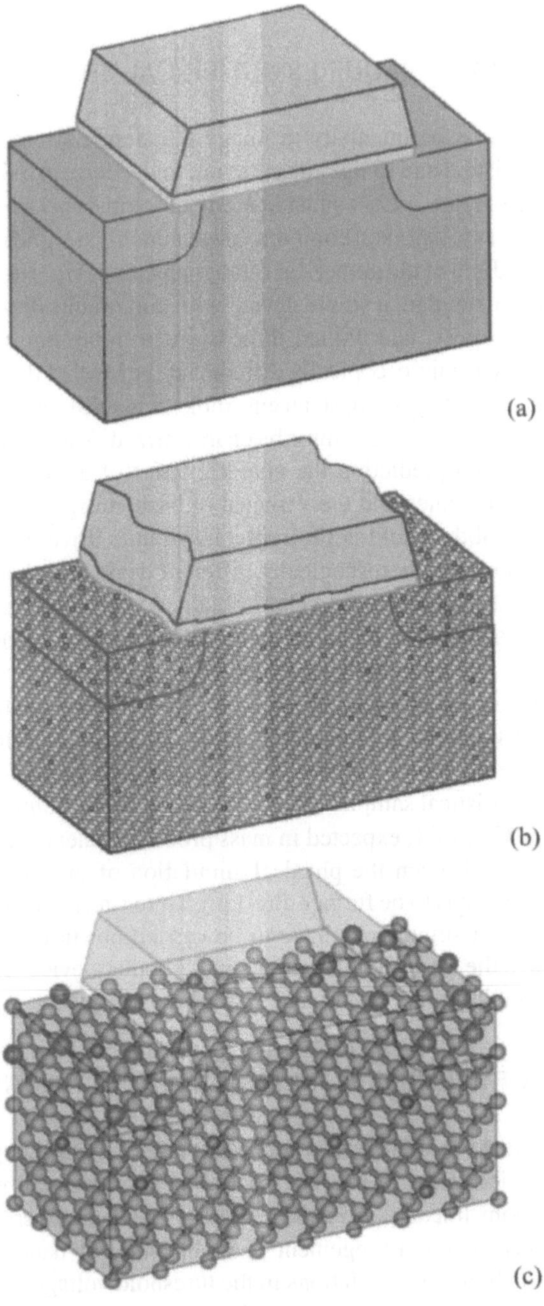

FIGURE 7.1. Transition from continuous towards "atomistic" device concepts. (a) The current approach to the understanding and modeling of CMOS devices assumes continuous ionized dopant charge and smooth boundaries and interfaces. (b) Sketch of a 20-nm MOSFET expected in mass production before 2009. There are less than 50 Si atoms along the channel. Random discrete dopants, atomic-scale interface roughness, and line-edge roughness introduce significant intrinsic parameter fluctuations. (c) Sketch of a 4-nm MOSFET expected in mass production in 2022. There are less than 10 Si atoms along the channel. The device becomes comparable to medium-sized molecules.

7.2. UNDERSTANDING REQUIRES NUMERICAL SIMULATION

It is still impossible experimentally to image the atomic structure of individual nano-CMOS transistors and to link the structure to the corresponding device characteristics. Therefore, the current understanding of intrinsic parameter variations in nano-CMOS devices is based on comprehensive numerical simulations. The problem shifts the paradigm of the numerical semiconductor device simulations. It is no longer sufficient to simulate a single device with continuous doping distribution, uniform oxide thickness, and unified dimensions to represent one macroscopic design. Each device is microscopically different at the level of dopant distribution, oxide thickness, and gate pattern, so an ensemble of macroscopically identical but microscopically different devices must be characterized. The aim of the numerical simulation shifts from predicting the characteristics of a single device towards estimating the mean values and the variance of basic design parameters, such as threshold voltage, subthreshold slope, transconductance, drive current, and so forth for a whole ensemble of microscopically different devices in the system. It must be emphasized that even the mean values obtained from, for example, statistical atomistic simulations are not identical to the values corresponding to continuous charge simulation. The simulation of a single device with random dopants, oxide thickness, and gate pattern variation requires a three-dimensional (3-D) solution with fine-grain discretisation. The requirement for statistical simulations transforms the problem into a four-dimensional (4-D) one, where the fourth dimension is the size of the statistical sample. Typical results of 3-D "atomistic" simulations of a 30 × 30-nm MOSFET, expected in mass production next year, are illustrated in Fig. 7.2. The result from the physical simulation of intrinsic fluctuations in ensembles of devices has to be further transferred into statistical circuit level models. Statistical circuit simulations have to be carried out in order to estimate at what scaling stage the intrinsic fluctuations in particular device architectures will become unacceptable from a circuit and systems point of view.

7.3. SOURCES OF INTRINSIC PARAMETER FLUCTUATIONS

Even if the "extrinsic" parameter variations associated with lithographic dimensions and layer thicknesses in modern CMOS technologies are well controlled, unavoidable random fluctuation in the relatively small number of dopants and their discrete microscopic arrangement in the channel of nano-MOSFETs will lead to significant "intrinsic" variations in the threshold voltage and drive current. This problem was pointed out in the early seventies and later experimentally confirmed for a wide range of fabricated and measured MOSFETs down to sub-0.1-μm dimensions.

7.3.1. Random Discrete Dopants

The MOSFET illustrated in Fig. 7.2 has, on average, less than 100 dopant atoms in the channel region. Their number varies from device to device following a

(a)

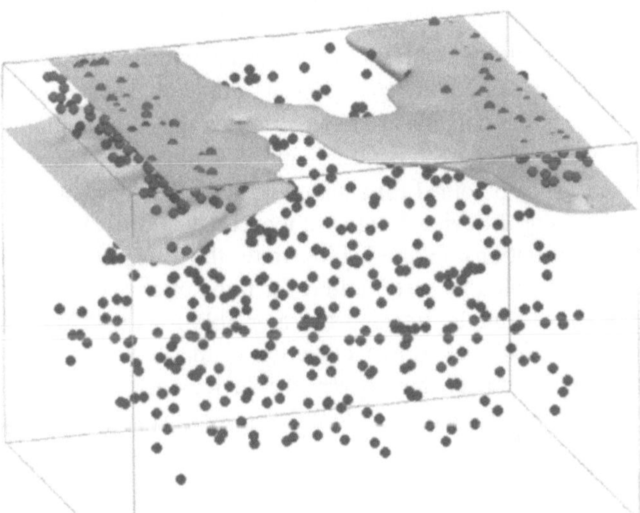

(b)

FIGURE 7.2. Results from the atomistic drift-diffusion simulation of a 30 × 30-nm MOSFET including quantum corrections. (a) Potential distribution indicating also the positions of the individual dopants; (b) one-electron equiconcentration contour indicating that the current starts to flow in the device though a random path determined by the discrete dopant distribution.

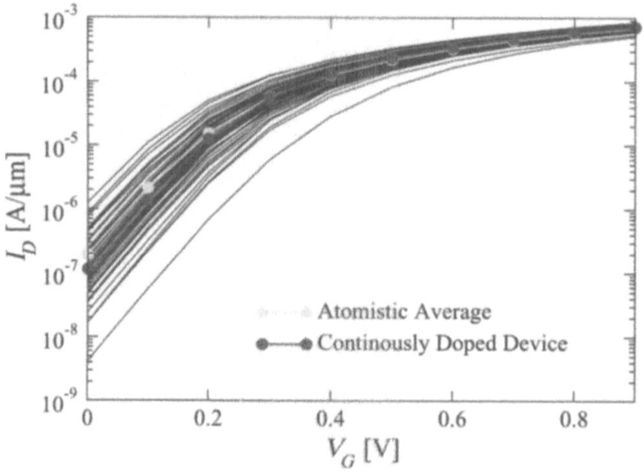

FIGURE 7.3. Current–voltage characteristics obtained from the simulation of 200 macroscopically identical but microscopically different 30 × 30-nm MOSFETs.

Poisson distribution, resulting in threshold voltage and current variations. Even if two transistors have the same number of dopants in the channel region, the fact that in the two devices the dopants are arrange in two different random patterns results in different device characteristics. Figure 7.3 illustrates the variation in the device characteristics obtained for a sample of 200, 30 × 30-nm MOSFETs with identical macroscopic device design but the expected differences in the microscopic doping distributions. Figure 7.4 illustrates the differences in the dopant and the corresponding potential distributions for three microscopically different devices from the above sample, which are responsible for the difference in the device characteristics. With the further reduction of the device dimensions, the number of dopants in the device that governs the device operation will be further reduced, which inevitably result in an intolerable increase in the intrinsic parameter variation.

FIGURE 7.4. Differences in the dopant and the potential distributions for three microscopically different devices from the sample used to generate the characteristics in Fig.7.3, which are responsible for the difference in the device characteristics.

There are two important aspects associated with the inclusion of random discrete dopants in drift-diffusion (DD) simulations: The first is how discrete dopants could be represented bearing in mind the continuum nature of the DD approach; the second is how the random position of the dopants will be chosen in any individual device that is to be simulated. In the Poisson equation associated with the DD approach, the charge in the system is presented as a charge density. Normally a discrete dopant is modeled in the continuum world of DD as a charge density produced by the spreading of the dopant charge in a volume, V, resulting in a doping density in this volume of $1/V$. Normally, the volume associated with the single point charge is the mesh spacing surrounding it.

The second problem that we must address is how to populate our device with random discrete dopants. The original method used in the simulator employs a simple rejection technique. In this approach, at each node of the discretization mesh representing a device, a decision is made to introduce a dopant, or not, based on the continuous doping distribution at this node using a rejection technique [1]. This was accomplished by generating a random number and a probability associated with each mesh point based on Eq. (7.1), where the probability is created from the volume surrounding the mesh node and the doping density. In Eq. (7.1), ρ is the probability of a charge being assigned, dx, dy and dz are the x, y and z dimensions, respectively, of the mesh cell, and N_d is the doping density associated with the mesh node.

$$\rho = (dx\,dy\,dz)N_d. \tag{7.1}$$

If the random number is greater than the generated probability, then a dopant is placed at the mesh node, otherwise only a very low background doping is introduced to avoid numerical instabilities. This method, while introducing discrete random dopants, artificially couples their position to the mesh nodes, and for a large mesh spacing, it introduces artificial correlation in the discrete dopant distribution.

A better approach to generating random discrete dopant positions has been outlined in Ref. 2. It allocates the dopants to sites of the Si crystal lattice covering the simulated device. This approach has become feasible due to the reduced size of modern MOSFETs, which are now below 100 nm. It becomes possible and practical, from a computational point of view, to cover the whole device with the actual crystalline lattice and to randomly populate the sites of the lattice with dopants.

In order to construct the silicon lattice, we first create a basis of eight silicon atoms, depicted in Fig. 7.5 where the darker spheres are part of the basis and the lighter spheres are the remaining silicon lattice. The positions of these atoms relative to a unit cube are (0, 0, 0), (0, 1/2, 1/2), (1/2, 0, 1/2), (1/2, 1/2, 0), (1/4, 1/4, 1/4), (1/4, 3/4, 3/4), (3/4, 1/4, 3/4) and (3/4, 3/4, 1/4). This cube determines the unit translational vector and its size replicates the lattice constant of Si, which is 0.543 nm. As this volume is replicated and translated by multiples of the translational vector it fills the entire simulation domain creating the lattice sites corresponding to a standard bi-cubic silicon lattice, depicted in Fig. 7.6. The

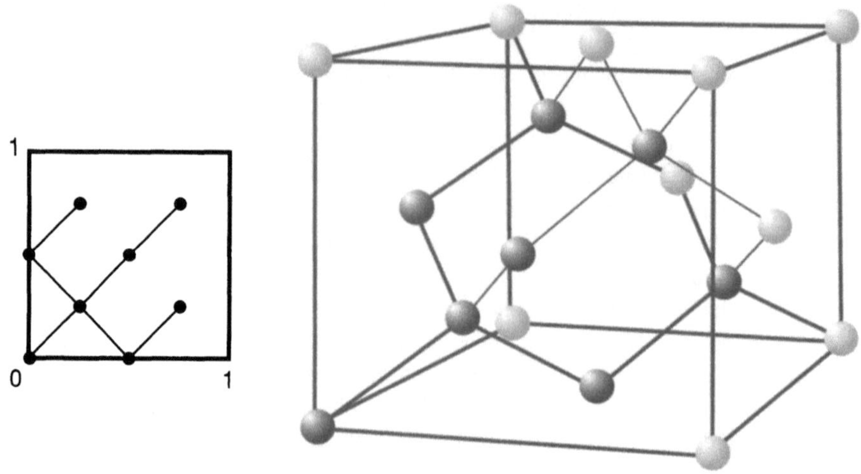

FIGURE 7.5. Silicon lattice basis consisting of eight silicon atoms at points (0,0,0), (0, 1/2, 1/2), (1/2, 0, 1/2), (1/2, 1/2, 0), (1/4, 1/4, 1/4), (1/4, 3/4, 3/4), (3/4, 1/4, 3/4) and (3/4, 3/4, 1/4). Offsets are normalized to a unit cube.

illustrated section of a generated lattice is generated by $5 \times 5 \times 5$ unit volumes (cells) and contains 1000 silicon atom sites.

The process by which Si atoms are replaced by dopants is similar to the method outlined previously. A simple rejection technique is used, but in this case instead of stepping over all of the mesh, this method steps over all of the possible lattice sites and selects whether a dopant atom is to be placed there using a rejection technique depending on a probability generated by the ratio of the doping concentration and the Si concentration at that site.

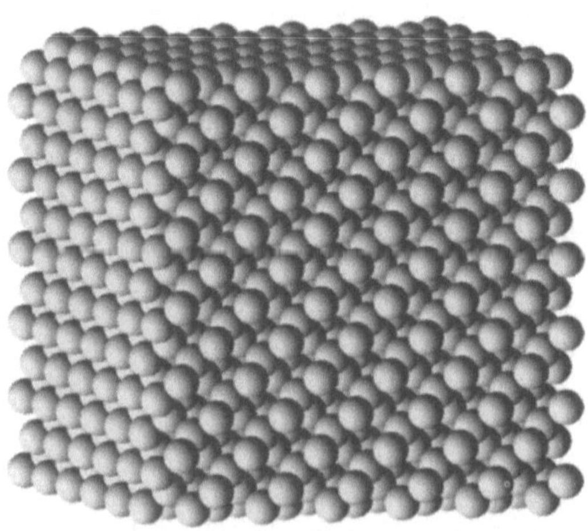

FIGURE 7.6. Section of a generated silicon lattice created by the replication of the basis shown in Fig. 7.5. This section is $5 \times 5 \times 5$ silicon sites and contains 1000 silicon atoms.

7.3.2. Line-Edge Roughness

Unfortunately, the random discrete dopants are not the only source of "intrinsic" parameter fluctuations in modern and future CMOS transistors. Line-edge roughness (LER) caused by tolerances inherent to materials and tools used in the lithography processes is yet another source of fluctuations that needs close attention. LER has caused little worry in the past since the critical dimensions of MOSFETs were orders of magnitude larger than the roughness. However, as the aggressive scaling continues into the decananometer regime, LER does not scale accordingly, becoming an increasingly larger fraction of the gate length. As shown in Fig. 7.7a, the edge roughness, which remains typically on the order of 5 nm almost independently of the type of lithography used in production or research, inevitably introduces variations from device to device. It will be increasingly difficult to

(a)

(b)

FIGURE 7.7. Electron distribution in a 50 × 200-nm MOSFET illustrating the effect of (a) LER and (b) LER with discrete random dopants.

reduce LER, which is limited by the molecular dimensions in the photoresist and, therefore, will be an increasingly important source of "intrinsic" parameter fluctuations in the future [3].

The method used to generate random junction patterns is based on a one-dimensional (1-D) Fourier synthesis that generates gate edges from a power spectrum corresponding to a Gaussian or exponential autocorrelation function. The parameters used to describe this gate edge are the correlation length, Λ, and the root mean square (rms) amplitude, Δ. To understand these parameters, we can think of the rms amplitude as the standard deviation of the x coordinate of the gate edge if we assume that the gate edge is parallel to the y direction; In most cases, the value quoted as LER is traditionally defined as three times the rms amplitude (i.e., 3Δ). The correlation length is obtained by fitting a particular type of autocorrelation function to the gate-edge line. The algorithm for generating a random line creates a complex array of N elements whose amplitudes are determined by the power spectrum obtained from either a Gaussian or exponential autocorrelation function. S_G, as shown in Eq. (7.2) is the power spectrum for a Gaussian autocorrelation function where $k = i(2\pi/Ndx)$, dx is the discrete spacing used for the line, and $0 \leq i \leq N/2$. Equation (7.3) shows a similar power spectrum S_E, this time for an exponential autocorrelation function.

$$S_G(k) = \sqrt{\pi}\Delta^2\Lambda e^{-k^2\Lambda^2/4}, \tag{7.2}$$

$$S_E(k) = \frac{2\Delta^2\Lambda}{1+k^2\Lambda^2}. \tag{7.3}$$

The phases of each of the elements is selected at random, which makes each line unique; however, only $(N/2)-2$ elements are independent and the rest are selected through symmetry operations, so that after an inverse Fourier transform, the resulting height function, $H(x)$, will be real. An example of these generated random lines is shown in Fig. 7.8, which shows two random lines: one generated with a Gaussian autocorrelation function and one with an exponential autocorrelation

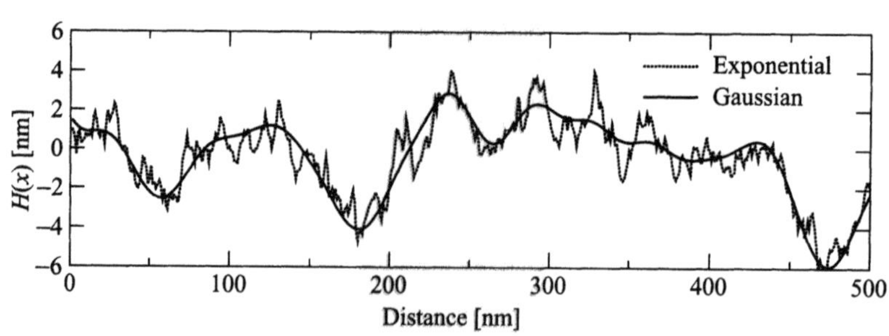

FIGURE 7.8. Two random lines generated from method outlined in Refs. [3 and 4]. One random line is generated using a Gaussian power spectrum whereas the second random line is generated from a exponential power spectrum. The exponential random line can be seen to contain higher-frequency components than the Gaussian random line.

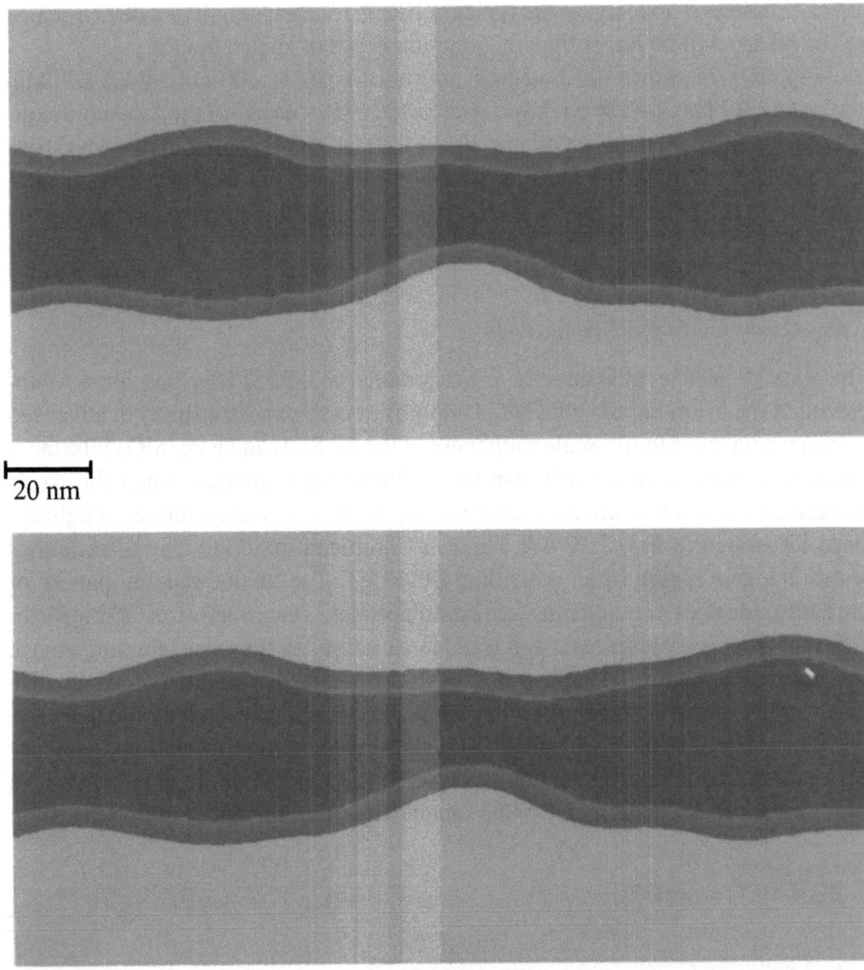

FIGURE 7.9. Gate and *p–n* junctions profiles from Taurus process simulation of a 35-nm channel-length MOSFET including LER after (a) ion implantation and (b) 30 s rapid thermal annealing (RTA) at 900°C.

function. In both cases, typical values of Δ and Λ are used. The line generated from the Gaussian autocorrelation function is smoother due to the lack of high-frequency components, which are characteristic of the corresponding exponential power spectrum.

An example of applied LER is shown in Fig. 7.9, taken from Ref. 4. This shows the gate-edge doping profile and the *p–n* junctions of a 35 × 200-nm MOSFET as a result of process simulation using Synopsys Taurus. In our simulated example, the shape of the surface *p–n* junction replicates the gate-edge profile and follows, through the depth of the device, a Gaussian doping profile. Figure 7.9 shows that even after rapid thermal annealing (RTA) the *p–n* junctions follow the shape of the gate edge, and this is assumed for all of our simulations. Such an approach is not as simple as it appears, as the typical LER correlation length of a device scaled below 50 nm will be larger than the junction depth of such a device.

Figure 7.7a shows the potential profile of a 30 × 200-nm MOSFET with applied LER. This MOSFET has a gate voltage, V_G, equal to the threshold voltage, V_T, at an applied drain voltage, $V_D = 0.01$ V. The potential in this MOSFET approximately follows the metallurgical *p–n* junction. Figure 7.7b shows the potential from the same device but with random discrete dopants also included in the simulation.

7.3.3. Oxide Thickness Fluctuations

The gate dielectric thickness in mass-production MOSFETs has already approached the 1-nm barrier with sub-1-nm physical thickness utilized in advanced research devices. Atomic-scale roughness of the Si/SiO_2 and gate/SiO_2 interfaces introduces significant intrinsic parameter fluctuations. Indeed, when the oxide thickness is only a few silicon atomic layers, the atomic-scale interface roughness steps illustrated in Fig. 7.10 will result in significant oxide thickness variations within the gate region of an individual MOSFET. The unique random pattern of the gate oxide thickness and interface landscape makes each nano-CMOS transistor different from its counterparts and leads to variations in the corresponding device characteristics. Figure 7.11 illustrates results of 3-D device simulation, which take into account the random pattern of the Si/SiO_2 interface in each individual device.

The random two-dimensional (2-D) surfaces used to represent the boundary between the oxide and the silicon or between the oxide and the gate material in our simulations are constructed using standard assumptions for the autocorrelation

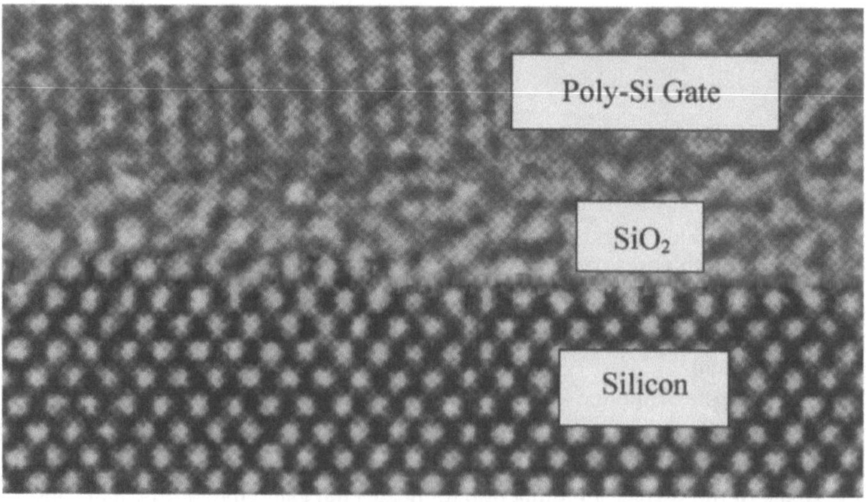

FIGURE 7.10. Atomic structure of the Si/SiO_2 interface in a state-of-the-art MOSFET with less than 1 nm oxide thickness. Random variations on the scale of one interatomic layer of silicon result in more than a 30% variation in the oxide thickness.

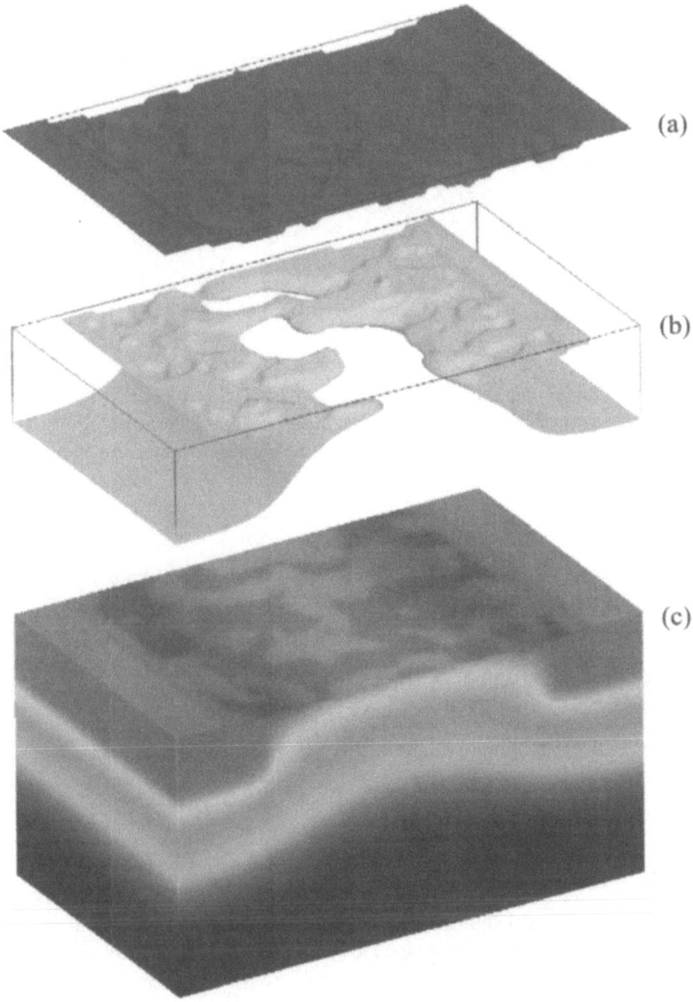

FIGURE 7.11. Results from the simulation of a 30 × 30-nm MOSFET that takes into account the atomic-scale variation of the gate dielectric thickness. (a) Random pattern of the Si/SiO_2 interface. (b) One-electron equiconcentration contour indication that the current starts to flow in the device though a random path determined by the random interface pattern. (c) Potential distribution. The surface potential fluctuations responsible for the intrinsic parameter fluctuations conform to the interface pattern.

function of the interface roughness. Generally, the interface is described as an exponential autocorrelation function with a given correlation length, Λ, and rms height, Δ [5]. The corresponding power spectrum can be obtained by 2-D Fourier transformation (in radial coordinates). In order to reconstruct the interface, we generate, in the Fourier domain, a complex $N \times N$ matrix. The magnitude of the elements of this matrix follows the power spectrum of the autocorrelation function, whereas the phase is selected at random. Several conditions [6] must be satisfied to ensure

FIGURE 7.12. Two-dimensional rough surface generated to act as a template for interface roughness. This 2-D surface was generated in a similar fashion to the random line shown in Fig. 7.8 using the inverse Fourier transform of a complex vector created from a Gaussian autocorrelation function.

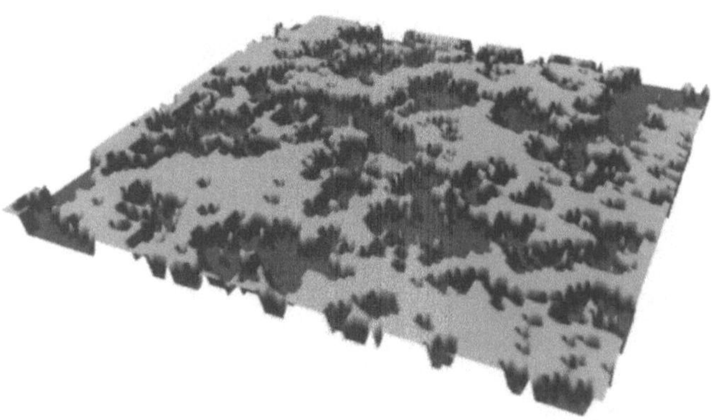

FIGURE 7.13. Two-dimensional digitized random surface created from the Gaussian random surface shown in Fig. 7.12. This surface can then be used to create a random interface by altering the material profile at the interface.

that the corresponding 2-D surface, obtained by the inverse Fourier transformation, represents a real function. The "analog" random 2-D surface obtained using this procedure and illustrated in Fig. 7.12 is then quantized in steps to take into account the discrete nature of the interface roughness steps associated with atomic layers in the crystalline silicon substrate [7]. The step height is approximately 0.3 nm for the (001) interface. The result of this digitization is illustrated in Fig. 7.13.

7.4. METHODOLOGY

In many cases, in order to investigate the effects of random dopant-induced fluctuations, it is sufficient to perform a classical simulation involving the solution of the DD approximation to the Boltzmann transport equation. This has been the basis for the majority of the work performed so far in this field [8,9]. The standard

DD approach requires the solution of Poisson's equation along with the current continuity equations for electrons, holes, or both. The DD simulator can be used to perform statistical investigations into threshold voltage fluctuations and lowering in MOSFETs scaled below 0.1 μm under arbitrary bias conditions. Short-channel effects such as drain-induced barrier lowering (DIBL) are naturally included in this approach, together with asymmetry in device characteristics due to dopant clustering at one end of the channel.

Our simulations are based on a 3-D "atomistic" DD simulator based on the decoupled Gummel procedure [10]. The solution of Poisson's equation uses a parallelized black/red Newton SOR (Successive Over Relaxation) solver, whereas a parallelised BiCGSTAB (BiConjugate Gradient Stabilised) solver has been implemented for the solution of the electron current continuity equation.

However, as devices are shrunk toward the nanometer scale, the effects associated with quantum mechanics start to dominate the device operation and must be accounted for within the simulations. To account for quantum effects, we implement the density gradient method [11], which is described in detail in the following section.

7.4.1. Density Gradient in Drift-Diffusion Simulations

The density gradient (DG) formalism can be derived from the equation of motion for the one-particle Wigner function:

$$\frac{\partial f(\mathbf{p},\mathbf{r},t)}{\partial t}+\mathbf{v}\cdot\nabla_r f(\mathbf{p},\mathbf{r},t)-\frac{2}{\hbar}V(\mathbf{r})\sin\left[\frac{\hbar\overleftarrow{\nabla}_{\mathbf{r}}\overrightarrow{\nabla}_{\mathbf{p}}}{2}\right]f(\mathbf{p},\mathbf{r},t)=0. \tag{7.4}$$

The quantum effects are included through the inherently nonlocal driving potential in the third term on the left-hand side, where it is understood that ∇_r acts only on V, and ∇_p acts only on the distribution function f. The operator within the sum may be written in terms of a power series, so that the transport equation for the Wigner distribution function can be written in the form of a modified Boltzmann transport equation (BTE) as

$$\frac{\partial f}{\partial t}+\mathbf{v}\cdot\nabla_r f-\frac{1}{\hbar}\nabla_r V\cdot\nabla_k f+\sum_{\alpha=1}^{\infty}\frac{\hbar^{(2\alpha-1)}(-1)^{\alpha+1}}{4^{\alpha}(2\alpha+1)!}(\nabla_r V\cdot\nabla_k f)^{2\alpha+1}$$
$$=\left(\frac{\partial f}{\partial t}\right)_{\text{coll}}, \tag{7.5}$$

where V represents the electrostatic potential. Expanding to first order in $\hbar$, so that only the first nonlocal quantum term is considered, has been shown to be sufficiently accurate to model nonequilibrium quantum transport and also for the inclusion of tunneling phenomena in particle-based Monte Carlo simulators. Although it is tempting to expand the nonlocal term to higher powers of Planck's constant, this may lead to spurious results since the equation of motion is not analytic in $\hbar$. The additional, nonclassical, quantum correction term may be viewed as a modification to the classical potential and acts like an additional quantum force term in particle simulations, similar in spirit to the Bohm interpretation. It should be noted that in the limit of slow spatial variations (e.g., for gutterlike potentials), the nonlocal terms disappear and the equation reduces to the classical BTE. However, in order to obtain a correction that may be used efficiently in device simulators, it is necessary to make an assumption regarding the carrier distribution function. The following form of the distribution is usually assumed:

$$f = \exp\left\{-\beta\left[\frac{\hbar^2 (k_i - \langle k_i \rangle)^2}{2m^*} + V(r) - E_f\right]\right\}, \quad \frac{\hbar^2 (k_i - \langle k_i \rangle)^2}{2m^*} \approx \frac{k_B T}{2} \tag{7.6}$$

for each independent spatial coordinate, i, along with

$$\frac{\partial^2 V}{\partial x^2} \approx -k_B T \frac{\partial^2 (\ln n)}{\partial x^2}, \tag{7.7}$$

where $\beta = 1/k_B T$.

The quantum-corrected BTE may be written in one dimension, without loss of generality, as

$$\frac{\partial f}{\partial t} + v \cdot \nabla_r f + \frac{1}{\hbar} F^{qc} \cdot \nabla_k f = \left(\frac{\partial f}{\partial t}\right)_{\text{coll}}, \tag{7.8}$$

where

$$F_x^{\text{qc}} = -\frac{\partial}{\partial x}\left(V - \frac{\hbar^2}{12m^*}\frac{\partial^2 (\ln n)}{\partial x^2}\right). \tag{7.9}$$

We may, in turn, write the carrier distribution in equilibrium as

$$n(x) = N_c \exp\left\{-\beta\left[V(x) + V_{\text{qc}}(x) - E_f\right]\right\} = n_{\text{cl}}(x) \exp\left[-\beta V_{\text{qc}}(x)\right], \tag{7.10}$$

where

$$V_{\text{qc}}(x) = -\frac{\hbar^2}{12m^*}\frac{\partial^2 (\ln n)}{\partial x^2}$$

and n_{cl} represents the classical carrier density without the quantum corrections. The density gradient approximation in a DD context may be derived in a manner similar to that for deriving the DD approximation from the BTE, making the following assumptions:

1. The electron temperature is equal to the lattice temperature and the electron temperature gradient is zero.

2. The $\mathbf{v} \cdot \nabla \mathbf{v}$ term is assumed to be small compared to other terms and can be neglected.

Thus, using these approximations the current may be written as

$$J = qn\mu_n E + qD_n \frac{\partial}{\partial x}\left(ne^{-\beta V_{qc}}\right). \tag{7.11}$$

By expanding the term $e^{-\beta V_{qc}}$ and taking the lowest-order component, we obtain what is termed the density gradient approximation:

$$J = qn\mu_n E + qD_n \frac{\partial n}{\partial x} - \frac{qn\mu_n \hbar^2}{6m^*}\frac{\partial}{\partial x}\left(\frac{\partial^2 \sqrt{n}/\partial x^2}{\sqrt{n}}\right). \tag{7.12}$$

The additional quantum potential term, which is proportional to the second derivative of the root of the density, may be taken as an additional term in the quasi-Fermi level so that the electronic equation of state becomes similar to that of an ideal gradient gas for typical low-density, high-temperature semiconductors, which includes an additional term that is dependent on the gradient of the carrier density:

$$2b_n \frac{\nabla^2 \sqrt{n}}{\sqrt{n}} = \phi_n - \psi + \frac{k_B T}{q} \ln\left(\frac{n}{n_i}\right), \tag{7.13}$$

where $b_n = \hbar^2/4q\, m_n^* r$; all other symbols have their usual meaning and r is a dimensionless parameter. In situations with strong quantum confinement (when only a single subband is occupied), the parameter r is considered equal to 1. However, as more subbands become filled (e.g., due to increase in temperature), statistical averaging causes r to change, asymptotically approaching 3. Throughout, we assume that r is equal to 3. This results in a quantum potential correction term in the standard DD flux:

$$F_n = -n\mu_n \nabla\psi + D_n \nabla n - 2n\mu_n \nabla\left(b_n \frac{\nabla^2\sqrt{n}}{\sqrt{n}}\right) \tag{7.14}$$

or

$$F_n = -n\mu_n \nabla\left(\psi + 2b_n \frac{\nabla^2 \sqrt{n}}{\sqrt{n}}\right) + D_n \nabla n = -n\mu_n \nabla\psi_{\text{eff}} + D_n \nabla n, \tag{7.15}$$

where ψ_{eff} is the effective quantum potential.

For our solution of the DG-corrected DD approximation, we use a modified Gummel approach in which the Poisson equation [Eq. (7.16)] and DG equation [Eq. (7.17)] are solved self-consistently for the electrostatic potential and the quantum-corrected electron density:

$$\nabla \cdot (\varepsilon \nabla \psi) = -q\left(p - n + N_D^+ - N_A^-\right), \tag{7.16}$$

$$2b_n \frac{\nabla^2 \sqrt{n}}{\sqrt{n}} = \phi_n - \psi + \frac{k_B T}{q} \ln\left(\frac{n}{n_i}\right). \tag{7.17}$$

The effective quantum potential is then calculated from

$$\psi_{\text{eff}} = \psi + 2b_n \frac{\nabla^2 \sqrt{n}}{\sqrt{n}} = \phi_n + \frac{k_B T}{q} \ln\left(\frac{n}{n_i}\right) \tag{7.18}$$

and used as the driving potential for the current continuity equation,

$$\nabla \cdot J_n = 0, \tag{7.19}$$

where

$$J_n = -qn\mu_n \nabla \psi_{\text{eff}} + qD_n \nabla n, \tag{7.20}$$

which is solved using a standard Sharfetter–Gummel discretisation based on the effective quantum potential.

The system of Eqs. ((7.16) and (7.17)) and Eq. (7.19) are solved self-consistently until convergence.

7.4.2. Boundary Conditions for Density Gradient

All differential equations require boundary conditions before they can be solved numerically and the DG equation for the quantum potential, ψ_q as given by Eq. (7.21), is no exception:

$$\psi_q = 2b_n \frac{\nabla^2 \sqrt{n}}{\sqrt{n}}. \tag{7.21}$$

When using the quantum potential for MOSFET simulations, we must consider three distinct boundary conditions for ψ_q: Dirichlet (electrodes), Si/SiO_2 interfaces, and Neumann boundary conditions for the flux. Each of these boundaries will be considered next.

7.4.2.1. Dirichlet Boundary Conditions

At the electrodes, the quantum potential, ψ_q, may be assumed to be zero since these regions are usually heavily doped ($\sim 10^{20}$ cm^{-3}) and the electron density can be assumed to be uniform in thermal equilibrium at the electrode interface as in classical DD simulations. In this region, the boundary conditions are given by

$$\psi_q = 0 \tag{7.22}$$

for the quantum potential.

7.4.2.2. Neumann Boundary Conditions

At the Neumann boundary condition in the classical DD simulations we have the boundary condition that the outward normal of the current density flux must be zero, which gives the following boundary condition for the classical DD simulations:

$$\nabla \psi \cdot \mathbf{n} = 0, \tag{7.23}$$

where ψ denotes the classical electrostatic potential and $\mathbf{n}$ denotes the outward normal vector to the surface. In the DG simulations, the current density flux is given by Eq. (7.14) and, therefore, in analogy to the boundary condition for the above classical potential, we may define the following boundary condition for the quantum potential, ψ_q:

$$\nabla \psi_q \cdot \mathbf{n} = 0. \tag{7.24}$$

7.4.2.3. Si/SiO_2 Interface Boundary Conditions

The Si/SiO_2 boundary conditions for the quantum potential in the DG approach require careful consideration, and in our simulations, we use the approach of Jin et al. [12]. The conventional DG model assumes an extremely small carrier (electron or hole) density at the interface [13–15], effectively assuming an infinite Si/SiO_2 potential barrier. However, this boundary condition overestimates slightly the quantum confinement effect since the actual potential is finite. This overestimation will become more apparent if we consider high-κ dielectrics [16] in which the barrier offset (conduction or valence) between Si and the dielectric is not as large as for SiO_2.

The effect of the finite barrier height and the associated electron wavefunction penetration into the oxide can be modeled in the following way within the DG formalism following Ref. 12. The electron density penetration into the oxide as a function of distance, x, from the Si/SiO_2 interface can be approximated as [17]

$$n(x) = n(0) \exp\left(\frac{-2x}{x_p}\right), \tag{7.25}$$

where $n(0)$ is the electron density at the interface and x_p, given by

$$x_p = \frac{\hbar}{\sqrt{2m^*_{\text{ox}}\Phi_B}}, \tag{7.26}$$

is the characteristic penetration depth obtained from the Wentzel–Kramers–Brillouin (WKB) approximation [18]. Here, m^*_{ox} is the electron effective mass within the oxide and Φ_B (3.1 eV in our simulations) is the potential barrier of the oxide. Using Eq. (7.25), the outward normal component of $b_n \nabla \sqrt{n}$ at the Si/SiO_2

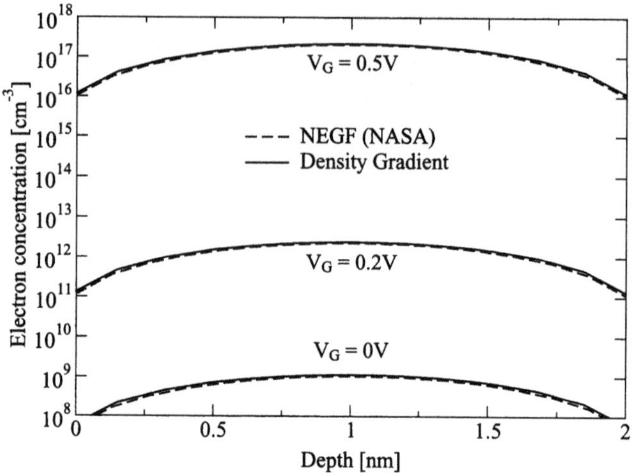

FIGURE 7.14. Calibration of the DG against NEGFs.

interface can be written as

$$\mathbf{n} \cdot b_n \nabla \sqrt{n} = -\left(\frac{b_{ox}}{x_p}\right)\sqrt{n}, \tag{7.27}$$

where b_{ox} is given by

$$b_{ox} = \frac{\hbar^2}{12qm_{ox}}. \tag{7.28}$$

Typically within DG simulations such as ours, the value of m^*_{ox} is chosen to match the results that are obtained from fully quantum mechanical simulations such as Poisson–Schrödinger or nonequilibrium Green's functions (NEGFs) [19]. We have calibrated our simulator against NEGFs simulations provided by NASA (see Fig. 7.14) and obtain effective mass values of $m^*_{ox} = 0.22$ and $m^*_{Si} = 0.3$.

7.5. PROBLEMS IN CLASSICAL SIMULATIONS

The inclusion of discrete random dopants in classical DD simulations introduces several problems when mesh-based discretization is used to solve Poisson's equation [20]. The use of Boltzmann or Fermi–Dirac statistics leads to a "trapping" of carriers by the sharply resolved potential wells or peaks associated with the mesh-resolved Coulomb potential of individual acceptors or donors. The effect of this is that an area of a device being simulated "atomistically" using discrete random dopants will have an artificially lower conductance compared to the case of a continuously doped simulation [21, 22]. This leads to problems, particularly in the case of MOSFET simulations where the transistors will have an artificially increased access resistance associated with the introduction of "atomistic" source and drain doping and, therefore, reduced current. A second problem is the strong mesh dependence of the solution. Normally, a discrete dopant is introduced into the continuum world of the DD formalism as a charge density inversely proportional to the volume of the mesh cell. This implies that as the mesh is decreased, the charge density increases, which results in a better resolution of the singular Coulomb potential of a point charge associated with the dopant and in sharper peaks in the solution. As a result, more mobile charge becomes localized and this reduces the free charge participating in the transport and contributing to the device current. Therefore, a simulation of a MOSFET that has a finer mesh size will result, for example, in a larger source/drain access resistance.

Various solutions have been proposed to correct the problems associated with the inclusion of "atomistic" dopants. These include charge assignment schemes in which the charge associated with an individual atom is spread over the surrounding mesh cells [23], the splitting of the Coulomb potential into long- and short-range components [24], and the use of quantum corrections to alter the shape of the electron concentration profile [21, 22, 25, 26]. In the case of splitting the potential into long- and short-range components based on screening considerations, there is the possibility of "double counting" the effects of mobile charge screening and also this method uses a somewhat arbitrary cutoff point.

7.5.1. Charge Localization

In the case of n-type material in which the majority carriers are electrons, a donor-type dopant atom will produce a sharply resolved potential well in the conduction band edge, as seen by an electron. As a result, a significant amount of electron charge can become "trapped" or localized in the sharply resolved Coulomb potential well created by a discrete dopant assigned to a fine mesh. Such trapping is physically impossible since, in quantum mechanical terms, confinement keeps the ground electron state high in the well. Figure 7.15, which depicts a 1-D Poisson–Schrödinger solution, shows the Coulomb potential well corresponding to a charge plane and the bound energy states, illustrating this point. It is clear that a large fraction of the potential in the well cannot be followed by the electron concentration due to such quantization. This results in a quantum reduction of the electron concentration in the well when compared to the classical electron concentration, which can also be seen in Fig. 7.15.

The effects of charge localization can have a significant impact on the results of classical MOSEFT simulation. Here, the inclusion of discrete random dopants, particularly in the source/drain, can result in the increased access resistance of these regions. This can be understood by inspecting the potential across the channel obtained from the "atomistic" DD simulations of a classical n-MOSFET. Figure 7.16a shows a 2-D potential slice along the channel of a 35×35-nm square n-MOSFET. In this case, only the substrate is "atomistically" doped, whereas the source and drain regions are continuously doped. In this case, the electrons as the majority carrier would move over the potential barrier and through the valleys produced

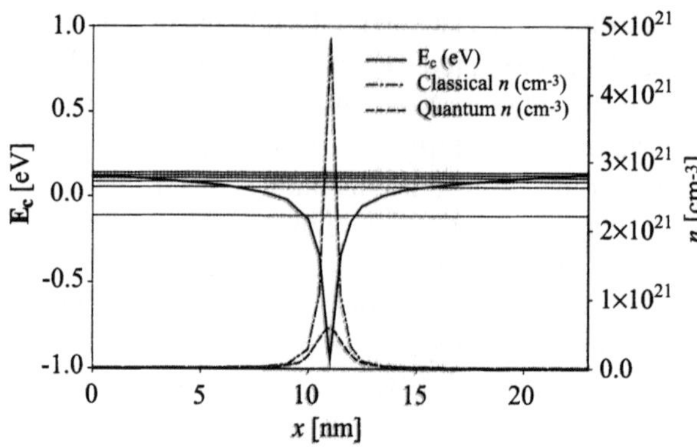

FIGURE 7.15. A 1-D Poisson–Schrödinger solution showing the Coulomb potential well and bound energy states. The classical electron concentration and the quantum electron concentration associated with this potential well are also shown.

by the potential spikes caused by the acceptors in the channel. There would be no charge localization in the source/drain regions to affect the access resistance and the device current.

Figure 7.16b shows the inverse situation. In this case, the substrate region is continuously doped, whereas the source and drain regions are now "atomistically" doped. As can be seen, there are now sharply resolved potential wells within the source/drain regions. When the carrier concentration is calculated, a high

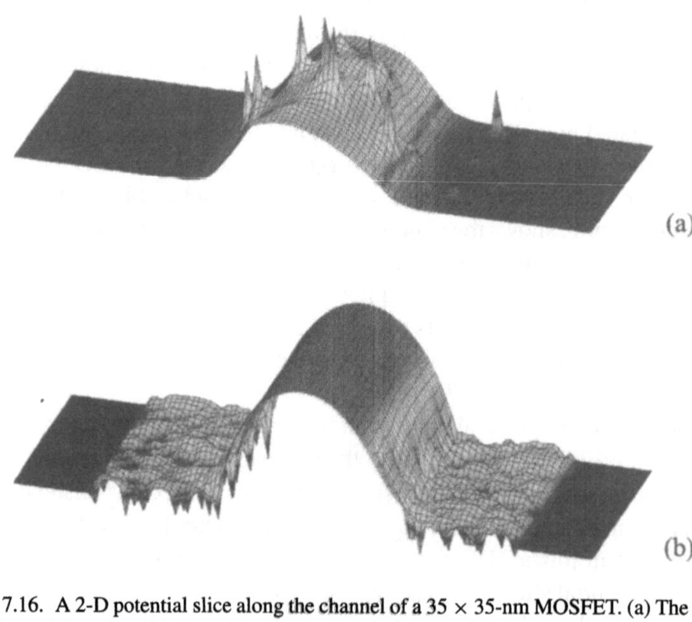

FIGURE 7.16. A 2-D potential slice along the channel of a 35 × 35-nm MOSFET. (a) The source and drain regions of the MOSFET are continuously doped, whereas the substrate is "atomistically" doped. (b) The source and drain regions of the MOSFET are "atomistically" doped, whereas the substrate region is continuously doped.

proportion of the majority carriers (in this case, electrons) will be trapped in these wells. The electrons in the base of these potential wells now have to overcome a much larger potential barrier in order to participate in the current. This results in an increased resistance in the source/drain regions and, in turn, leads to a reduced current in the device.

7.5.2. *"Atomistic" Resistor Study*

To investigate the overall increase in resistance associated with charge trapping, a simple device structure is needed so that the more complex effects associated with the design and operation of a MOSFET would be excluded. A simple resistor was chosen as the structure most suitable for this purpose. The resistor was modeled as a slab of n-doped silicon, which could be doped at various concentrations to study the effects of charge localization. For the simulations presented here, the doping concentration used was 10^{20}cm^{-3}, although the results were consistent at other doping levels. This particular concentration is representative of the doping concentrations in the source/drain regions of modern MOSFETs. Once the analytical value for the resistance was calculated based on the fixed value of mobility used for the simulations, the continuously doped resistor was simulated using the classical DD approach. A perfect match was found between the analytical value for the resistance and the results of the DD simulation.

In the case of the "atomistic" simulations, we chose to simulate and average the results from an ensemble of 200 "atomistic" devices as a statistical sample indicative of the effects caused by the inclusion of discrete random dopants. One possible solution to the problem of charge localization is to smear out the associated charge over a broader (preferably constant) area and so reduce the amount of the charge density at any given mesh point, thus reducing the depth of the associated potential well. In order to do this, various charge assignment schemes can be used, and in this chapter we compare two charge assignment methods: nearest grid point (NGP) and cloud-in-cell (CIC) [27]. The main principle of these two charge assignment schemes is that an individual dopant atom will make a contribution to the charge density assigned to the neighboring mesh points in the simulation. The amount of this contribution generally depends on the distance between the dopant position and the corresponding mesh point and results in a weighting between 0 and 1 determining the fraction of the doping density, $1/V$. The various charge assignment schemes vary in terms of the weighting factors and, therefore, in the positional dependence of the contribution to a particular mesh point. The simplest charge assignment scheme is NGP. In the NGP approach, all of the charge associated with a single dopant atom is assigned to the single closest mesh point. A weighting of 1 is given to the closet mesh point, and all other weightings are zero. A more complex scheme for charge assignment is the CIC approach. In this approach, the charge associated with a single dopant atom in three dimensions is spread across the surrounding eight mesh points. Here, the weighting depends on the position of the dopant within the surrounding box; the closer a dopant is to a mesh point, the larger the doping density assigned at that point. This method

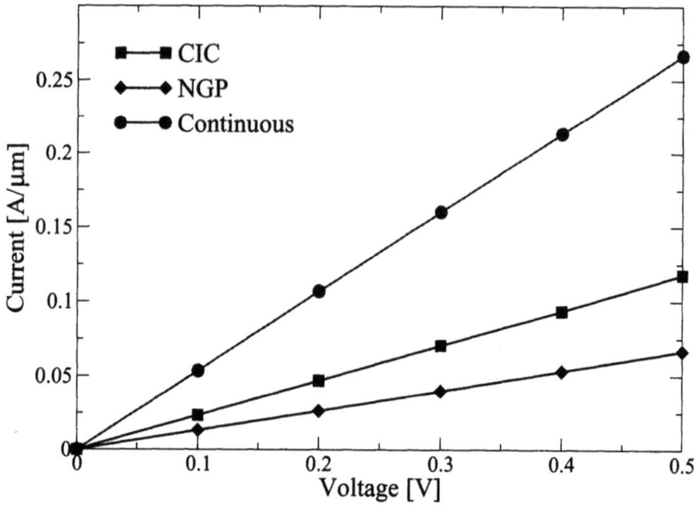

FIGURE 7.17. The I–V characteristics of a 30 × 20 × 20-nm resistor with a 1-nm mesh spacing showing the cases of a continuously doped resistor as well as the atomistic average of 200 devices for 2 different charge assignment schemes: NGP and CIC.

of charge assignment is an improvement over NGP, as the charge associated with a single dopant atom is spread over a region of eight nodes as opposed to one node. For a dopant close to the center of the cell, this gives the best possible improvement compared to NGP, assigning one-eighth of the doping density to each node of the cell. However, as a dopant moves closer to a single mesh point, this charge assignment scheme becomes more like NGP. CIC exhibits charge trapping, which increases with the reduction in mesh size; however, the effect is significantly reduced when compared with the NGP scheme.

Figure 7.17 depicts the I–V characteristics obtained from classical simulations of a 30 × 20 × 20-nm resistor for both continuous doping and for the two methods of charge assignment for a mesh spacing of 1 nm. It can be seen that the resistance of the device increases with the inclusion of discrete random dopants and that when a coarse method of charge assignment such as NGP is used, the increase in resistance is greatest. This is due to the sharply resolved potential wells associated with the high doping concentrations representing the discrete random dopants. When using the CIC assignment, the resistance is lowered in comparison to the NGP approach.

7.5.3. Quantum Corrections

An alternative approach to the use of charge assignment schemes for reducing charge localization is to introduce quantum corrections, in particular the use of the DG method. As has been discussed previously, this method has been used successfully to model confinement effects at the interface of a MOSFET and the corresponding shape of the electron distribution peaking away from the interface. If these quantum corrections are applied to the region that contains discrete random

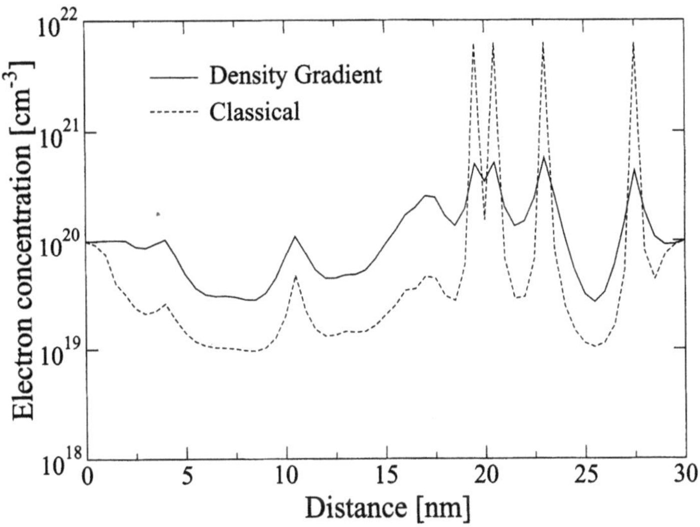

FIGURE 7.18. The electron concentration along the x direction of a $30 \times 20 \times 20$-nm resistor, showing both the classical electron concentration due to Boltzmann statistics and the quantum electron concentration generated by the inclusion of DG quantum corrections.

dopants, a marked reduction in the charge localization at impurity sites can be observed. In the case of a sharply resolved potential well, electrons are unable to follow the real electrostatic potential because their concentration is determined by the effective quantum potential, which introduces a force pushing the electrons out of the well. The result of this is that when DG is applied in conjunction with the DD system of equations, it smoothes out the unnaturally large variations in the electron concentration around the dopants and more carriers are able to contribute to the current flow in the simulation.

The impact of the inclusion of quantum confinement effects in simulations by use of the DG approximation is illustrated in Fig. 7.18. Quantum confinement in the sharp potential wells smooths the overall electron concentration by reducing the sharper peaks, eliminating the charge localization. It can also be seen that this effect raises the concentration in the surrounding area, allowing more charge to participate in the simulation instead of being trapped in the potential wells. Because less of the mobile charge is trapped within the Coulomb potential wells, the overall resistance of the simulated slab is reduced.

Figure 7.19 illustrates the impact on the I–V characteristics of the introduction of DG quantum corrections. A significant improvement in the "atomistically" simulated resistance can be observed with the introduction of the quantum corrections for both CIC and NGP charge assignment schemes.

7.5.4. *Mesh Sensitivity*

The amount of charge that becomes localized is strongly dependent on the mesh size used in the simulation. If we consider the case of NGP, the charge density at a given mesh point is equal to $1/V$. If we then reduce the mesh size, a subsequent

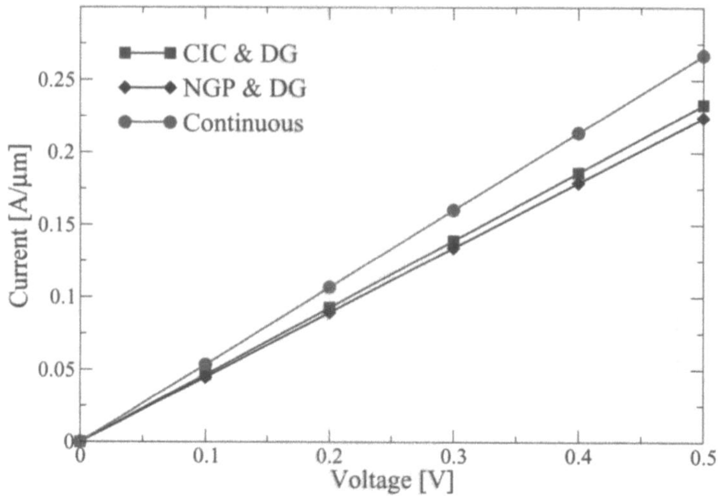

FIGURE 7.19. The I–V characteristics of a $30 \times 20 \times 20$-nm resistor with a 1-nm mesh spacing showing the cases of a continuously doped resistor as well as the atomistic average of 200 devices for 2 different charge assignment schemes: (NGP and CIC), with the inclusion of DG quantum corrections.

reduction in volume causes the charge density at that point to increase. The fine mesh in combination with the charge density increase results in a more sharply resolved Coulomb potential well, leading to a greater charge localization.

As expected, the singularity in the electrostatic potential increases with the reduction of the mesh size. At the same time, the effective quantum potential is much less sensitive to the mesh size, showing no differences below the 0.5-nm mesh spacing. The effective potential in the middle of the Coulomb well, which is approximately 40 mV, corresponds roughly to the energy level of the electron ground state and is in good agreement with the ground state of a hydrogenic model of an impurity in Si. The effects of the mesh size on the overall resistance of the simulated slab is shown in Fig. 7.20. For an "atomistically" doped resistor, the use of a NGP charge assignment scheme results in a progressively larger increase in the resistance of the slab as the mesh spacing is reduced. With the inclusion of DG quantum corrections, not only is most of the charge localization removed from the simulation but also the mesh dependence of the solution is greatly reduced below the 0.5-nm mesh spacing.

7.5.5. DG Corrections for Holes

It has been sufficient to outline the problems associated with charge trapping by only considering the case of electrons being trapped within the source and drain regions of a MOSFET, which leads to a increase in the series resistance of the device. However, for the accurate "atomistic" simulation of a MOSFET, it is also important to consider the trapping of holes in the channel region when "atomistic" acceptors are considered. This trapping leads to a change in both the size and shape

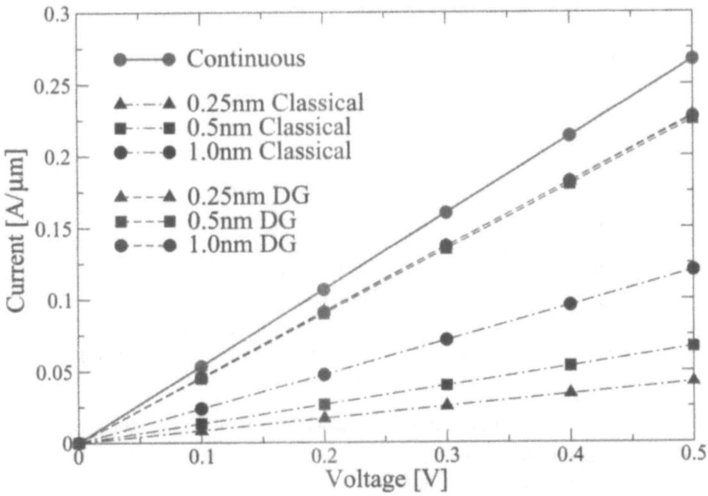

FIGURE 7.20. The I–V characteristics of a 30 × 20 × 20-nm resistor for both continuous and "atomistic" doping using the NGP charge assignment scheme showing the average of 200 "atomistically" doped resistors for mesh sizes of 1 nm, 0.5 nm, and 0.25 nm in both classical and quantum-corrected simulations.

of the depletion region. This, in turn, alters the characteristics of the simulated device by reducing, for example, the threshold voltage.

To overcome this problem, DG corrections for holes also have to be considered in the MOSFET. Typically, in the DG DD simulation of n-channel MOSFETs, it is sufficient to solve, self-consistently, the Poisson equation and the electron current continuity equation, eventually adding DG corrections for electrons. However, both the trapping of electrons in the source/drain regions and the trapping of holes in the substrate have to be avoided in the "atomistic" simulation and, therefore, the DG corrections have been introduced for the two types of carriers even when simulating unipolar MOSFETs.

Figure 7.21 shows the hole concentration in the substrate depletion region of a 50 × 50-nm MOSFET. It is clear that, in the classical case, the holes become localized around individual dopant atoms, producing high concentration spikes. With the inclusion of DG quantum corrections for holes, these high-concentration regions are reduced and the resulting concentration in the surrounding regions is increased. This, in turn, modifies the shape of the depletion region under the gate and alters the threshold voltage.

7.6. *AB INITIO* COULOMB SCATTERING IN MONTE CARLO SIMULATIONS

Previous simulation studies of random dopant fluctuation effects, similar to the studies of effects associated with trapped charges, have been predominantly

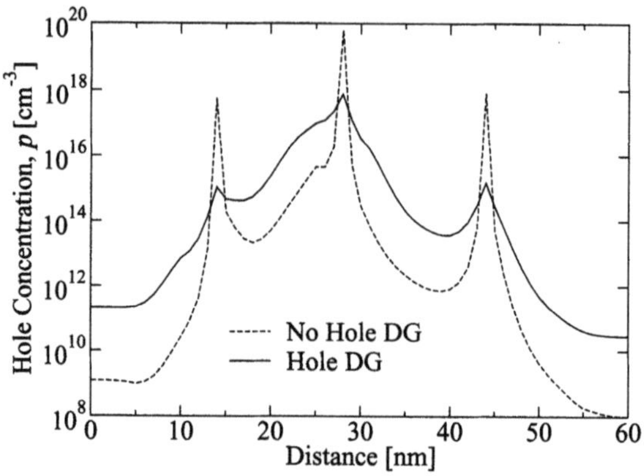

FIGURE 7.21. The hole concentration along the channel of a 50 × 50-nm transistor showing both the classical hole concentration due to Boltzmann statistics and the quantum hole concentration generated by the inclusion of DG quantum corrections.

performed using 3-D DD simulators [20, 28–30]. Such simulations only capture the electrostatic effects associated with random discrete dopant distributions in providing an estimate for the variations in the threshold voltage and the drive current. However, the DD simulations cannot capture the complex effects associated with the variation in carrier transport from device to device, associated with the different numbers and configuration of ionized dopants within the channel that act as Coulomb scattering centers in a non-self-averaging manner. Therefore, the published results for the variation in the drive current obtained from DD simulation most probably underestimate the real magnitude of the intrinsic parameter variations.

In our Monte Carlo (MC) simulations, the impurity scattering is removed from the conventional scattering rate tables and introduced through the real space trajectories of the electrons in the mesh-resolved potential of the individual discrete dopants. This requires a correction to be applied at short range to the mesh-calculated force to account for the aliasing in the mesh solution of Poisson's equation. The *ab initio* approach reproduces the dopant concentration dependence of the mobility in Si [31].

Although *ab initio* Coulomb scattering from ionized impurities through the real space carrier trajectories has previously been included in 3-D ensemble MC simulations [32–35], little statistical analysis has been performed with such simulators. In particular, no results have been published comparing the impact of the variation of numbers and position of Coulomb scattering centers in the channel on the current fluctuations, with their electrostatic impact on the current percolation paths within the device. In this section, a careful comparison between DD and MC simulations of MOSFETs with different discrete dopant configurations is presented. The relative importance of the electrostatic and scattering effects when studying random dopant-induced intrinsic parameter variations is investigated and highlighted.

7.6.1. Simulation Methodology

The variation in current associated with a random distribution of dopant atoms within the active region of a 30 × 30-nm MOSFET was studied in comparison with simulation of a uniformly doped device of the same design. An ensemble of 50 microscopically different but macroscopically identical devices are randomly generated and simulated using the DD method. These devices have continuously doped source and drain regions, whereas the channel region is described atomistically. In the DD simulations, field-dependent and concentration-dependent mobility models are employed, although in the atomistic DD simulations the mobility values used are taken from mobility values stored at each mesh node as a result of simulation of the continuous device. MC simulation is then performed in the frozen-field approximation, where the source-to-drain bias is suitably low to ensure that nonequilibrium transport effects are minimized and the two simulation methods are equivalent. The MC and DD simulations of the atomistic devices, in comparison with their respective results for the continuous device, allow the variation in current to be determined for each atomistic device and comparison between DD and MC results to be made.

7.6.2. Percentage Change in Current

Figure 7.22 shows the percentage change in current calculated from both DD and MC simulations. From the DD results it can be seen that the atomistic description

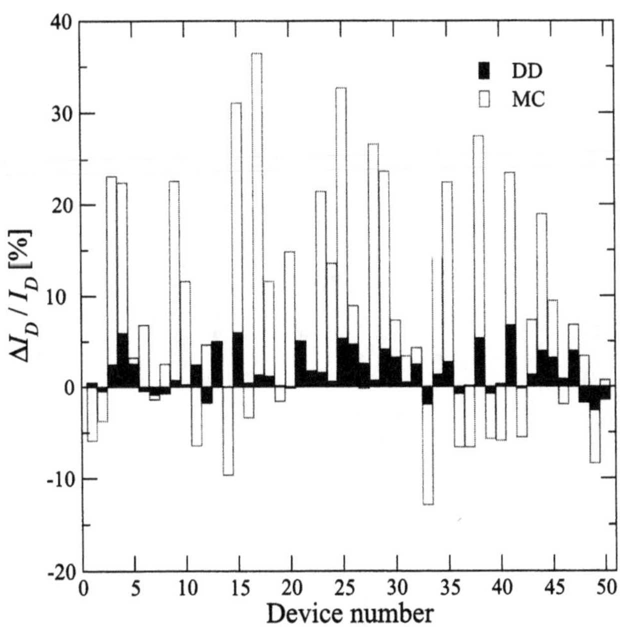

FIGURE 7.22. Histogram of the percentage change in current for each of the 50 randomly configured devices as calculated from DD and MC simulations. MC largely reproduces the trend of DD simulation, but with a much greater range in values.

increases the current on average or, alternatively, reduces the threshold voltage, due to the presence of current percolation paths. This is a purely electrostatic effect. The trend of current increase is also reproduced in the MC simulations. A minority of devices show reduced current in the DD simulation associated with a larger than average number of donors in the channel, raising the electrostatic barrier. The MC simulations also show current reduction in a minority of the devices with reduced current. In most cases, the sign and magnitude of the current variation in the MC simulations closely follow the DD results, although there is not a 100% correlation. In device 11 for example, the DD simulation predicts an increase in current, whereas MC predicts a reduction.

A plot of the current variation obtained using DD simulations against the current variation obtained using MC simulations is presented in Fig. 7.23. The corresponding correlation coefficient is 0.59, showing a certain degree of correlation. The cause of the randomization that reduces the correlation lies in the random positioning of the acceptor atoms within the channel. This affects the MC simulations through the resulting position-dependent scattering and more realistic position-dependent mobility. This is best highlighted with reference to devices 3, 5, and 11 from the ensemble of 50 devices. For these devices, the percentage variation in current calculated by both simulation methods is tabulated in Table 7.3. Devices 3, 5, and 11 show a similar percentage increase in current as predicted by DD. In contrast, MC simulation shows a percentage increase that is much higher

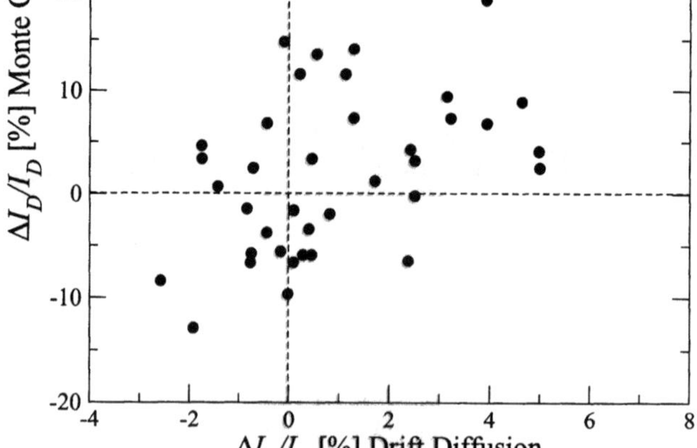

FIGURE 7.23. Scatterplot of the percentage current variation calculated via DD against percentage current variation calculated via MC. The correlation coefficient for this distribution is calculated as 0.59.

TABLE 7.3. Percentage variation in current due to atomistic channel for three microscopically different devices from the ensemble of 50.

Device no.	3	5	11
$\Delta I/I$ [%] (DD)	2.43	2.51	2.38
$\Delta I/I$ [%] (MC)	23.21	3.21	−6.43

Results calculated from Drift Diffusion (DD) and Monte Carlo (MC) simulation are shown for comparison. MC results show significant variation compared with the DD results which show very similar increased current.

for device 3 compared to device 5, whereas device 11 shows a substantial reduction in current. These three devices are analysed in detail next.

7.6.2.1. Device 3

In Fig. 7.24, the electron concentration calculated by DD and MC simulations at the Si/SiO_2 interface is shown throughout the channel for device 3. Within the DD simulations, each acceptor sits at the center of a circular charge exclusion region. The number of acceptors, and hence the resulting electrostatic impact, is similar in each of the three devices, leading to similar levels of current increase. However, the comparison of the electron concentration from DD and the equivalent MC simulation reveals some differences. The MC simulations result in a larger reduction in the electron concentration behind the close grouping of three acceptors near the source end. This is associated with the scattering of electrons injected into the channel by these acceptors, blocking the propagation into the opposite side. It can be seen that the corresponding charge exclusion regions are now, in general, not circular. Scattering from ionized impurities has a larger effect nearer the source due to the lower kinetic energy of the electrons. At the drain, where carriers have higher energy, impurity scattering is less effective and, with reference again to the MC results in Fig. 7.24, the charge exclusion regions more closely resemble the

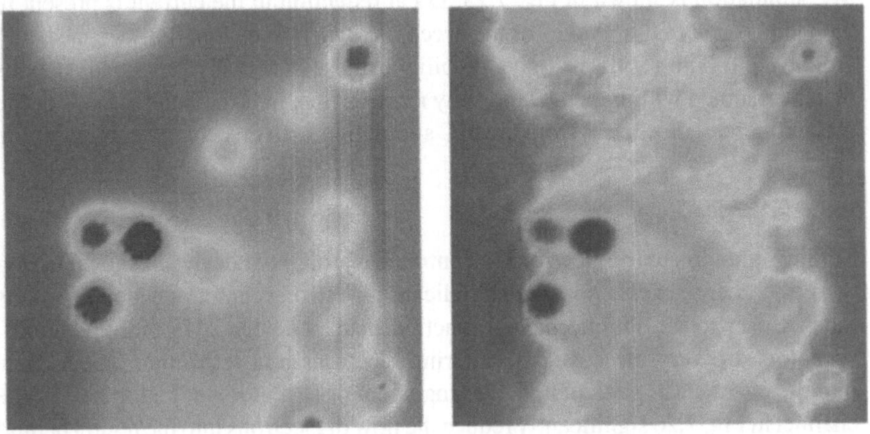

FIGURE 7.24. Electron concentration throughout the channel at the Si/SiO_2 interface resulting from DD (left) and MC (right) simulation for device 3.

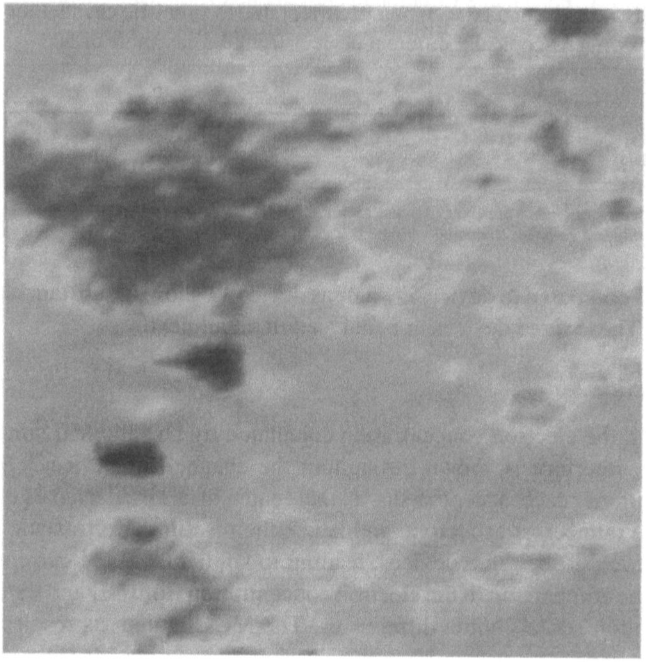

FIGURE 7.25. Average current density within the channel at the Si/SiO_2 interface for device 3. The reduction in carrier velocity and density around the acceptors is reflected in the lower current density, which extends beyond the scattering center, whereas current flow is unimpeded in the regions with few acceptors.

DD result. Within this device, acceptors are grouped together, presenting areas of the channel devoid of scattering centers and allowing a largely unimpeded flow of current throughout the channel. The corresponding average current density from MC simulation is shown in Fig. 7.25. Local reduction in the current is present in the vicinity and in the "wake" of the acceptors, whereas charge flows consistently around them. This results in higher mobility in these effectively "undoped" regions than is assumed within the DD mobility model. This, in turn, is the reason for the greater increase in current from the MC simulation with this charge configuration.

7.6.2.2. Device 5

Turning attention to the electron concentration throughout device 5, shown in Fig. 7.26, the charge exclusion regions indicate that two acceptors occupy positions much closer to the source/channel junction than in device 3. In these positions, acceptors are very effective at scattering electrons and restricting charge flow into the channel. In particular, the arrangement of acceptors in the top half of the channel in Fig. 7.26 significantly reduce the flow of electrons into the region behind them, creating an extended charge exclusion region. Analyzing the current density for this device in Fig. 7.27, it is clear that current flow is predominantly in the

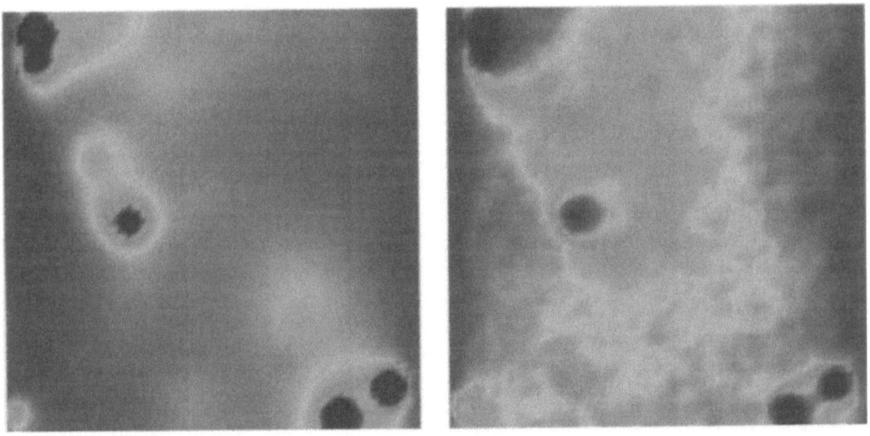

FIGURE 7.26. Electron concentration throughout the channel at the Si/SiO_2 interface resulting from DD (left) and MC (right) simulation for device 5 from the ensemble of 50.

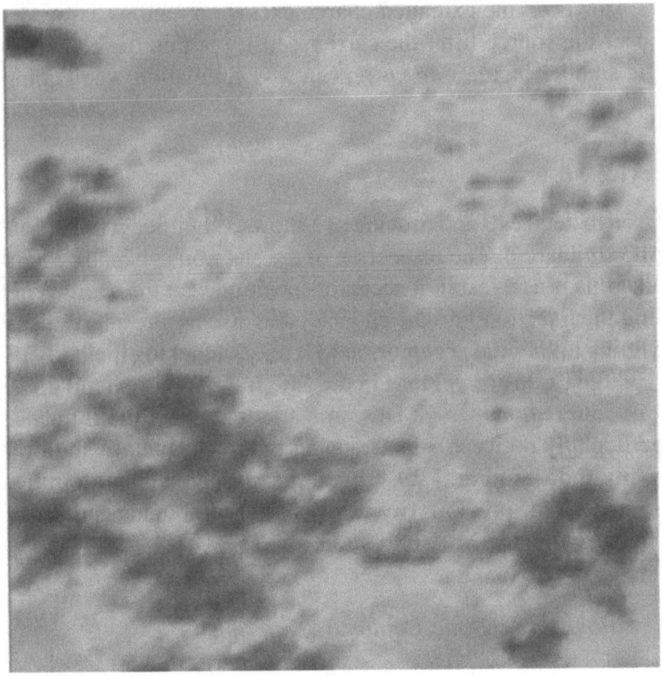

FIGURE 7.27. Average current density within the channel at the Si/SiO_2 interface for device 5. Similar to Fig. 7.25, current flows predominantly in regions absent of acceptors.

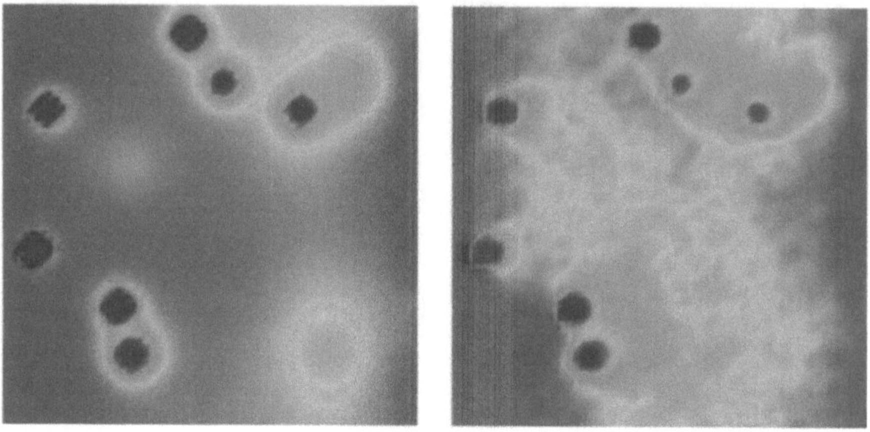

FIGURE 7.28. Electron concentration throughout the channel at the Si/SiO_2 interface resulting from DD (left) and MC (right) simulation for device 11 from the ensemble of 50.

area around this grouping where there are no impeding acceptors. This increased mobility increases the current beyond that estimated by DD in the same manner as device 3. However, the extended charge exclusion regions associated with the acceptors near the source limit the carrier density within the channel; this is the cause of the smaller increase in current in MC simulation when compared with device 3.

7.6.2.3. Device 11

In comparison to devices 3 and 5, device 11 shows a significant reduction in current from the MC simulation. The inspection of the electron concentration in Fig. 7.28 indicates that this is as a result of acceptors positioned at the source/channel junction, limiting the injection into the channel, and also as a result of the positioning of the remaining influential acceptors across the channel to efficiently impede current flow. Charge exclusion regions in the MC simulations extend greatly beyond the acceptors into regions where DD simulations shows little influence from the dopants, particularly noticeable in the center of the channel. In addition, from the current density shown in Fig. 7.29, current flow has no path with little or no scattering present, as in the previous two devices. The combined result is the observed reduction in current in the MC simulation.

7.6.3. Conclusions on Ab Initio Coulomb Scattering

The inclusion of position-dependent ionized impurity scattering, associated with the unique arrangement of dopant atoms, in MC simulation of an ensemble of 30-nm n-MOSFET devices shows increased current variation compared to the simulation of the same ensemble performed using the DD approximation. The DD results show an average increase in current of 1.5% with a standard

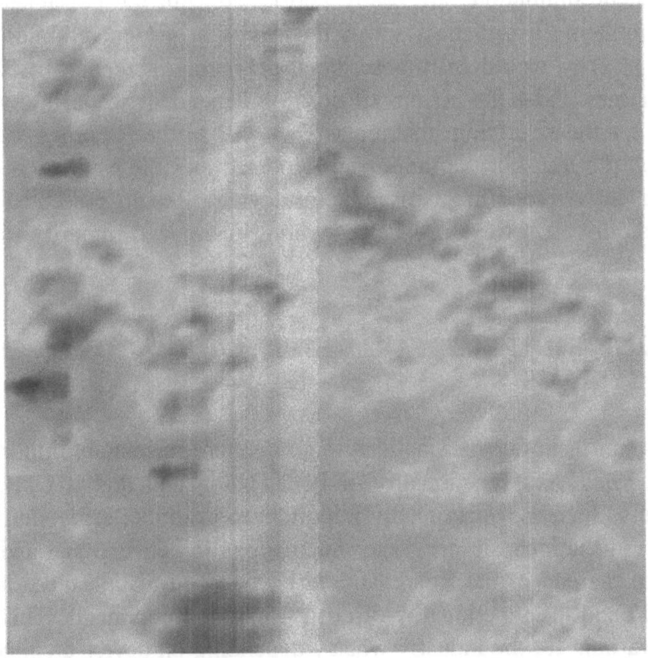

FIGURE 7.29. Average current density within the channel at the Si/SiO interface for device 11. The well-spaced acceptors across the channel prevent the kind of resistance-free current paths that are seen in Figs. 7.25 and 7.27 and minimize the current flow throughout the channel.

deviation of 2.3%. Such simulations only consider the electrostatic landscape created by the charge configurations. By comparison, MC results show an average current increase of 7.8% with a much larger standard deviation of 11.6%. In general, the trend in the DD results are reproduced within the MC results, but the incorporation and the sensitivity of the position-dependent impurity scattering to the unique arrangements of dopants within the MC simulations reduces the direct correlation. The comparison of the current variation obtained by both methods gives a correlation coefficient of 0.59. Although the electrostatic impact of a particular configuration of donors may be similar, differences in the position-dependent scattering between such similar configurations may have large effects. By affecting both the mobility and the electron concentration throughout the channel, the ionized impurity scattering plays an important role in determining the magnitude of the random dopant-induced device parameter variations.

It is noted, however, that the MC simulations reported here are performed in the frozen-field approximation to allow efficient simulation of an ensemble of devices. It is reasonable to expect that a self-consistent MC simulation with *ab initio* scattering should yield differences in the steady-state electron concentration, since the electron concentration is otherwise representative of the initial DD solution, which is insensitive to the position-dependent scattering. Such self-consistent simulations are easily performed with the developed code, although simulating a statistical set of devices is at present prohibitively time-consuming. They would fully describe the dynamic screening of dopants by mobile charges under the actions of position-dependent scattering, which will, in turn, alter the scattering potentials. This will further improve the accuracy with which the transport variations are modeled and the estimation of the statistical current variation in an ensemble of devices. Results presented here highlight the requirement for a proper accounting of ionized impurity scattering in nanoscale devices.

7.7. IMPACT OF INTRINSIC PARAMETER FLUCTUATION ON CIRCUITS AND SYSTEMS

In circuits, intrinsic fluctuations lead to component mismatch, limiting the performance, yield, and functionality of both analog and digital CMOS circuits and systems [36, 37]. It is crucially important to examine, using simulation, the magnitude of the intrinsic parameter fluctuations in conventional and novel devices up to the end of the Roadmap and beyond and to link these fluctuations to the yield and the functionality of critical circuit components. This will issue an early warning for the industry when—depending on device architecture—the level of intrinsic parameter fluctuation becomes unacceptable for particular applications. Simplified analyses, based mainly on analytical models, have been used to evaluate dopant fluctuation effects in circuits [38]. However, they suffer from a lack of accuracy, being unable to take into account 3-D and quantum mechanical effects and other sources of fluctuations. Therefore, they could be misleading.

On the other hand, numerical simulation is a powerful tool for describing fundamental physical mechanisms inside semiconductor devices and is able to predict device characteristics several generations in advance of production. However, such studies are largely limited to the device level. It is crucially important to transfer all the fluctuation information obtained by numerical simulation into compact models, so the impact of intrinsic parameter fluctuations can be investigated at the circuit level based on an accuracy similar to that obtained by 3-D "atomistic" simulations at a device level.

The physical analysis of the atomistic simulation results shows that random dopant fluctuations have the following consequences: First, in weak inversion, due to lack of inversion layer screening, they inflict dramatically drain current and subthreshold slope variation; second, the variation in number and location of dopant atoms in the channel region cause well-documented threshold voltage fluctuation; third, in the saturation region, such "atomistic" variation will cause bulk charge effect fluctuation, and particularly due to lack of inversion layer in the pinch-off area, the electrical field fluctuation at this area will become more pronounced; finally, it will also introduce DIBL effect variation through effective doping profile fluctuation.

7.7.1. Statistical Compact Modeling

It is desirable to integrate random doping effects into an industry-standard compact model, BSIM3v3 [39] in order to take advantage of its excellent acceptance in today's Electronic Design Automation (EDA) tools. Although BSIM3v3 does not include atomistic fluctuation directly, it does contain a number of empirical parameters aimed at modeling device performance variation caused by different foundry process conditions. In principle, it is possible to use the model parameters dedicated to such process differences to reproduce the above four dopant fluctuation effects. We have found that BSIM3v3 is flexible enough to comprehend atomistic fluctuation, and the critical step is to map atomistic fluctuations into a particular parameter group.

Seven critical parameters were chosen to map and describe random doping effects in BSIM3v3. The parameter V_{off} was initially introduced to handle the difference between threshold voltage in the subthreshold region and strong inversion region. It has a large effect on subthreshold drain current. The N_{factor} was introduced to better describe the subthreshold slope in case of any uncertainty in the calculation of the depletion capacitance. When combined, these two parameters can reflect random doping effects in the subthreshold region. N_{ch}, which is the channel-doping concentration, can be chosen to reflect threshold voltage variation caused by atomistic fluctuations. Initially, the parameter A_0 was introduced to account for channel-length dependence of the bulk charge effects at different process conditions. In this work, A_0 has been chosen to map atomistic fluctuations in bulk charge effects. A_1 and A_2 were originally introduced to control saturation voltage behavior. In this work, they are used to map random doping effects caused by electrical field fluctuation at the pinch-off region. D_{sub} was introduced to handle channel-length dependence of DIBL effects at different process conditions and is used here similarly to A_0 to describe atomistic-fluctuation-induced DIBL variation.

Based on the above analysis, a two-stage statistical compact model parameter extraction strategy is developed. During the first stage, a combination of local optimization and group extraction strategy is employed to extract the complete set of BSIM3v3 parameters for a typical device with continuous doping profiles. This process involves three main steps, each extracts a different subset of parameters: long channel, short channel, and high field. Gate and drain characteristics of different channel-length devices are needed as extraction targets at this stage.

In second-stage extraction, except for the seven critical parameters, all other parameter value remains fixed. A large ensemble of devices, each with a different discrete doping pattern, is involved at this stage. If device characteristics were measured or simulated for each member of the ensemble over its complete operating range, the datasets would become unwieldy (and prohibitively expensive, computationally, if produced by the 3-D "atomistic" simulator). We find that most fluctuation information is reflected in the gate characteristics at low and high drain bias. Therefore, only gate characteristics are used for second-stage extraction. Two extraction substeps are involved at this stage: N_{ch}, V_{off}, and N_{factor} considered as the first-order random doping effect parameters are extracted from the gate characteristics at low drain bias at the first substep. A_0, A_1, A_2 and, D_{sub} considered as second-order parameters targeting high-field fluctuation effects are extracted from gate characteristics at high drain bias at the final substep.

7.7.2. Extraction Results

The above two-stage parameter extraction strategy has been successfully applied on a well-scaled 35-nm gate-length device by using the commercial parameter extraction tool Aurora [40]. Figure 7.30 shows results from the first-stage parameter extraction.

An ensemble of 200 microscopically different 35-nm devices is created using the calibrated 3-D "atomistic" DD device simulator. The simulated characteristics of this ensemble are used as data sources for the second-stage parameter extraction. Figure 7.31 shows the extraction results for three typical 35-nm atomistic devices in the ensemble. The overall I_{off} and I_{on} fluctuation is shown in Fig. 7.32. The collection of BSIM compact models replicates, with a high degree of accuracy, the characteristics of the physically simulated devices. The averages and the standard deviations are practically the same.

Figure 7.33 shows scatterplots between two mapped parameters. There are two sets of parameters that have strong correlation: $A_1 - A_2$ and $A_0 - D_{sub}$. This is probably because A_1 and A_2 are combined in BSIM3v3 to control saturation behavior, and A_0 and D_{sub} are both related to effective channel length in the model equations. The mapped parameters are not completely statistically independent. This means that the use of traditional MC SPICE simulation can create a device that may not exist in reality. In order to avoid this, a statistical model card library is

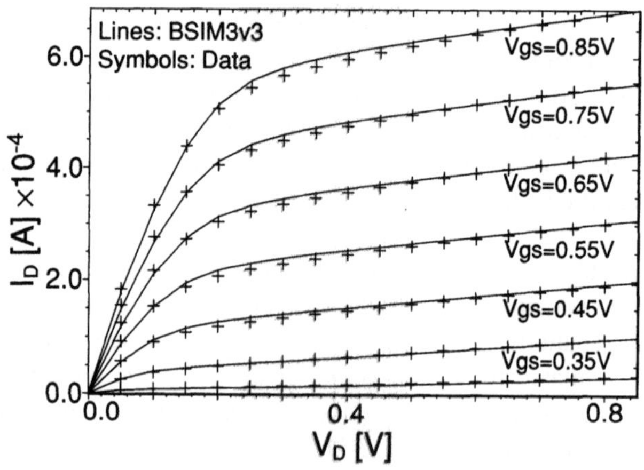

FIGURE 7.30. Compact model calibrated result for a typical 35-nm gate-length device; the symbols denote the experiment results.

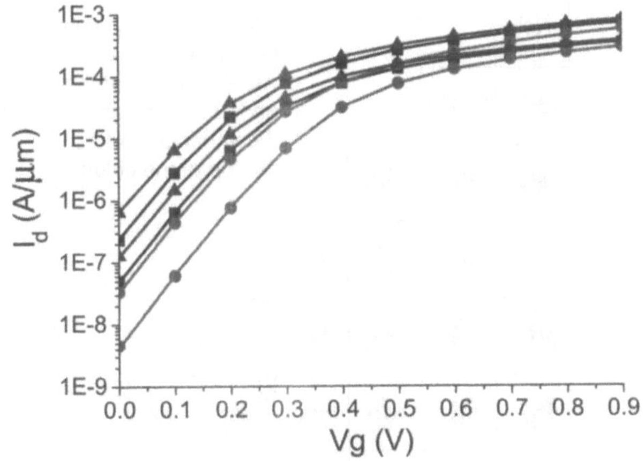

FIGURE 7.31. Statistical parameter extraction result for three typical 35-nm gate-length devices; the symbols represent "atomistic" simulation results, and the high and low drain bias are 0.85 V and 0.1 V, respectively.

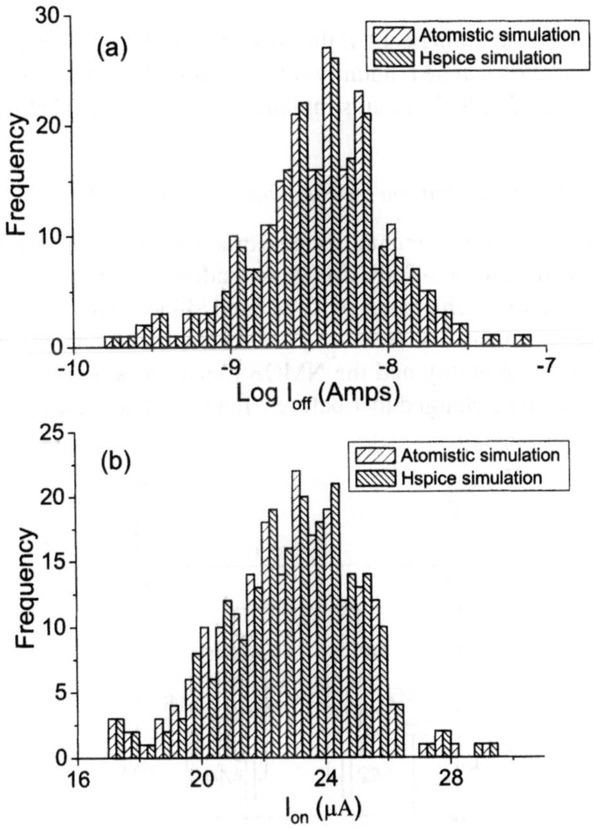

FIGURE 7.32. Distribution of (a) I_{off} and (b) I_{on} over an ensemble of 200 atomistic devices. The I_{off} bias condition is $V_g = 0$ and $V_{\text{ds}} = 0.85$ V; the I_{on} bias condition is $V_g = 0.85$ V, and $V_{\text{ds}} = 0.85V$.

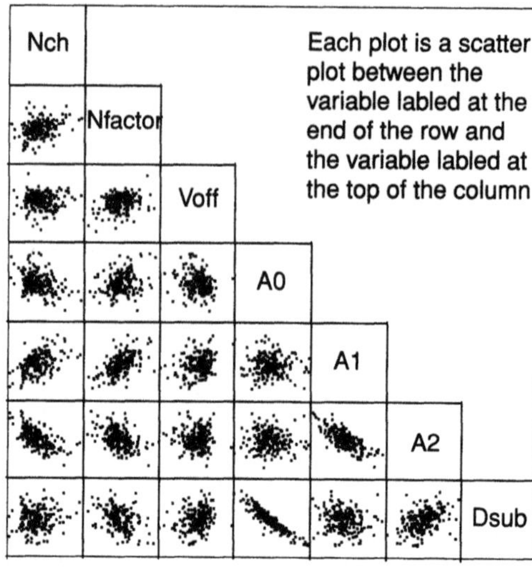

FIGURE 7.33. Scatterplots between two mapped parameters.

built based on the extraction result of the ensemble of 200 devices, and the devices in a circuit simulation can be randomly selected from this library. This guarantees that the devices used in MC circuit simulations represent "real" atomistic effects.

7.7.3. Impact of Intrinsic Parameter Fluctuation on 6-T SRAM

Above 35 nm, physical gate-length device extraction results are employed in statistical circuit analysis of a 6-T SRAM (static random access memory) cell. Figure 7.34 shows the circuit's schematics, with M2 and M4 as driver transistors, and M5 and M6 as access transistors. Because PMOS load transistors M1 and M3 have generally low driving ability and the NMOS transistor is not good at passing 1, the bit line needs to be charged to 1 before a read operation. Design optimization

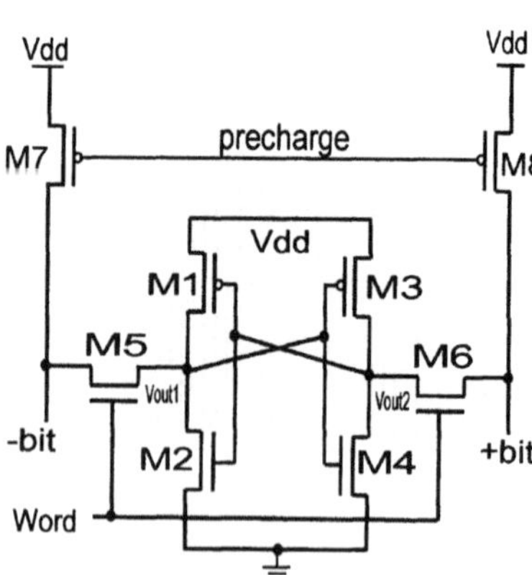

FIGURE 7.34. Circuit schematic of a CMOS SRAM.

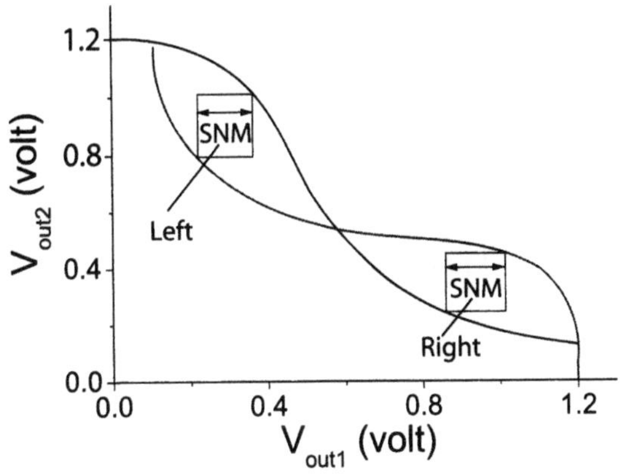

FIGURE 7.35. The static transfer characteristics and SNM of a normal SRAM case.

of an SRAM circuit concentrates on pulling the bit line from 1 to 0, and the cell is most vulnerable to noise at the initiation of this operation.

In SRAM cell design, the width/length ratios of the load transistors and access transistors are often as close to 1.0 as possible. The ratio of the driver transistor's W/L to the access transistor's W/L is called the *cell ratio*; it determines the cell stability as well as cell size [41]. Initially, a cell ratio is set to 1. In order to have enough samples for statistical analysis, 200 circuit simulations are carried out by using HSPICE and it takes about 10 min of CPU time on a SunBlade 1500 workstation.

Graphically, the static noise margin (SNM) is the side length of the maximum possible square in the static transfer characteristics (Fig. 7.35). Because of device mismatch, the left side of the SNM is not necessarily equal to the right side, therefore the smaller value will be chosen as the SNM of the SRAM. Due to random doping effects, large fluctuations occur in static transfer characteristics (Fig. 7.36a). An extreme example is shown in Fig. 7.36b, and in this case, as no square exists at the left side of the static transfer characteristics, the SNM is reduced to zero and the SRAM cannot operate correctly even under ideal conditions. Figure 7.36c further shows the read behavior of this extreme case. Before the read, 0 is stored on node out1 and 1 is stored on node out2. Although we ignore circuit noise, the state of the cell begins to flip on initiation of the read operation. After 100 ps, the state of node out1 is changed from 0 to 1. For a cell ratio of 1, due to random doping effects nearly 10% of cells will malfunction even under noise-free condition.

Increasing the cell ratio has two benefits for improved SNM behavior. First, a larger cell ratio directly improves cell stability—reflected in the mean value of SNM, μ. Second, a larger W/L ratio will reduce the magnitude of the characteristic fluctuations caused by random doping effects, which is partly reflected in the normal standard deviation σ of SNM, σ/μ. Figure 7.37a clearly shows these benefits from a larger cell ratio.

Although each individual transistor in the SRAM cell has a statistically identical characteristic fluctuation distribution, their contributions to the total SNM

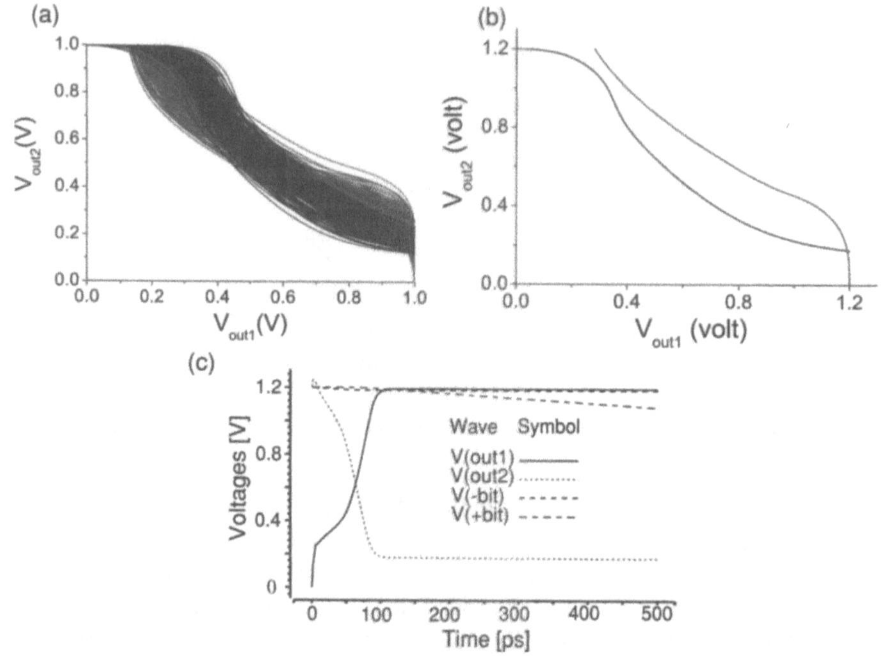

FIGURE 7.36. (a) Static transfer characteristics over an ensemble of 200 SRAM. (b) Static transfer characteristics of a extreme case. (c) Read behavior of an extreme case; the cell ratio is 1.

variation are different. For a cell ratio of 3, because of the larger W/L value, the driver transistors have a smaller absolute magnitude of characteristic fluctuations caused by random doping. However, Fig. 7.37b shows that SNM is most sensitive to driver transistor variation, which will contribute about 70% of total SNM fluctuation, and is less sensitive to access and load transistor variations.

The normalized standard deviation of SNM caused by power supply instability and ignoring and fluctuations due to intrinsic parameter variations is shown in Fig. 7.37c, in which the cell ratio is 3. For these devices, SNM fluctuations caused by random doping effects are of the same level as the fluctuations caused by ±20% supply instability.

As a guideline, $\mu - 6\sigma$ is required to exceed approximately 4% of the supply voltage to achieve 90% yield for 1Mbit SRAMs [37]. If we consider only the fluctuations caused by random doping effects (see Table 7.4), the cell ratio should be at least 3, but if other intrinsic fluctuation sources are taken into account, a larger cell area will be required in order to achieve reasonable yield. This implies that SRAM may not gain all the benefits of further bulk CMOS scaling, from the SNM point of view.

Depending on which type of sense amplifier is employed, there are two modes of read operation for an SRAM cell: voltage or current mode. Although the issues associated with read-time fluctuations are not as critical as SNMs for SRAM

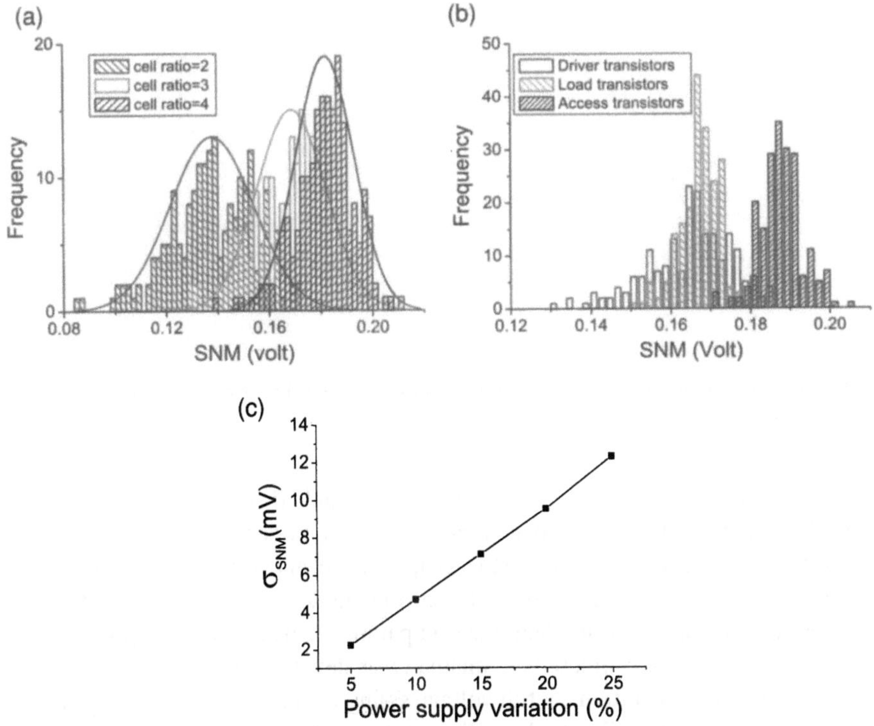

FIGURE 7.37. (a) Distribution of SNM; (b) SNM distributions due to random doping effects in different types of transistors; (c) normalized standard deviation of SNM due to power supply variation.

operation, they will determine the memory access speed and thus affect system performance. In the statistical circuit simulations, a 0.1-pF bit capacitance is assumed. In order to have a sufficient noise margin, we also assume that the threshold for the sense amplifier is 0.6 V for the voltage mode and that the read time is roughly the time taken for the bit-line voltage to drop to the sense threshold. In the current mode, the voltage swing is not critical for the read operation; the peak current is used as a probe to detect read-time fluctuations. Figure 7.38 shows the impact of random dopant variation on both read modes.

For larger cell ratios, cell pull-down resistance becomes lower, which will help to improve general read access performance. Compared to the SNM case, the fluctuation behavior of read operations are less sensitive to the cell ratio. Roughly,

TABLE 7.4. Static noise margins of different cell ratio SRAMs.

Cell ratio	Mean of SNM, μ (mV)	SD of SNM, σ (mV)	$\mu-6\sigma$ (mV)	σ/μ
2	137.69	16.28	40	11.8%
3	168.57	13.39	88.2	7.94%
4	181.59	11.32	113.6	6.23%

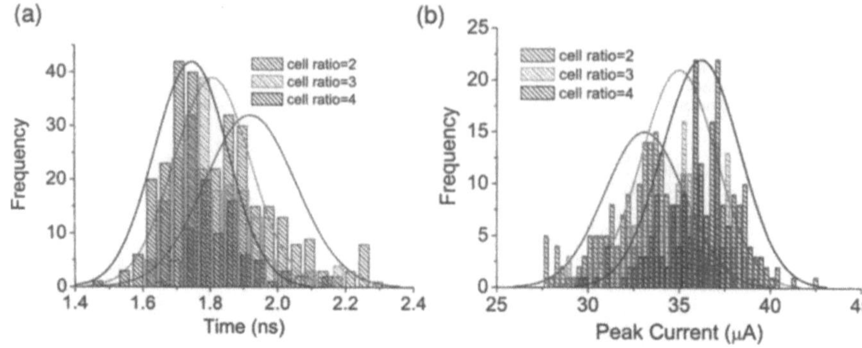

FIGURE 7.38. (a) Read-time distribution for the voltage mode; (b) peak current distribution for the current mode.

random dopant effects will cause a 40% performance difference between fastest and slowest memory accesses. In general, the current mode is superior in all aspects of voltage swing and the sensitivity to bit capacitance, and as the impacts of random dopant effects on both modes are similar, current mode would still be a good choice for the read operation even when intrinsic parameter fluctuations begin to play a greater role in device characteristic mismatch as devices shrink.

During write operations, a full voltage swing on a bit line is often required to override the previous cell data. In reality, such signals are produced by peripheral circuitry. In order to clearly illustrate the impact of random dopant effects on the cell itself, the peripheral circuit is excluded in the circuit simulation and an ideal complementary write signal is directly applied on the bit lines.

The switch point voltage is defined as the bit line voltage which will cause cell data to begin to change under a write operation. It is another important parameter in cell design, which, together with the SNM, will determine cell stability. Typical write behavior is shown in Fig. 7.39a, in which quasi-static operation is considered in order to clearly show the switch point voltage.

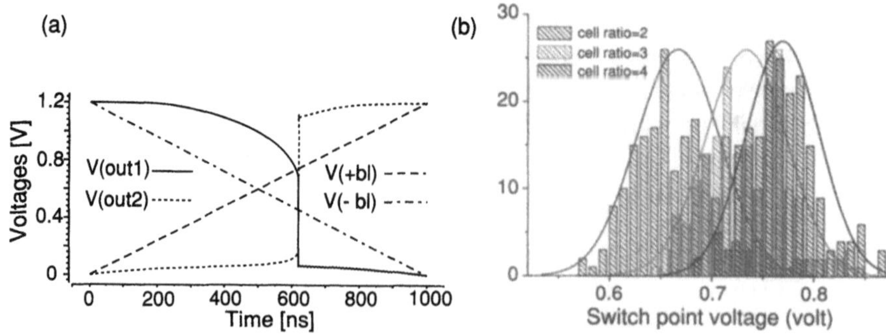

FIGURE 7.39. (a) HSPICE simulation result for a typical writing operation; the cell ratio is 3. (b) Distribution of switch point voltage (+ bl).

A larger cell ratio will give a higher switch point voltage (shown in Fig. 7.39b), which results in better noise immunity, in concert with the earlier results for SNM. When the cell ratio is increased from 2 to 4, the magnitude of the relative differences between the highest and lowest switch point voltages is also improved. Such quantitative results allow the circuit designer to trade off these benefits against the requirements of circuit speed, area and power dissipation for a given system application.

7.7.4. Conclusions on Fluctuation in Circuits and Systems

An effective methodology is presented to introduce "atomistic" fluctuations into an industry-standard BSIM compact model, integrating naturally with present EDA tools. The proposed methodology will allow analysis of the effects of a range of "atomistic" phenomena—both singly and in combination—on circuits and systems. As an example, the methodology has been successfully applied in 6-T SRAM; the result shows that the read/write variation caused by random dopant fluctuations will degrade overall SRAM speed, but SNM is a fundamental limitation for further bulk SRAM scaling.

7.8. CONCLUSIONS

The intrinsic parameter fluctuations introduced by atomic-scale effects in present- and future-generation CMOS devices and the corresponding device characteristics mismatch are already starting to affect the functionality and reliability not only of analog but also digital circuits and systems. The designer should be acutely aware of the forthcoming problems. Shift to fault tolerant design, including redundancy, self-organizing, and reconfigurable architectures and intensive on-chip testing will be required to combat the increasing levels of intrinsic parameter fluctuations that will accompany the scaling in the next two decades.

REFERENCES

1. A. Asenov, A.R. Brown, J.H. Davies and S. Saini, *IEEE Trans. Computer-Aided Design of Integrated Circuits and Systems* **18**, 1558 (1999).
2. D.J. Frank, Y. Taur, M. Ieong and H.-S.P. Wong, *Symposium on VLSI Circuits Digest of Technical Papers*, p. 171, (1999).
3. S. Kaya, A.R. Brown, A. Asenov, D. Magot and T. Linton, *Proc. Simulation of Semiconductor Processes and Devices*, (Athens, Greece, 2001), p. 78.
4. A. Asenov, Savas Kaya and Andrew R. Brown, *IEEE Transactions on Electron Devices* **50**, 1254 (2003).
5. S.M. Goodnick, D.K. Ferry, C.W. Wilmsen, Z. Liliental, D. Fathy and O.L. Krivack, *Phys. Rev. B.* **32**, 8171 (1985).
6. R.M. Feenstra, M.A. Lutz, F. Stern, K. Ismail, P. Mooney, F.K. LeGoues, C. Stanis, J.O. Chu and B.S. Meyerson, *J. Vac. Sci. Technol. B. Microelectron. Process. Phenom.* **13**, 1608 (1995).

7. M. Niva, T. Kouzaki, K. Okada, M. Udagawa and R. Sinclair, *Jpn. J. Appl. Phys.* **33**, 388 (1994).
8. D. Vasileska, W.J. Gross and D.K. Ferry, *Extended Abstracts of the 6th International Workshop on Computational Electronics*, IEEE Cat. No. 98EX116, (Osaka, Japan, 1998), p. 259.
9. A. Asenov, *IEEE Trans. Electron Dev.* **45**, 2505 (1998).
10. H.K. Gummel, *IEEE Trans. Electron Dev.* **11**, 455 (1964).
11. M.G. Ancona, *Phys. Rev. B* **42**, 1222 (1990).
12. S. Jin, Y.J. Park and H.S. Min, *Journal of Semiconductor Technology and Science* **4**, 32 (2004).
13. J.R. Watling, A.R. Brown, A. Asenov and D.K. Ferry, *Proc. Simulation of Semiconductor Processes and Devices*, (Athens, Greece, 2001), p. 82.
14. Z. Yu, R.W. Dutton, D.W. Yergeau and M.G. Ancona, *Proc. Simulation of Semiconductor Processes and Devices*, (Athens, Greece, 2001), p. 1.
15. J.R. Watling, A.R. Brown, A. Asenov, A. Svizhenko and M.P. Anantram, *Proc. Simulation of Semiconductor Processes and Devices*, (Kobe, Japan, 2002), p. 267.
16. G.D. Wilk, R.M. Wallace, J.M. Anthony, *J. Appl. Phys.* **89**, 5243 (2001).
17. S. Jin, Y.J. Park and H.S. Min, *Proc. Simulation of Semiconductor Processes and Devices*, (Cambridge, MA, USA, 2003), p. 263.
18. A.T. Fromhold, Jr. Quantum Mechanics for Applied Physics and Engineering, (Dover Publications, Inc., New York, 1981).
19. A. Svizhenko, M.P. Anantram, T.R. Govindan, B. Biegel and R. Venugopal, *J. Appl. Phys.* **91**, 2343 (2002).
20. T. Ezaki, T. Ikezawa, A. Notsu, K. Tanaka and M. Hane, *Proc. Simulation of Semiconductor Processes and Devices*, (Kobe, Japan, 2002), p. 91.
21. G. Roy, A.R. Brown, A. Asenov and S. Roy, *J. Comp. Elec.* **2**, 323 (2003).
22. G. Roy, A.R. Brown, A. Asenov and S. Roy, *Superlattices and Microstructures* **34**, 327 (2003).
23. A. Asenov, M. Jaraiz, S. Roy, G. Roy, F. Adamu-Lema, A.R. Brown, V. Moroz and R. Gafiteanu, *Proc. Simulation of Semiconductor Processes and Devices*, (Kobe, Japan, 2002), pp. 87.
24. N. Sano, K. Matsuzawa, M. Mukai and N Nakayama, *International Electron Device Meeting (IEDM) Digest Tech. Papers*, p. 275 (2000).
25. Z. Qin and S.T. Dunham, *Proc. Mater. Res. Soc. Symp.* **717**, C3.8 (2002).
26. Z. Qin and S.T. Dunham, *Phys. Rev B* **68**, 245201 (2003).
27. R.W. Hockney and J.W. Eastwood, Computer Simulation using Particles (McGraw-Hill, New York, 1981).
28. A. Asenov, R. Balasubramaniam, A.R. Brown and J.H. Davies, *IEEE Trans. Electron Dev.* **50**, 839 (2003).
29. P.A. Stolk, F.P. Widdershoven and D.B. M Klaassen, *IEEE Trans. Elec. Dev.* **45**, 1960 (1998).
30. T. Ezaki, T. Ikezawa and M. Hane, *International Electron Device Meeting (IEDM) Digest Tech. Papers*, p. 311, 2002.
31. C. Alexander, J.R. Watling, A.R. Brown and A. Asenov, *Superlattices and Microstructures* **34**, 319 (2003).
32. W.J. Gross, D. Vasileska and D.K. Ferry, *IEEE Trans. Elec. Dev. Lett.* **20**, 463 (1999).
33. C.J. Wordelman and U. Ravaioli, *IEEE Trans. Elec. Dev.* **47**, 410 (2000).
34. S. Barraud, P. Dollfus, S. Galdin and P. Hesto, *Solid State Electronics* **46**, 1061 (2002).
35. S.M. Ramey and D.K. Ferry, *IEEE Trans. Nanotechnology* **2**, 193 (2003).
36. H.P. Tuinhout, *Proc. 32th European Solid-State Device Research Conference* (Florence, Italy, 2002) p. 95.
37. P.A. Stolk, H.P. Tuinhout, R. Duffy, E. Augendre, L.P. Bellefroid, M.J.B. Bolt, J. Croon, C.J.J. Dachs, F.R.J. Huisman, A.J. Moonen, Y.V. Ponomarev, R.F.M. Roes, M. Da Rold, E. Sevinck, K.N. Sreerambhatla, R. Surdeanu, R.M.D.A. Velghe, M. Vertregt, M.N.

Webster, N.K.J. van Winkelhoff, A.G.A. Zegers-Van Duijnhoven, *International Electron Device Meeting (IEDM) Digest Tech. Papers*, p. 215 (2001).

38. K. Takeuchi, R. Koh, and T. Mogami, *IEEE Trans. Elec. Dev.* **48**, 1995 (2001).
39. BSIM software, University of California, Berkeley, CA, USA, http://www-device.eecs.berkeley.edu/~bsim3/.
40. Aurora User's Manual, Synopsys Inc., Mountain View, CA, USA, 2002.
41. E. Seevinck, F.J. List, and J. Lohstroh, *IEEE J. Solid-State Circuits* **22**, 748 (1987).

8

Lattice Polarons and Switching in Molecular Nanowires and Quantum Dots

A.S. Alexandrov*

8.1. INTRODUCTION

Conducting electrons in inorganic and organic matter interact with vibrating ions. If phonon frequencies are sufficiently low, the local deformation of ions, caused by the electron itself, creates a potential well, which traps the electron even in a perfect crystal lattice. This *self-trapping* phenomenon was predicted by Landau [1]. It was studied in greater detail by Pekar [2], Fröhlich [3], Feynman [4], Rashba [5], Devreese [6], and other authors in the effective mass approximation for the electron placed in a continuous polarizible medium, which leads to a so-called *large* or *continuous* polaron. Large-polaron wavefunctions and corresponding lattice distortions spread over many lattice constants (see Fig. 8.1). The trapping is never complete in the perfect lattice. Due to finite phonon frequencies, ion polarizations follow polaron motion if the motion is sufficiently slow. Hence, large polarons with a low kinetic energy propagate through the lattice as free electrons but with an enhanced effective mass.

When the electron–phonon interaction (e-ph) energy E_p is larger than the electron energy bandwidth, all electrons in the Bloch bands of the crystal are "dressed" by phonons. In this strong-coupling regime, $\lambda = E_p/D > 1$, the finite bandwidth $2D$ becomes important, so the continuous approximation cannot be applied. It cannot be applied in quantum (molecular) dots at any coupling either, because there are no Bloch bands, only discrete energy levels in molecules. The carriers in those cases are described as *small* or *lattice* polarons. The main features of

* Department of Physics, Loughborough University, Loughborough, United Kingdom a.s.alexandrov@lboro.ac.uk

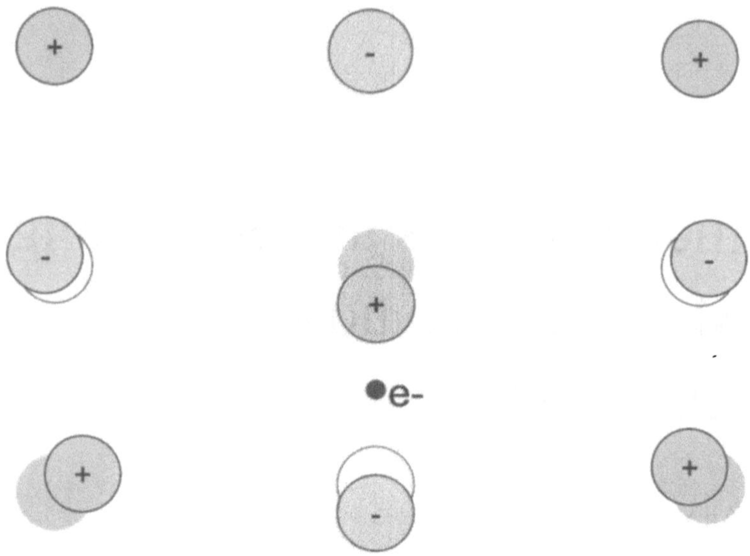

FIGURE 8.1. An electron shifts the equilibrium position of Na^+ and Cl^- ions in the ionic lattice of NaCl and forms a large (or intermediate)-radius polaron.

small polarons were understood by Tjablikov [7], Yamashita and Kurosawa [8], Sewell [9], Holstein [10] and his school, Lang and Firsov [11], Eagles [12], and other researchers and described in several review papers and textbooks [13]. A polaron shift of the atomic level and an exponential reduction of the bandwidth (Section 8.3) at large values of λ are among those features. The shift can be easily understood using a toy model of an electron localized on site **n** and interacting with a single ion vibrating near site **m** in the direction connecting **n** and **m** (see Fig. 8.2). The vibration part of the Hamiltonian in our model is

$$H_{\mathrm{ph}} = -\frac{1}{2M}\frac{\partial^2}{\partial x^2} + \frac{kx^2}{2}, \tag{8.1}$$

where M is the ion mass, $k = M\omega^2$ is the spring constant, and x is the ion displacement (here and throughout we take the Planck constant $\hbar = 1$). The electron potential energy due to the Coulomb interaction with the ion is approximately

$$V = V_0\left(1 - \frac{x}{a}\right), \tag{8.2}$$

where $V_0 = -Ze^2/a$ is the Coulomb energy in a rigid lattice (an analog of the crystal field potential) and a is the average distance between sites. Hence, the Hamiltonian of the model is given by

$$H = E_a\hat{n} + fx\hat{n} - \frac{1}{2M}\frac{\partial^2}{\partial x^2} + \frac{kx^2}{2}, \tag{8.3}$$

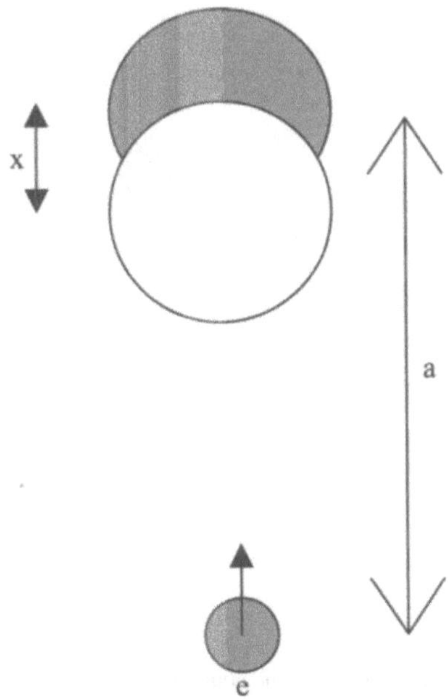

FIGURE 8.2. Localized electron shifts the equilibrium position of an ion and lowers its atomic energy level.

where E_a is the atomic level at site **n** in the rigid lattice, which includes the crystal field, $f = Ze^2/a^2$ is the Coulomb force, and $\hat{n} = c^\dagger c$ is the occupation number operator on site **n** expressed in terms of the electron annihilation c and creation $c^\dagger$ operators. This Hamiltonian is readily diagonalized using a displacement transformation of the vibration coordinate x:

$$x = y - \frac{\hat{n} f}{k}. \tag{8.4}$$

The transformed Hamiltonian has no electron–phonon coupling,

$$\tilde{H} = (E_a - E_p)\hat{n} - \frac{1}{2M}\frac{\partial^2}{\partial y^2} + \frac{ky^2}{2}, \tag{8.5}$$

where we used $\hat{n}^2 = \hat{n}$ because of the Fermi statistics. It describes a small polaron at the atomic level shifted by the polaron level shift $E_p = f^2/2k$ and entirely decoupled from ion vibrations. The ion vibrates near a new equilibrium, shifted by f/k, with the "old" frequency ω. As a result of the local ion deformation, the total energy of the whole system decreases by E_p since a decrease of the electron energy by $-2E_p$ overruns an increase of the deformation energy E_p.

The lattice deformation also strongly affects the interaction between electrons. At large distances, polarons repel each other in ionic crystals, but their Coulomb repulsion is substantially reduced due to the ion polarization. Nevertheless, two

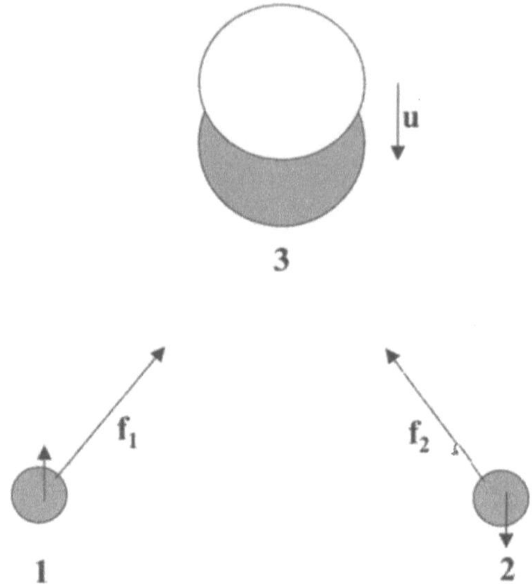

FIGURE 8.3. Two localized electrons shift the equilibrium position of the ion (**3**). As a result, two electrons on neighboring sites **1** and **2** attract each other.

large polarons can be bound into a *large* bipolaron by an exchange interaction even with no additional e-ph interaction but the Fröhlich one [14–18].

When a short-range deformation potential and molecular e-ph interactions (e.g., of the Jahn–Teller type [19]) are taken into account together with the long-range Fröhlich interaction, they can overcome the Coulomb repulsion [20]. The resulting interaction becomes attractive at a short distance of about a lattice constant. Then two small polarons readily form a bound state (i.e., a *small* bipolaron) [21, 22], because their band is narrow. Consideration of particular lattice structures shows that small bipolarons are mobile even when the electron–phonon coupling is strong and the bipolaron binding energy is large [20] (Section 8.4). Here, we encounter a novel electronic state of matter, a charged Bose liquid of electron molecules with double elementary charge 2e, qualitatively different from normal Fermi liquids in ordinary metals and from the Bardeen–Cooper–Schrieffer (BCS) superfluids in conventional superconductors.

The origin of the attractive force between two small polarons can be readily understood from the same model as in Fig.8.2, but with two electrons on neighbor sites **1** and **2** interacting with an ion in between **3** (see Fig. 8.3). For generality, we now assume that the ion is a three-dimensional oscillator described by a displacement vector $\mathbf{u}$, rather than by a single-component displacement x as in Fig. 8.2. The vibration part of the Hamiltonian in the model is

$$H_{\text{ph}} = -\frac{1}{2M}\left(\frac{\partial}{\partial \mathbf{u}}\right)^2 + \frac{ku^2}{2}. \tag{8.6}$$

Electron potential energies due to the Coulomb interaction with the ion are approximately

$$V_{1,2} = V_0\left(1 - \frac{\mathbf{u}\cdot\mathbf{e}_{1,2}}{a}\right), \tag{8.7}$$

where $\mathbf{e}_{1,2}$ are unit vectors connecting sites **1** and **2** and site **3**. Hence, the Hamiltonian of the model is given by

$$H = E_a(\hat{n}_1 + \hat{n}_2) + \mathbf{u}\cdot(\mathbf{f}_1\hat{n}_1 + \mathbf{f}_2\hat{n}_2) - \frac{1}{2M}\left(\frac{\partial}{\partial\mathbf{u}}\right)^2 + \frac{ku^2}{2}, \tag{8.8}$$

where $\mathbf{f}_{1,2} = Ze^2\mathbf{e}_{1,2}/a^2$ is the Coulomb force and $\hat{n}_1$ and $\hat{n}_2$ are occupation number operators at every site. This Hamiltonian is also readily diagonalized by the same displacement transformation of the vibronic coordinate $\mathbf{u}$ as above:

$$\mathbf{u} = \mathbf{v} - \frac{\mathbf{f}_1\hat{n}_1 + \mathbf{f}_2\hat{n}_2}{k}. \tag{8.9}$$

The transformed Hamiltonian has no e-ph coupling,

$$\tilde{H} = (E_a - E_p)(\hat{n}_1 + \hat{n}_2) + V_{\text{ph}}\,\hat{n}_1\hat{n}_2 - \frac{1}{2M}\left(\frac{\partial}{\partial\mathbf{v}}\right)^2 + \frac{kv^2}{2}, \tag{8.10}$$

and describes two small polarons at their atomic levels shifted by the polaron level shift $E_p = f_{1,2}^2/2k$, which are entirely decoupled from ion vibrations. As a result, the lattice deformation caused by two electrons leads to an effective interaction between them, V_{ph}, which should be added to their Coulomb repulsion, V_c:

$$V_{\text{ph}} = -\frac{\mathbf{f}_1\cdot\mathbf{f}_2}{k}. \tag{8.11}$$

When V_{ph} is negative and large compared to positive V_c, the full interaction becomes negative.

Experimental evidence for an exceptionally strong electron-phonon interaction in high-temperature superconductors, colossal magnetoresistance oxides, and organic conductors is now overwhelming, so that the multipolaron problem is of a great importance [23–30]. Recent interest in polarons extends, of course, well beyond physical descriptions of materials. The field is a testing ground for analytical techniques, such as the $1/\lambda$ diagrammatic expansion [30], quantum statistical mean-field models [31], different unitary transformations [32], dynamical mean-field theory [33], and numerical techniques, such as the numerical diagonalization of vibrating clusters [25, 34, 35], advanced variational methods [36–40], and advanced quantum Monte Carlo (QMC) simulations [41, 42].

Small polarons with their attractive correlations are quite feasible also in molecular nanowires and quantum dots (MQDs) used as the "transmission lines"

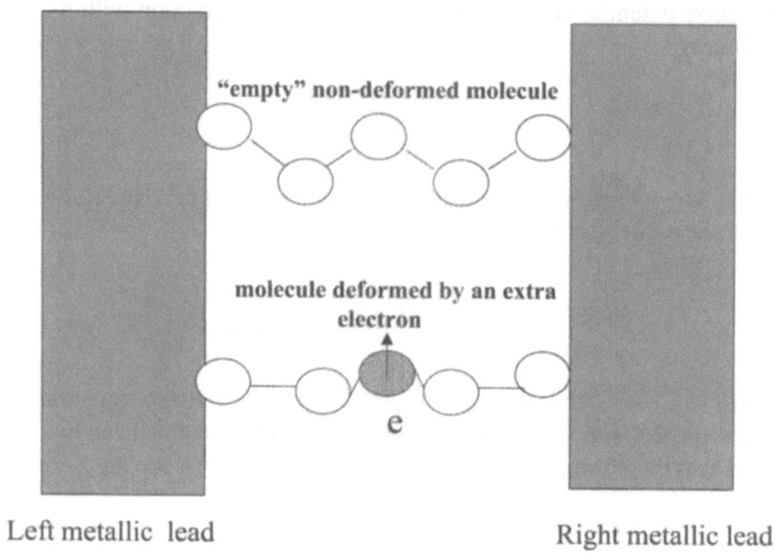

FIGURE 8.4. An ionic molecular chain attached to metallic leads. When electrons tunnel from the lead to the molecule, they shift equilibrium positions of the ions and form small polarons. The "lifetime" of electrons in the molecule should be long enough compared to the inverse vibron frequency to allow for the formation of small polarons.

[43, 44] and active molecular elements [45, 46] in molecular-scale electronics. In such devices, a molecule is placed between two metallic leads (see Fig. 8.4). When electrons from the lead tunnel to the molecular orbits, they deform the molecule and tunnel via molecular sites as small polarons.

It has been experimentally demonstrated that the low-bias conductance of molecules is dominated by resonant tunneling through coupled electronic and vibration levels [47]. Conductance peaks due to electron–vibron interactions has been seen in C_{60} [48]. Different aspects of the e-ph/vibron interaction effect on the tunneling through molecules and quantum dots (QDs) have been studied by several authors [49–55]. In particular, Glazman and Shekhter and, later, Wingreen et al. [49] presented the exact resonant-tunneling transmission probability fully taking into account the e-ph interaction on a nondegenerate resonant site. Phonons produced transmission side bands but did not affect the integral transmission probability. Li et al. [50] studied the conductance of a double-degenerate (due to spin) QD with Coulomb repulsion and the e-ph interaction. Their numerical results also showed the side-band peaks and the main peak related to the Coulomb repulsion, which was decreased by the e-ph interaction. Kang [51] studied the boson (vibron)-assisted transport through a double-degenerate QD coupled to two *superconducting* leads and found multiple peaks in the current–voltage (I–V) curves, which originated from the singular BCS density of states and the phonon side bands.

Whereas a correlated transport through mesoscopic systems with repulsive electron–electron interactions received considerable attention in the past and continues to be the focus of intense investigations [56], much less has been known about the role of attractive correlations in MQDs. Recently, we have proposed a negative$-U$ Hubbard model of a d-fold degenerate QD [57]. We argued that the *attractive* electron correlations caused by a strong e-ph (vibron) interaction in the molecule provide a molecular switching effect when the $I-V$ characteristics show two branches with high and low current for the same voltage. Different switching phenomena were observed in a number experimental studies with complex [46] and simple molecules [58].

Here, we review key features of polarons and bipolarons and the theory of the *correlated* transport in degenerate MQDs fully taking into account both Coulomb and e-ph interactions [59]. In the framework of a simple *negative–U* Hubbard model, we show how the attractive correlations cause the switching behavior of MQDs. Finally, we show that the phonon side bands significantly modify the switching behavior in comparison to the negative–U Hubbard model [57]. Nevertheless, the switching effect is robust. It shows up when the effective interaction of polarons is attractive and the state of the dot is multiply degenerate, $d > 2$.

8.2. STRONG- AND WEAK-COUPLING CONTINUOUS POLARONS

8.2.1. Variational Approach

To approach the multipolaron problem, let us first discuss a single electron interacting with the lattice deformation in the continuous approximation, as studied by Pekar [2]. In his model, a free electron interacts with the dielectric polarizable continuum, described by the static ϵ_0 and the optical (high-frequency) dielectric constants ϵ. This is the case for carriers interacting with optical phonons in ionic crystals under the condition that the size of the self-trapped state is large compared with the lattice constant; thus, the lattice discreteness is irrelevant.

Describing the ionic crystal as a polarizable dielectric continuum, one should keep in mind that only the ionic part of the total polarization contributes to the polaron state. The interaction of a carrier with valence electrons responsible for the optical properties is taken into account via the Hartree–Fock periodic potential and included in the band mass m. Therefore, only ion displacements contribute to the self-trapping. Following Pekar, we minimize the sum $E(\psi)$ of the electron kinetic energy and the potential energy due to the self-induced polarization field:

$$E(\psi) = \int d\mathbf{r} \left[\psi^*(\mathbf{r}) \left(-\frac{\nabla^2}{2m} \right) \psi(\mathbf{r}) - \mathbf{P}(\mathbf{r}) \cdot \mathbf{D}(\mathbf{r}) \right], \tag{8.12}$$

where

$$\mathbf{D}(\mathbf{r}) = e\nabla \int d\mathbf{r}' \frac{|\psi(\mathbf{r}')|^2}{|\mathbf{r} - \mathbf{r}'|} \tag{8.13}$$

is the electric field of an electron in the state with the wavefunction $\psi(\mathbf{r})$ and $\mathbf{P}$ is the ionic part of the lattice polarization. Here and throughout, we set $\hbar = c = k_B = 1$ unless specified otherwise. Minimizing $E(\psi)$ with respect to $\psi^*(\mathbf{r})$ at *fixed* $\mathbf{P}$ and $\int d\mathbf{r}\,|\psi(\mathbf{r})|^2 = 1$, one arrives at the equation of motion

$$\left(-\frac{\nabla^2}{2m} - e\int d\mathbf{r}'\,\mathbf{P}(\mathbf{r}')\cdot\nabla'\frac{1}{|\mathbf{r}'-\mathbf{r}|}\right)\psi(\mathbf{r}) = E_0\psi(\mathbf{r}), \tag{8.14}$$

where E_0 is the electron part of the ground-state state energy. The ion part of the total polarization is given by the definition of the susceptibility χ_0 and χ:

$$\mathbf{P} = (\chi_0 - \chi)\mathbf{D}. \tag{8.15}$$

The dielectric susceptibilities χ_0 and χ are expressed through the static and high-frequency dielectric constants, respectively $[\chi_0 = (\epsilon_0 - 1)/4\pi\epsilon_0$, $\chi = (\epsilon - 1)/4\pi\epsilon]$, to obtain

$$\mathbf{P} = \frac{\mathbf{D}}{4\pi\kappa} \tag{8.16}$$

with $\kappa^{-1} = \epsilon^{-1} - \epsilon_0^{-1}$. Then the equation of motion is

$$\left(-\frac{\nabla^2}{2m} - \frac{e^2}{4\pi\kappa}\int d\mathbf{r}'\int d\mathbf{r}''\,|\psi(\mathbf{r}'')|^2\nabla'\frac{1}{|\mathbf{r}'-\mathbf{r}''|}\cdot\nabla'\frac{1}{|\mathbf{r}'-\mathbf{r}|}\right)\psi(\mathbf{r}) = E_0\psi(\mathbf{r}). \tag{8.17}$$

Differentiating by parts with the use of the equation

$$\nabla^2\frac{1}{r} = -4\pi\delta(\mathbf{r}), \tag{8.18}$$

we obtain

$$\left(-\frac{\nabla^2}{2m} - \frac{e^2}{\kappa}\int d\mathbf{r}'\frac{|\psi(\mathbf{r}')|^2}{|\mathbf{r}'-\mathbf{r}|}\right)\psi(\mathbf{r}) = E_0\psi(\mathbf{r}). \tag{8.19}$$

The solution of this nonlinear integro-differential equation can be calculated with the help of a variational minimization of the functional

$$J(\psi) = \frac{1}{2m}\int d\mathbf{r}\,|\nabla\psi(\mathbf{r})|^2 - \frac{1}{2ma_B}\int d\mathbf{r}\,d\mathbf{r}'\frac{|\psi(\mathbf{r})|^2|\psi(\mathbf{r}')|^2}{|\mathbf{r}'-\mathbf{r}|}, \tag{8.20}$$

where $a_B = \kappa/me^2$ is the effective Bohr radius. The simplest choice of the normalized trial function is

$$\psi(\mathbf{r}) = Ae^{-r/r_p}, \tag{8.21}$$

with

$$A = \frac{1}{\sqrt{\pi r_p^3}}. \tag{8.22}$$

Substituting the trial function into the functional yields

$$J(\psi) = T + \frac{1}{2}U, \tag{8.23}$$

where the kinetic energy is

$$T = \frac{1}{2mr_p^2}. \tag{8.24}$$

To calculate the potential energy U, we first integrate over the angle θ between $\mathbf{r}$ and $\mathbf{r}'$;

$$\int_{-\pi}^{\pi} \frac{\sin\theta \, d\theta}{|\mathbf{r} - \mathbf{r}'|} = \frac{2\Theta(r - r')}{r} + \frac{2\Theta(r' - r)}{r'} \tag{8.25}$$

to obtain

$$U = -\frac{32\pi^2}{ma_B} \int_0^\infty dr\, r^2 \psi(r)^2 \int_r^\infty dr'\, r' \psi(r')^2. \tag{8.26}$$

Here, $\Theta(x) = 1$ for $x > 0$ and zero otherwise. As a result, we find

$$U = -\frac{5}{8ma_B r_p} \tag{8.27}$$

and the function to be minimized with respect to r_p is

$$J(\psi) = \frac{1}{2mr_p^2} - \frac{5}{16ma_B r_p}. \tag{8.28}$$

As a result, for the polaron radius we obtain

$$r_p = \frac{16a_B}{5} \tag{8.29}$$

and the ground-state energy $E_0 = T + U$ is

$$E_0 = -0.146\frac{1}{ma_B^2}. \tag{8.30}$$

This can be compared with the ground-state energy of the hydrogen atom, $-0.5m_e e^4$, where m_e is the free-electron mass. Their ratio is $0.3m/(m_e\kappa^2)$. In polar solids $4 < \kappa \simeq 20$. Then the large-polaron binding energy is below 0.25 eV if $m \simeq m_e$. The potential energy in the ground state is

$$U = -4T = \frac{4}{3}E_0. \tag{8.31}$$

The lowest photon energy $\nu_{\min}$ needed to excite a polaron into the electron band is

$$\nu_{\min} = |E_0|. \tag{8.32}$$

The ion configuration does not change during the photoexcitation of a polaron. A lower activation energy, W_T, is necessary, however, if the self-trapped state disappears together with the polarization well due to a thermal fluctuation

$$W_T = |E_0| - U_d, \tag{8.33}$$

where U_d is the deformation energy. In ionic crystals,

$$U_d = \frac{1}{2}\int d\mathbf{r}\, \mathbf{P}(\mathbf{r}) \cdot \mathbf{D}(\mathbf{r}), \tag{8.34}$$

which for the ground state is

$$U_d = \frac{2}{3}|E_0|. \tag{8.35}$$

Therefore, the thermal activation energy is

$$W_T = \frac{1}{3}|E_0|. \tag{8.36}$$

The ratio of four characteristic energies for the large strong-coupling polaron is given by

$$W_T : U_d : \nu_{min} : |U| = 1 : 2 : 3 : 4. \tag{8.37}$$

Different trial functions yield practically the same ground-state energy with numerically different polaron radii. In particular, with Pekar's choice

$$\psi(r) = A\left(1 + \frac{r}{r_p} + \beta r^2\right) e^{-r/r_p}, \tag{8.38}$$

one obtains $A = 0.12/r_p^{3/2}$, $\beta = 0.45/r_p^2$, and the polaron radius

$$r_p = 1.51 a_B. \tag{8.39}$$

The ground-state energy is

$$E_0 = -0.164 \frac{1}{m a_B^2}. \tag{8.40}$$

8.2.2. *Effective Mass of a Continuous Strong-Coupling Polaron*

Any polaron in a perfect crystal can move because of the translational symmetry. The lattice polarization is responsible for the polaron mass enhancement. Within the continuum approximation, the evolution of the lattice polarization $\mathbf{P}(\mathbf{r}, t)$ is described by the harmonic oscillator subjected to an external force, $\sim\mathbf{D}/\kappa$:

$$\omega^{-2} \frac{\partial^2 \mathbf{P}(\mathbf{r}, t)}{\partial t^2} + \mathbf{P}(\mathbf{r}, t) = \frac{\mathbf{D}(\mathbf{r}, t)}{4\pi\kappa}, \tag{8.41}$$

where ω is the optical phonon frequency. If during the characteristic time of the lattice relaxation ($\simeq \omega^{-1}$) the polaron moves a distance much less than the polaron radius, the polarization practically follows the polaron motion. Therefore, for a slow motion with velocity

$$v \ll \omega a_B, \tag{8.42}$$

the first term in Eq. (8.41), responsible for the retardation, is a small perturbation. Then

$$\mathbf{P}(\mathbf{r}, t) \approx \frac{1}{4\pi\kappa} \left(\mathbf{D}(\mathbf{r}, t) - \omega^{-2} \frac{\partial^2 \mathbf{D}(\mathbf{r}, t)}{\partial t^2} \right). \tag{8.43}$$

The total energy of the crystal with an extra electron,

$$E = E(\psi) + 2\pi\kappa \int d\mathbf{r} \left[\mathbf{P}(\mathbf{r}, \mathbf{t})^2 + \omega^{-2} \left(\frac{\partial \mathbf{P}(\mathbf{r}, t)}{\partial t} \right)^2 \right], \tag{8.44}$$

is determined in such a way that it gives Eq. (8.41) if minimized with respect to $\mathbf{P}$. We note that the first term of the lattice contribution is the deformation energy U_d, discussed in the previous subsection. The lattice part of the total energy depends on the polaron velocity and contributes to the effective mass. Replacing the static wavefunction $\psi(\mathbf{r})$ in all expressions for $\psi(\mathbf{r} - \mathbf{v}t)$ and neglecting a contribution to the total energy of higher order than v'^2, one obtains

$$E = E_0 + U_d + \frac{m^* v'^2}{2}, \tag{8.45}$$

where

$$m^* = -\frac{1}{12\pi\omega^2\kappa}\int d\mathbf{r}\,\mathbf{D}(\mathbf{r})\cdot\nabla^2\mathbf{D}(\mathbf{r}) \tag{8.46}$$

is the polaron mass. The use of the equations

$$\nabla^2\mathbf{D} = -4\pi e\nabla|\psi(\mathbf{r})|^2 \tag{8.47}$$

and

$$\nabla\cdot\mathbf{D} = -4\pi e|\psi(\mathbf{r})|^2 \tag{8.48}$$

yields

$$m^* = \frac{4\pi e^2}{3\omega^2\kappa}\int d\mathbf{r}\,|\psi(\mathbf{r})|^4. \tag{8.49}$$

Calculating the integral with the trial function, one obtains

$$m^* \simeq 0.02\alpha^4 m, \tag{8.50}$$

where α is the dimensionless constant, defined by Fröhlich as

$$\alpha = \frac{e^2}{\kappa}\sqrt{\frac{m}{2\omega}}. \tag{8.51}$$

To conclude our discussion of the strong-coupling large polaron, let us determine the condition of its existence. The polaron radius should be large compared with the lattice constant, $r_p \gg a$, to justify the effective mass approximation for the electron. Then the value of α should not be very large:

$$\alpha \ll \sqrt{\frac{D}{z\omega}}, \tag{8.52}$$

where $D \simeq z/2ma^2$ is half of the bare-electron bandwidth and, z is the lattice nearest-neighbor number. On the other side, the continuum (classical) approximation for the lattice polarisation is justified if the number of phonons taking part in the polaron formation is large. This number is of order U_d/ω. The total energy of the immobile polaron and deformed lattice is expressed as

$$E = -0.109\alpha^2\omega, \tag{8.53}$$

and $U_d = 0.218\alpha^2\omega$. Then α is bounded below by the condition $U_d/\omega \gg 1$, which yields

$$\alpha^2 \gg 5. \tag{8.54}$$

The adiabatic ratio D/ω is of the order 10–100. In fact, in many transition metal oxides with narrow bands and high optical phonon frequency, this ratio is below 10, which makes Eqs. (8.54) and (8.55) incompatible. Therefore, the continuous strong-coupling polaron is difficult to realize in transition metal oxides and also in molecules, where the "bandwidth" is zero.

8.2.3. *Weak-Coupling (Fröhlich) Polaron*

Fröhlich [3] and other workers applied the second quantisation form of the electron–lattice interaction to describe a weak-coupling large polaron when $\alpha \leq 1$, so the quantum nature of lattice polarization becomes important. The electron potential energy in a crystal field distorted by phonons is

$$V(\mathbf{r}) = \sum_{\mathbf{l}} v(\mathbf{r} - \mathbf{R}_{\mathbf{l}}), \tag{8.55}$$

where the interaction of an electron with a single ion is described by the potential $v(\mathbf{r})$. The distance of ions $\mathbf{u}_{\mathbf{l}} = \mathbf{R}_{\mathbf{l}} - \mathbf{l}$ from the equilibrium positions $\mathbf{l}$ is small compared with the lattice constant a, which allows us to expand $v(\mathbf{r} - \mathbf{R}_{\mathbf{l}})$ near equilibrium:

$$v(\mathbf{r} \quad \mathbf{R}_{\mathbf{l}}) \simeq v(\mathbf{r} - \mathbf{l}) - \mathbf{u}_{\mathbf{l}} \cdot \nabla v(\mathbf{r} - \mathbf{l}). \tag{8.56}$$

The lattice part of the Hamiltonian can be diagonalized with harmonic phonons such that the ion displacement is a linear combination of their annihilation $d_{\mathbf{q}}$ and creation $d_{\mathbf{q}}^{\dagger}$ bosonic operators:

$$\mathbf{u}_{\mathbf{l}} = \sum_{\mathbf{q}} \frac{\mathbf{e}_{\mathbf{q}}}{\sqrt{2NM\omega_{\mathbf{q}}}} d_{\mathbf{q}} e^{i\mathbf{q}\cdot\mathbf{l}} + \text{h.c.} \tag{8.57}$$

where $\mathbf{q}$ is the phonon momenta in the first Brillouin zone, $\omega_{\mathbf{q}}$ the phonon frequency, M ionic mass, $\mathbf{e}_{\mathbf{q}}$ is a unit polarization vector, and N is the number of ions (sites) in a crystal. Using the Fourier transform,

$$v(\mathbf{r}) = \sum_{\mathbf{k}} v_{\mathbf{k}} e^{i\mathbf{k}\cdot\mathbf{r}}; \tag{8.58}$$

the e-ph interaction in the second quantization is written as

$$H_{\text{e-ph}} = \frac{1}{\sqrt{2N}} \sum_{\mathbf{q}} \gamma(\mathbf{q}) \omega_{\mathbf{q}} d_{\mathbf{q}} e^{i\mathbf{q}\cdot\mathbf{r}} + \text{h.c.}, \tag{8.59}$$

where a dimensionless matrix element is given by

$$\gamma(\mathbf{q}) = -i \frac{N \mathbf{e}_{\mathbf{q}} \cdot \mathbf{q}}{M^{1/2} \omega_{\mathbf{q}}^{3/2}} v_{\mathbf{q}}. \tag{8.60}$$

In ionic crystals, the interaction $v(\mathbf{r})$ is the Coulomb one with the Fourier component $v_{\mathbf{q}} \simeq 4\pi/\Omega\kappa q^2$ (Ω is the crystal volume). Then the coupling with longitudinal ($\mathbf{e}_{\mathbf{q}} \parallel \mathbf{q}$) ionic plasmons is

$$\frac{\gamma(\mathbf{q})\omega}{2N} \equiv V_{\mathbf{q}} = -\frac{i\omega}{q} \left(\frac{4\pi\alpha}{\Omega\sqrt{2m\omega}} \right)^{1/2}, \tag{8.61}$$

where α is the Fröhlich constant and $\omega = \sqrt{4\pi N e^2/M\Omega\kappa}$ is the ionic plasma frequency. The complete Hamiltonian, including the quantized deformation energy, has the form

$$H = -\frac{\nabla^2}{2m} + \sum_{\mathbf{q}} \left(V_{\mathbf{q}} d_{\mathbf{q}} e^{i\mathbf{q}\cdot\mathbf{r}} + \text{h.c.} \right) + \sum_{\mathbf{q}} \omega_{\mathbf{q}} \left(d_{\mathbf{q}}^{\dagger} d_{\mathbf{q}} + \frac{1}{2} \right). \tag{8.62}$$

The quantum states of the noninteracting electron and phonons are classified with the electron momentum $\mathbf{k}$ and with the phonon occupation numbers $\langle d_{\mathbf{q}}^{\dagger} d_{\mathbf{q}} \rangle \equiv n_{\mathbf{q}} = 0, 1, 2, \ldots, \infty$. For zero temperature, the unperturbed state is the vacuum $|0\rangle$ of phonons and the electron plane wave

$$|\mathbf{k}, 0\rangle = \frac{1}{\sqrt{\Omega}} e^{i\mathbf{k}\cdot\mathbf{r}} |0\rangle. \tag{8.63}$$

While the coupling is weak one can apply the perturbation theory. The interaction couples the state Eq. (8.52) with the energy $k^2/2m$ and states of a single phonon of momentum $\mathbf{q}$ and the electron of momentum $\mathbf{k} - \mathbf{q}$ with the energy $(\mathbf{k} - \mathbf{q})^2/2m + \omega$:

$$|\mathbf{k} - \mathbf{q}, 1_{\mathbf{q}}\rangle = \frac{1}{\sqrt{\Omega}} e^{i(\mathbf{k}-\mathbf{q})\cdot\mathbf{r}} |1_{\mathbf{q}}\rangle. \tag{8.64}$$

The corresponding matrix element is

$$\langle \mathbf{k} - \mathbf{q}, 1_{\mathbf{q}} | H_{\text{e-ph}} | \mathbf{k}, 0 \rangle = V_{\mathbf{q}}^{*}. \tag{8.65}$$

There is no diagonal matrix elements of $H_{\text{e-ph}}$. Then the renormalized energy $\tilde{E}_{\mathbf{k}}$ in the lowest second order is

$$\tilde{E}_{\mathbf{k}} = \frac{k^2}{2m} - \sum_{\mathbf{q}} \frac{|V_{\mathbf{q}}|^2}{(\mathbf{k}-\mathbf{q})^2/2m + \omega - k^2/2m}. \tag{8.66}$$

As in Section 8.2.2, we consider a slow electron such as

$$k < q_p, \tag{8.67}$$

where

$$q_p = \min\left(\frac{m\omega}{q} + \frac{q}{2}\right) = \sqrt{2m\omega}. \tag{8.68}$$

In this case, there is no imaginary part of $\tilde{E}_{\mathbf{k}}$, which means that the momentum is conserved. On substituting the expression for $V_{\mathbf{q}}$, changing the sum over $\mathbf{q}$ to integrals, and taking the upper limit for q to infinity, one obtains

$$\tilde{E}_{\mathbf{k}} = \frac{k^2}{2m} - \frac{\alpha\omega}{\pi} \int_{-1}^{1} dx \int_0^{\infty} \frac{dy}{y^2 - 2yxk/q_p + 1}. \tag{8.69}$$

The integral over q converges because the coupling constant ($\sim 1/q$) decreases with q. Therefore, long-wave optical phonons contribute mainly to the polaron self-energy. That is not the case for molecular or acoustical phonons when all states of the Brillouin zone contribute (see Section 8.3.1). Evaluating the integrals one arrives at

$$\tilde{E}_{\mathbf{k}} = \frac{k^2}{2m} - \frac{\alpha\omega q_p}{k} \arcsin\left(\frac{k}{q_p}\right), \tag{8.70}$$

which for a very slow motion $k \ll q_p$ yields

$$\tilde{E}_{\mathbf{k}} \simeq -\alpha\omega + \frac{k^2}{2m^*}. \tag{8.71}$$

Here, the first term is the polaron binding energy $-E_p$. The effective mass of the polaron is enhanced as

$$m^* = \frac{m}{1-\alpha/6} \simeq m\left(1 + \frac{\alpha}{6}\right). \tag{8.72}$$

This is due to a phonon cloud accompanying a slow polaron. The number of virtual phonons N_{ph} in the cloud is given by taking the expectation value of the phonon number operator

$$N_{\text{ph}} = \left\langle \sum_{\mathbf{q}} d_{\mathbf{q}}^{\dagger} d_{\mathbf{q}} \right\rangle, \tag{8.73}$$

where *bra* and *ket* refer to the perturbed state

$$|\,\rangle = |0\rangle + \sum_{\mathbf{q}'} \frac{V_{\mathbf{q}'}^{*}}{k^2/2m - (\mathbf{k}-\mathbf{q}')^2/2m - \omega} |1_{\mathbf{q}'}\rangle. \tag{8.74}$$

For a polaron at rest ($\mathbf{k} = 0$), one obtains

$$N_{\text{ph}} = \sum_{\mathbf{q}} \frac{|V_{\mathbf{q}}|^2}{(\omega + q^2/2m)^2}. \tag{8.75}$$

The value of the integral is

$$N_{\text{ph}} = \frac{\alpha}{2}. \tag{8.76}$$

Therefore, the Fröhlich coupling constant measures the cloud "thickness" directly. One can also calculate the lattice charge density induced by the electron. The electrostatic potential $e\phi(\mathbf{r})$ is given by the average of the interaction term of the Hamiltonian,

$$e\phi(\mathbf{r}) = \left\langle \sum_{\mathbf{q}} V_{\mathbf{q}} e^{i\mathbf{q}\cdot\mathbf{r}} d_{\mathbf{q}} + \text{h.c.} \right\rangle, \tag{8.77}$$

and the charge density $\rho(\mathbf{r})$ is related to the electrostatic potential by Poisson's equation,

$$\nabla\phi = -4\pi\rho. \tag{8.78}$$

As a result, one obtains

$$\rho(\mathbf{r}) = -\frac{1}{2\pi e} \sum_{\mathbf{q}} \frac{q^2 |V_{\mathbf{q}}|^2 \cos[\mathbf{q}\cdot\mathbf{r}]}{\omega + q^2/2m}. \tag{8.79}$$

Integrating over $\mathbf{q}$ yields

$$\rho(\mathbf{r}) = -\frac{e q_p^3}{4\pi\kappa} \frac{e^{-q_p r}}{q_p r}. \tag{8.80}$$

The mean extension of the phonon cloud, which can be taken as the radius of a weak-coupling polaron, is

$$r_p = q_p^{-1} \tag{8.81}$$

and the total induced charge is

$$Q = \int d\mathbf{r}\, \rho(\mathbf{r}) = -\frac{e}{\kappa}. \tag{8.82}$$

8.3. SMALL (LATTICE) POLARON

8.3.1. Holstein Model

When the coupling with phonons increases, the polaron radius decreases and becomes of the order of the lattice constant. Then all momenta of the Brillouin zone contribute to the polaron wavefunction and the effective mass approximation cannot be applied. This regime occurs if the characteristic potential energy E_p due to the local lattice deformation is compared or larger than the half-bandwidth D. The strong-coupling regime with the dimensionless coupling constant

$$\lambda \equiv \frac{E_p}{D} \geq 1 \tag{8.83}$$

is called a *small* polaron. In general, E_p is expressed as

$$E_p = \frac{1}{2N} \sum_{\mathbf{q}} |\gamma(\mathbf{q})|^2 \omega_{\mathbf{q}} \tag{8.84}$$

for any type of phonons involved in the polaron cloud. For the Fröhlich interaction with optical phonons, one obtains

$$E_p \simeq \frac{q_D e^2}{\pi \kappa}, \tag{8.85}$$

where $q_d = (6\pi^2/\Omega)^{1/3}$ is the Debye momentum. For example, with parameters appropriate for high-T_c copper oxides, $\epsilon_0 \gg \epsilon_\infty \simeq 5$ and $q_D \simeq 0.7$ Å^{-1}, one obtains $E_p \simeq 0.6$ eV. The exact value of λ_c when the large polaron collapses into a small one depends on the lattice structure, phonon frequency dispersion, and the radius of the e-ph interaction. However, in any case the transition occurs around $\lambda_c \simeq 1$. *Small* polarons are expected to be the carriers in high-T_c oxides, which are strongly polarizible doped semiconductors with rather narrow electron bands, and in molecular nanowires.

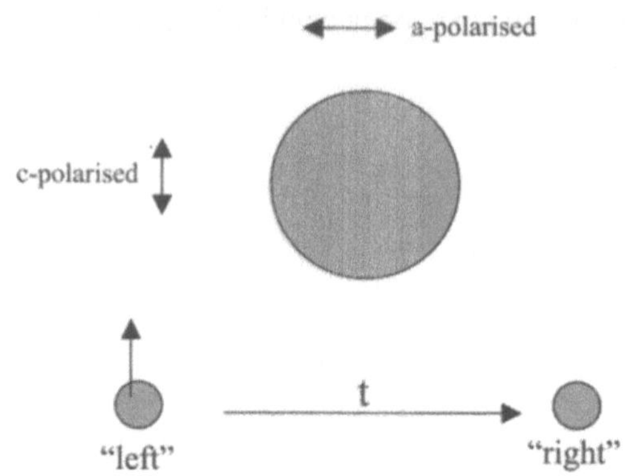

FIGURE 8.5. The electron tunnels between sites 1 ("left") and 2 ("right") with the amplitude t and interacts with c-axis or a-axis polarized vibrational modes of the ion, placed in between.

The main features of small polarons are revealed in the simplest Holstein model [10] consisting of only two sites and one electron tunneling between them (see Fig. 8.5).

Now, following Holstein [10], we introduce the two-site Hamiltonian, describing an electron tunneling between sites 1 ("left") and 2 ("right") with amplitude t and interacting with a vibrational mode of an ion, placed at some distance in between (Fig. 8.5):

$$H = t(c_1^\dagger c_2 + c_2^\dagger c_1) + H_{\text{ph}} + H_{\text{e-ph}}, \tag{8.86}$$

where $H_{\text{e-ph}}$ depends on the polarization of vibrations, and we take $E_a = 0$. If the ion vibrates along the perpendicular direction to the hopping (in the c direction), we have

$$H_{\text{e-ph}} = f_c x(c_1^\dagger c_1 + c_2^\dagger c_2), \tag{8.87}$$

and we have

$$H_{\text{e-ph}} = f_a x(c_1^\dagger c_1 - c_2^\dagger c_2) \tag{8.88}$$

if the ion vibrates along the hopping (a direction).

The wavefunction of the electron and the ion is a linear superposition of two terms describing the electron on the "left" and on the "right" site, respectively:

$$\psi = \left[u(x)c_1^\dagger + v(x)c_2^\dagger\right]|0\rangle, \tag{8.89}$$

where $|0\rangle$ is the vacuum state describing a rigid lattice without the extra electron. Substituting ψ into the Schrödinger equation, $H\psi = E\psi$, we obtain two coupled equations for the amplitudes:

$$(E - f_{a,c}x - H_{ph})u(x) = tv(x), \quad (8.90)$$

$$(E \pm f_{a,c}x - H_{ph})v(x) = tu(x). \quad (8.91)$$

There is the exact solution for the c-axis polarization when a change in the ion position leads to the same shift of the electron energy on the left and on the right sites:

$$u(x) = u\chi_n(x), \quad (8.92)$$

$$v(x) = v\chi_n(x), \quad (8.93)$$

where u and v are constants and

$$\chi_n(x) = \left(\frac{M\omega}{\pi(2^n n!)^2}\right)^{1/4} H_n\left[\left(x - \frac{f_c}{k}\right)(M\omega)^{1/2}\right] \exp\left[-M\left(x - \frac{f_c}{k}\right)^2 \omega/2\right], \quad (8.94)$$

is the harmonic oscillator wavefunction. There are two ladders of levels given by

$$E_n^{\pm} = -E_p \pm t + \omega(n + 1/2) \quad (8.95)$$

with $E_p = f_c^2/2k$. Here,

$$H_n(\xi) = (-1)^n e^{\xi^2} \frac{d^n e^{-\xi^2}}{d\xi^n} \quad (8.96)$$

are the Hermite polynomials, and $n = 0, 1, 2, 3, \ldots$. Hence, the c-axis single-ion deformation leads to the polaron level shift but without any renormalization of the hopping integral t. In contrast, a-polarized vibrations with the opposite shift of the electron energy on the left and on the right sites strongly renormalize the hopping integral. There is no simple general solution of the Holstein model in this case, but one can find it in two limiting cases: *nonadiabatic*, when $t \ll \omega$, and *adiabatic*, when $t \gg \omega$.

8.3.2. Nonadiabatic Small Polaron

In the nonadiabatic regime, the ion vibrations are fast and the electron hopping is slow. Hence, one can apply a perturbation theory in powers of t to solve

$$\begin{pmatrix} E - f_a x - H_{ph} & -t \\ -t & E + f_a x - H_{ph} \end{pmatrix} \begin{bmatrix} u(x) \\ v(x) \end{bmatrix} = 0. \quad (8.97)$$

We take $t = 0$ in zero order and obtain a twofold *degenerate* ground state $[u^{l,r}(x), v^{l,r}(x)]$, corresponding to the polaron localized on the left (l) or on the right (r) sites:

$$u^l(x) = \exp\left[-\frac{M\omega}{2}\left(x + \frac{f_a}{k}\right)^2\right],$$
$$v^l(x) = 0, \tag{8.98}$$

and

$$u^r(x) = 0,$$
$$v^r(x) = \exp\left[-\frac{M\omega}{2}\left(x - \frac{f_a}{k}\right)^2\right] \tag{8.99}$$

with the energy $E_0 = -E_p + \omega/2$, where $E_p = f_a^2/2k$. The eigenstates are found as linear superpositions of two unperturbed states:

$$\begin{bmatrix} u(x) \\ v(x) \end{bmatrix} = \alpha \begin{bmatrix} u^l(x) \\ 0 \end{bmatrix} + \beta \begin{bmatrix} 0 \\ v^r(x) \end{bmatrix}. \tag{8.100}$$

Here, the coefficients α and β are independent of x. The conventional secular equation for E is obtained, multiplying the first row by $u^l(x)$ and the second row by $v^r(x)$ and integrating over the vibration coordinate, x, each of two equations of the system. The result is

$$\det\begin{pmatrix} E - E_0 & \tilde{t} \\ \tilde{t} & E - E_0 \end{pmatrix} = 0 \tag{8.101}$$

with the renormalised hopping integral

$$\frac{\tilde{t}}{t} = \frac{\int_{-\infty}^{\infty} dx\, u^l(x) v^r(x)}{\int_{-\infty}^{\infty} dx\, |u^l(x)|^2}. \tag{8.102}$$

The corresponding eigenvalues, $E_\pm$, are

$$E_\pm = \frac{\omega}{2} - E_p \pm \tilde{t}. \tag{8.103}$$

The hopping integral splits the degenerate level, as in the rigid lattice, but an effective "bandwidth" $2\tilde{t}$ is significantly reduced compared with the bare one:

$$\tilde{t} = t\, \exp\left(-\frac{2E_p}{\omega}\right). \tag{8.104}$$

This polaron band narrowing originates in a small overlap integral of two displaced oscillator wavefunctions $u^l(x)$ and $v^r(x)$.

8.3.3. Adiabatic Small Polaron

In the adiabatic regime, when $t \gg \omega$, the electron tunneling is fast compared to the ion motion. Hence, one can apply the Born–Oppenheimer adiabatic approximation taking the wavefunction in the form

$$\begin{bmatrix} u(x) \\ v(x) \end{bmatrix} = \chi(x) \begin{bmatrix} u_a(x) \\ v_a(x) \end{bmatrix}. \tag{8.105}$$

Here, $u_a(x)$ and $v_a(x)$ are the electron wavefunctions obeying the Schrödinger equation with the *frozen* ion deformation x; that is,

$$\begin{pmatrix} E_a(x) - f_a x & -t \\ -t & E_a(x) + f_a x \end{pmatrix} \begin{bmatrix} u_a(x) \\ v_a(x) \end{bmatrix} = 0. \tag{8.106}$$

The lowest energy level is found as

$$E_a(x) = -\sqrt{(f_a x)^2 + t^2}. \tag{8.107}$$

$E_a(x)$, together with $kx^2/2$, plays the role of a potential energy term in the equation for the "vibration" wavefunction, $\chi(x)$:

$$\left[-\frac{1}{2M}\frac{\partial^2}{\partial x^2} + \frac{kx^2}{2} - \sqrt{(f_a x)^2 + t^2}\right]\chi(x) = E\chi(x). \tag{8.108}$$

Terms with the first and second derivatives of the electron wavefunctions $u_a(x)$ and $v_a(x)$ are small compared with the corresponding derivatives of $\chi(x)$ in the adiabatic approximation, so they can be neglected in Eq. (8.108). As a result, we arrive at the familiar double-well potential problem, where the potential energy $U(x) = kx^2/2 - \sqrt{(f_a x)^2 + t^2}$ has two symmetric minima, separated by a barrier. Minima are located approximately at

$$x_m = \pm\frac{f_a}{k}. \tag{8.109}$$

in the strong-coupling limit, $E_p \gg t$ and the potential energy near the bottom of each potential well is about

$$U(x) = -E_p + \frac{k(|x| - f_a/k)^2}{2}. \tag{8.110}$$

If the barrier were impenetrable, there would be the ground-state energy level $E_0 = -E_p + \omega/2$, the same for both wells. The tunneling under the barrier results in a splitting of this level $2\tilde{t}$, which corresponds to a polaron band in the lattice. It can be estimated using the quasi-classical approximation as

$$\tilde{t} \propto \exp\left[-2\int_0^{x_m} p(x)\,dx\right], \tag{8.111}$$

where $p(x) = \sqrt{2M[U(x) - E_0]} \approx (Mk)^{1/2}|x - f_a/k|$ is the classical momentum.

Calculating the integral, one finds the exponential reduction of the "bandwidth":

$$\tilde{t} \propto \exp\left(-\frac{2E_p}{\omega}\right), \tag{8.112}$$

which is the same as in the nonadiabatic regime. Holstein also found corrections to this expression up to terms of the order of $1/\lambda^2$, which allowed him to estimate the preexponential factor as

$$\tilde{t} \approx \sqrt{E_p\omega}\exp\left(-\frac{2E_p}{\omega}\right). \tag{8.113}$$

The term in front of the exponent differs from the t of the nonadiabatic case. It is thus apparent that the perturbation approach covers only part of the entire lattice polaron region, $\lambda \gtrsim 1$. The upper limit of applicability of the perturbation theory is given by $t < \sqrt{E_p\omega}$. For the remainder of the region, the adiabatic approximation is more appropriate.

8.3.4. Electron–phonon and Coulomb Interactions in Wannier Representation

Let us generalize the Holstein results for an infinite lattice. In the strong-coupling regime, it is convenient to transform the Bloch states to the site (Wannier) states using the canonical linear transformation of the electron operators:

$$c_i = \frac{1}{\sqrt{N}}\sum_{\mathbf{k}} e^{i\mathbf{k}\cdot\mathbf{m}} c_{\mathbf{k}s}, \tag{8.114}$$

where $i = (\mathbf{m}, s)$ includes both site $\mathbf{m}$ and spin s quantum numbers. In the new representation, the electron kinetic energy takes the form

$$H_e = \sum_{i,j} t(\mathbf{m} - \mathbf{n})\delta_{ss'} c_i^\dagger c_j, \tag{8.115}$$

where

$$t(\mathbf{m}) = \frac{1}{N}\sum_{\mathbf{k}} E_{\mathbf{k}} e^{i\mathbf{k}\cdot\mathbf{m}}$$

is the "bare" hopping integral. Here, $j = (\mathbf{n}, s')$ and $E_\mathbf{k}$ is the Bloch band dispersion in the rigid lattice.

The e-ph interaction and the Coulomb correlations acquire simple forms in the Wannier representation if their matrix elements in the momentum representation depend only on the momentum transfer $\mathbf{q}$:

$$H_{\text{e-ph}} = \sum_{\mathbf{q},i} \omega_\mathbf{q} \hat{n}_i \left[u_i(\mathbf{q},) d_\mathbf{q} + \text{h.c.} \right], \tag{8.116}$$

$$H_{\text{e-e}} = \frac{1}{2} \sum_{i \neq j} V_c(\mathbf{m} - \mathbf{n}) \hat{n}_i \hat{n}_j. \tag{8.117}$$

Here,

$$u_i(\mathbf{q}) = \frac{1}{\sqrt{2N}} \gamma(\mathbf{q}) e^{i\mathbf{q}\cdot\mathbf{m}} \tag{8.118}$$

and

$$V_c(\mathbf{m}) = \frac{1}{N} \sum_\mathbf{q} V_c(\mathbf{q}) e^{i\mathbf{q}\cdot\mathbf{m}} \tag{8.119}$$

are the matrix elements of the e-ph and Coulomb interactions, respectively, in the Wannier representation for electrons and $\hat{n}_i = c_i^\dagger c_i$ is the density operator. The dimensionless e-ph coupling constant $\gamma(\mathbf{q})$ is defined using the Bloch states $\psi_\mathbf{k}(\mathbf{r})$ as

$$\gamma(\mathbf{q}) = -\frac{N}{M^{1/2}\omega_\mathbf{q}^{3/2}} \int d\mathbf{r} \left(\mathbf{e}_\mathbf{q} \cdot \nabla v(\mathbf{r}) \right) \psi_\mathbf{k}^*(\mathbf{r}) \psi_{\mathbf{k}-\mathbf{q}}(\mathbf{r}). \tag{8.120}$$

Taking the interaction matrix elements depending only on the momentum transfer, one neglects the terms in the e-ph and Coulomb interactions, which are proportional to the overlap integrals of the Wannier orbitals on different sites. This approximation is justified for narrow-band materials, whose bandwidth $2D$ is less than the characteristic value of the crystal field. As a result, in the Wannier representation the Hamiltonian is

$$H = \sum_{i,j} t(\mathbf{m} - \mathbf{n}) \delta_{ss'} c_i^\dagger c_j + \sum_{\mathbf{q},i} \omega_\mathbf{q} \hat{n}_i \left[u_i(\mathbf{q}) d_\mathbf{q} + \text{h.c.} \right] + \frac{1}{2} \sum_{i \neq j} V_c(\mathbf{m} - \mathbf{n}) \hat{n}_i \hat{n}_j + \sum_\mathbf{q} \omega_\mathbf{q} \left(d_\mathbf{q}^\dagger d_\mathbf{q} + \frac{1}{2} \right). \tag{8.121}$$

This Hamiltonian should be treated as a "bare" one for metals, where the matrix elements and phonon frequencies are ill-defined. To the contrary, the bare phonons (vibrons) $\omega_\mathbf{q}$ and the electron band structure $E_\mathbf{k}$ are well defined in doped semiconductors and molecules. Here, the effect of carriers on the dynamic matrix is small and the carrier density is much less than the atomic one. Depending on the particular phonon mode, the interaction constant $\gamma(\mathbf{q})$ has different q dependences. For example, in the long-wavelength limit ($q \ll \pi/a$),

$$\gamma(\mathbf{q}) \propto \frac{1}{q} \tag{8.122}$$

$$\propto \text{constant} \tag{8.123}$$

$$\propto \frac{1}{\sqrt{q}} \tag{8.124}$$

for optical, molecular ($\omega_{\mathbf{q}} \propto$ constant), and acoustic ($\omega_{\mathbf{q}} \propto q$) phonons, respectively. We can transform the e-ph interaction further using the site representation also for phonons. Replacing the Bloch functions in the definition of $\gamma(\mathbf{q})$ by their Wannier series yields

$$\gamma(\mathbf{q}) = -\frac{1}{M^{1/2}\omega_{\mathbf{q}}^{3/2}} \sum_{\mathbf{n}} e^{-i\mathbf{q}\cdot\mathbf{n}} \mathbf{e}_{\mathbf{q}} \cdot \nabla_{\mathbf{n}} v(\mathbf{n}). \tag{8.125}$$

This result is obtained by neglecting the overlap integrals of the Wannier orbitals on different sites and by assuming that the single-ion potential $v(\mathbf{r})$ varies over the distance, which is much larger than the radius of the orbital. Then using the displacement operators, one arrives at

$$H_{\text{e-ph}} = \sum_{\mathbf{m},\mathbf{n},s} \hat{n}_{\mathbf{m}s} \mathbf{u}_{\mathbf{n}} \cdot \nabla_{\mathbf{n}} v(\mathbf{m}-\mathbf{n}). \tag{8.126}$$

The site representation of $H_{\text{e-ph}}$ is particularly convenient for the interaction with dispersionless modes, when $\omega_{\mathbf{q}} = \omega$ and $\mathbf{e}_{\mathbf{q}} = \mathbf{e}$ are q independent. Introducing the phonon site operators

$$d_{\mathbf{n}} = \frac{1}{\sqrt{N}} \sum_{\mathbf{q}} e^{i\mathbf{q}\cdot\mathbf{n}} d_{\mathbf{q}}, \tag{8.127}$$

in this case we obtain

$$\mathbf{u}_{\mathbf{n}} = \frac{\mathbf{e}}{\sqrt{2M\omega}} (d_{\mathbf{n}} + d_{\mathbf{n}}^{\dagger}),$$

$$H_{\text{ph}} = \omega \sum_{\mathbf{n}} \left(d_{\mathbf{n}}^{\dagger} d_{\mathbf{n}} + \frac{1}{2} \right),$$

and

$$H_{\text{e-ph}} = \omega \sum_{\mathbf{n},\mathbf{m},s} g(\mathbf{m}-\mathbf{n}) (\mathbf{e} \cdot \mathbf{e}_{\mathbf{m}-\mathbf{n}}) \hat{n}_{\mathbf{m}s} (d_{\mathbf{n}}^{\dagger} + d_{\mathbf{n}}), \tag{8.128}$$

where

$$g(\mathbf{m}) = \frac{1}{\omega\sqrt{2M\omega}} \frac{dv(m)}{dm}$$

is a dimensionless *force* acting between the electron on site **m** and the displacement of ion **n** and $\mathbf{e}_{\mathbf{m}-\mathbf{n}} \equiv (\mathbf{m}-\mathbf{n})/|\mathbf{m}-\mathbf{n}|$ is the unit vector in the direction from the electron **m** to the ion **n**. The "real-space" representation is convenient in modeling the e-ph interaction in complex lattices. Atomic orbitals of an ion adiabatically follow its motion. Therefore, the electron does not interact with the displacement of the ion, whose orbital it occupies; that is, $g(0) = 0$.

8.3.5. Polaron Band

The kinetic energy is smaller than the interaction energy as long as $\lambda > 1$. Hence, a self-consistent approach to the multipolaron problem is possible with the "$1/\lambda$" expansion technique [30], which treats the kinetic energy as a perturbation. The technique is based on the fact, known for a long time, that there is an analytical exact solution of a *single*-polaron problem in the strong-coupling limit $\lambda \to \infty$. Following Lang and Firsov [11], we apply the canonical transformation e^S to diagonalize the Hamiltonian. The diagonalization is exact if $t(\mathbf{m}) = 0$ (or $\lambda = \infty$):

$$\tilde{H} = e^S H e^{-S}, \tag{8.129}$$

where

$$S = -\sum_{\mathbf{q},i} \hat{n}_i \left[u_i(\mathbf{q})d_{\mathbf{q}} - \text{h.c.}\right] \tag{8.130}$$

is such that $S^\dagger = -S$. The electron and phonon operators are transformed as

$$\tilde{c}_i = c_i \exp\left[\sum_{\mathbf{q}} u_i(\mathbf{q})d_{\mathbf{q}} - \text{h.c.}\right], \tag{8.131}$$

and

$$\tilde{d}_{\mathbf{q}} = d_{\mathbf{q}} - \sum_i \hat{n}_i u_i^*(\mathbf{q}), \tag{8.132}$$

respectively. The Lang–Firsov canonical transformation shifts the ions to new equilibrium positions. In a more general sense, it changes the boson vacuum. As a result, the transformed Hamiltonian takes the following form:

$$\tilde{H} = \sum_{i,j} \hat{\sigma}_{ij} c_i^\dagger c_j - E_p \sum_i \hat{n}_i + \sum_{\mathbf{q}} \omega_{\mathbf{q}} \left(d_{\mathbf{q}}^\dagger d_{\mathbf{q}} + \frac{1}{2}\right) + \frac{1}{2}\sum_{i \neq j} v_{ij}\hat{n}_i\hat{n}_j, \tag{8.133}$$

where

$$\hat{\sigma}_{ij} = t(\mathbf{m}-\mathbf{n})\delta_{ss'} \exp\left(\sum_{\mathbf{q}}[u_j(\mathbf{q}) - u_i(\mathbf{q})]d_{\mathbf{q}} - \text{h.c.}\right) \tag{8.134}$$

is the renormalized hopping integral depending on the phonon operators and

$$v_{ij} \equiv v(\mathbf{m}-\mathbf{n}) = V_c(\mathbf{m}-\mathbf{n}) - \frac{1}{N}\sum_{\mathbf{q}} |\gamma(\mathbf{q})|^2 \omega_{\mathbf{q}} \cos[\mathbf{q}\cdot(\mathbf{m}-\mathbf{n})] \tag{8.135}$$

is the interaction of polarons comprising their Coulomb repulsion and the interaction via a local lattice deformation. In the extreme infinite-coupling limit $\lambda \to \infty$, we can neglect the hopping term of the transformed Hamiltonian. The rest has analytically determined eigenstates and eigenvalues. The eigenstates $|\tilde{N}\rangle = |n_i, n_{\mathbf{q}}\rangle$ are sorted by the polaron $n_{\mathbf{m}s}$ and phonon $n_{\mathbf{q}}$ occupation numbers. The energy levels are

$$E = -(\mu + E_p)\sum_i n_i + \frac{1}{2}\sum_{i\neq j} v_{ij} n_i n_j + \sum_{\mathbf{q}} \omega_{\mathbf{q}}\left(n_{\mathbf{q}} + \frac{1}{2}\right), \tag{8.136}$$

where $n_i = 0, 1$ and $n_{\mathbf{q}} = 0, 1, 2, 3, \ldots, \infty$.

The Hamiltonian in zero order with respect to the hopping describes localized polarons and independent phonons, which are vibrations of ions relative to new equilibrium positions, which depend on the polaron occupation numbers. The phonon frequencies remain unchanged in this limit. The middle of the electron band falls by the polaron level shift E_p due to a potential well created by lattice deformation.

Now let us discuss the $1/\lambda$ expansion. First, we restrict the discussion to a single-polaron problem with no polaron–polaron interaction. The finite hopping term leads to the polaron tunneling because of degeneracy of the zero-order Hamiltonian with respect to the site position of the polaron. To see how the tunneling occurs, we apply the perturbation theory using $1/\lambda$ as a small parameter $[D = zt(a)]$. The proper Bloch set of N-degenerate zero-order eigenstates with the lowest energy $(-E_p)$ of the unperturbed Hamiltonian is

$$|\mathbf{k}, 0\rangle = \frac{1}{\sqrt{N}}\sum_{\mathbf{m}} c^{\dagger}_{\mathbf{m}s} \exp(i\mathbf{k}\cdot\mathbf{m})|0\rangle, \tag{8.137}$$

where $|0\rangle$ is the vacuum. By applying the textbook perturbation theory, one readily calculates the perturbed energy levels. Up to the second order in the hopping integral, they are given by

$$E(\mathbf{k}) = -E_p + \epsilon_{\mathbf{k}} - \sum_{\mathbf{k}', n_{\mathbf{q}}} \frac{|\langle \mathbf{k}, 0| \sum_{i,j} \hat{\sigma}_{ij} c^{\dagger}_i c_j |\mathbf{k}', n_{\mathbf{q}}\rangle|^2}{\sum_{\mathbf{q}} \omega_{\mathbf{q}} n_{\mathbf{q}}}, \tag{8.138}$$

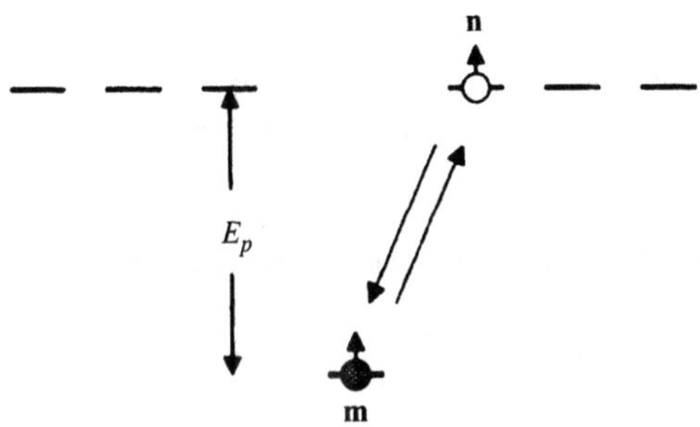

FIGURE 8.6. "Back and forth" virtual transitions of the polaron without any transfer of the lattice deformation from one site to another. These transitions shift the middle of the band down without any real charge delocalization.

where $|\mathbf{k}', n_{\mathbf{q}}\rangle$ are the exited states of the unperturbed Hamiltonian with one electron and at least one real phonon. The second term in this equation, which is linear with respect to the bare hopping $t(\mathbf{m})$, describes the polaron–band dispersion,

$$\epsilon_{\mathbf{k}} = \sum_{\mathbf{m}} t(\mathbf{m})\, e^{-g^2(\mathbf{m})} \exp(-i\mathbf{k}\cdot\mathbf{m}), \tag{8.139}$$

where

$$g^2(\mathbf{m}) = \frac{1}{2N}\sum_{\mathbf{q}} |\gamma(\mathbf{q})|^2[1 - \cos(\mathbf{q}\cdot\mathbf{m})] \tag{8.140}$$

is the *band-narrowing factor* at zero temperature. The third term, quadratic in $t(\mathbf{m})$, yields a **k**-*independent* correction to the polaron level shift of the order of $1/\lambda^2$. The origin of this correction, which could be much larger than the first-order contribution is understood in Fig. 8.6. The polaron localized in potential well of depth E_p on the site **m** hops onto a neighboring site **n** with no deformation around and comes back. As any second-order correction, this transition shifts the energy down by an amount of about $-t^2(\mathbf{m})/E_p$. It has little to do with the polaron effective mass and the polaron tunneling mobility because the lattice deformation around **m** does not follow the electron. The electron hops back and forth many times (about e^{g^2}), waiting for a sufficient lattice deformation to appear around the site **n**. Only after the deformation around **n** is created does the polaron tunnel onto the next site together with the deformation.

8.3.6. *From Continuous to Small Holstein and Small Fröhlich Polarons: QMC Simulation*

The narrowing of the band and the polaron effective mass strongly depend on the radius of the e-ph interaction [20]. Let us compare the small Holstein polaron (SHP) formed by the short-range e-ph interaction and a small polaron formed by the long-range (Fröhlich) interaction, which we refer as the small Fröhlich polaron (SFP). We use the real-space representation of $H_{\text{e-ph}}$ [Eq. (8.116)], introducing a normal coordinate at site **n** as

$$\xi_{\mathbf{n}} = \sum_{\mathbf{q}} (2NM\omega_{\mathbf{q}})^{-1/2} e^{i\mathbf{q}\cdot\mathbf{n}} d_{\mathbf{q}} + \text{h.c.} \tag{8.141}$$

and a "force" between the electron at site **m** and the normal coordinate $\xi_{\mathbf{n}}$,

$$f(\mathbf{m}) = N^{-1} \sum_{\mathbf{q}} \gamma(\mathbf{q})(M\omega_{\mathbf{q}}^3)^{1/2} e^{i\mathbf{q}\cdot\mathbf{m}}. \tag{8.142}$$

Then we write the e-ph interaction as

$$H_{\text{e-ph}} = \sum_{\mathbf{n},i} f(\mathbf{m} - \mathbf{n})\xi_{\mathbf{n}}\hat{n}_i. \tag{8.143}$$

For simplicity, we consider the interaction with a single-phonon branch and $\gamma(-\mathbf{q}) = \gamma(\mathbf{q})$. In general, there is no simple relation between the polaron level shift E_p and the exponent g^2 of the mass enhancement. This relation depends on the form of the e-ph interaction. Indeed, for dispersionless phonons ($\omega_{\mathbf{q}} = \omega$), we obtain

$$E_p = \frac{1}{2M\omega_0^2} \sum_{\mathbf{m}} f^2(\mathbf{m}) \tag{8.144}$$

and

$$g^2 = \frac{1}{2M\omega^3} \sum_{\mathbf{m}} \left[f^2(\mathbf{m}) - f(\mathbf{m}) f(\mathbf{m} + \mathbf{a}) \right], \tag{8.145}$$

where **a** is the primitive lattice vector. In the nearest-neighbor approximation, the effective mass renormalization is given by

$$\frac{m^*}{m} = e^{g^2},$$

where m is the bare band mass and $1/m^* = \partial^2\epsilon_{\mathbf{k}}/\partial k^2$ at $k \to 0$ is the inverse polaron mass.

If the interaction is short ranged, $f(\mathbf{m}) = \kappa\delta_{\mathbf{m},0}$ (the Holstein model), then $g^2 = E_p/\omega$. Here, κ is a constant. In general, we have $g^2 = \gamma E_p/\omega$ with the numerical coefficient

$$\gamma = \frac{1 - \sum_{\mathbf{m}} f(\mathbf{m}) f(\mathbf{m} + \mathbf{a})}{\sum_{\mathbf{n}} f^2(\mathbf{n})}, \tag{8.146}$$

which might be less than 1. To estimate γ, let us consider a one-dimensional chain model with the long-range Coulomb interaction between the electron on the chain

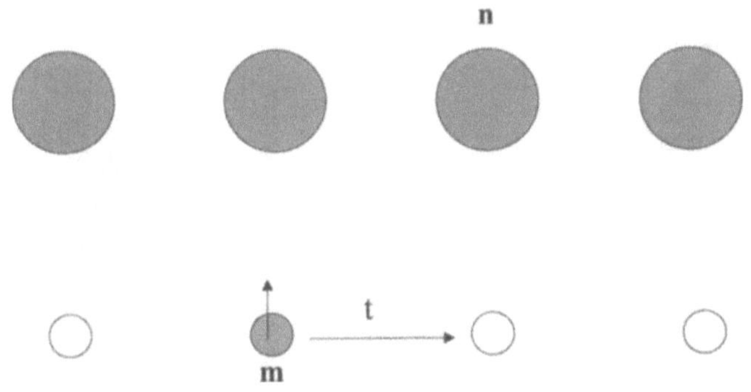

FIGURE 8.7. A one-dimensional model of the lattice polaron on the chain interacting with displacements of all ions of another chain.

($\times$) and ion vibrations of the chain ($\circ$), polarized in the direction perpendicular to the chains [60] (see Fig. 8.7). The corresponding force is given by

$$f(\mathbf{m}-\mathbf{n}) = \frac{\kappa}{(|\mathbf{m}-\mathbf{n}|^2+1)^{3/2}}. \tag{8.147}$$

Here, the distance along the chains $|\mathbf{m}-\mathbf{n}|$ is measured in units of the lattice constant a, the interchain distance is also a, and we take $a=1$. For this long-range interaction we obtain $E_p = 1.27\kappa^2/(2M\omega^2)$, $g^2 = 0.49\kappa^2/(2M\omega^3)$, and

$$g^2 = \frac{0.39E_p}{\omega}. \tag{8.148}$$

Thus, the effective mass renormalization is much smaller than in the Holstein model, roughly as $m^*_{\text{SFP}} \propto (m^*_{\text{SHP}})^{1/2}$.

Not only does the small-polaron mass strongly depend on the radius of the e-ph interaction but also does the range of the applicability of the analytical $1/\lambda$ expansion theory. The theory appears almost exact in a wide region of parameters for the Fröhlich interaction. The exact polaron mass in a wide region of the adiabatic parameter $\omega/t(a)$ and coupling was calculated with the continuous-time path-integral quantum Monte Carlo (QMC) algorithm [60]. This method is free from any systematic finite-size, finite-time-step, and finite-temperature errors and allows for an *exact* (in the QMC sense) calculation of the ground-state energy and the effective mass of the lattice polaron for any e-ph interaction. At large λ (>1.5), SFP was found to be much lighter than SHP, and the large Fröhlich polaron (i.e., at $\lambda < 1$) was *heavier* than the large Holstein polaron with the same binding energy (see Fig. 8.8). The mass ratio $m^*_{\text{FP}}/m^*_{\text{HP}}$ is a nonmonotonic function of λ. The effective mass of the Fröhlich polaron, $m^*_{\text{FP}}(\lambda)$, is well fitted by a single exponent, which is $e^{0.73\lambda}$ for $\omega = t(a)$ and $e^{1.4\lambda}$ for $\omega = 0.5\,t(a)$. The exponents are remarkably close to those obtained with the Lang–Firsov transformation: $e^{0.78\lambda}$ and $e^{1.56\lambda}$, respectively. Hence, in the case of the Fröhlich interaction, the transformation is perfectly accurate even in the moderate adiabatic regime; $\omega/t(a) \le 1$ for *any*

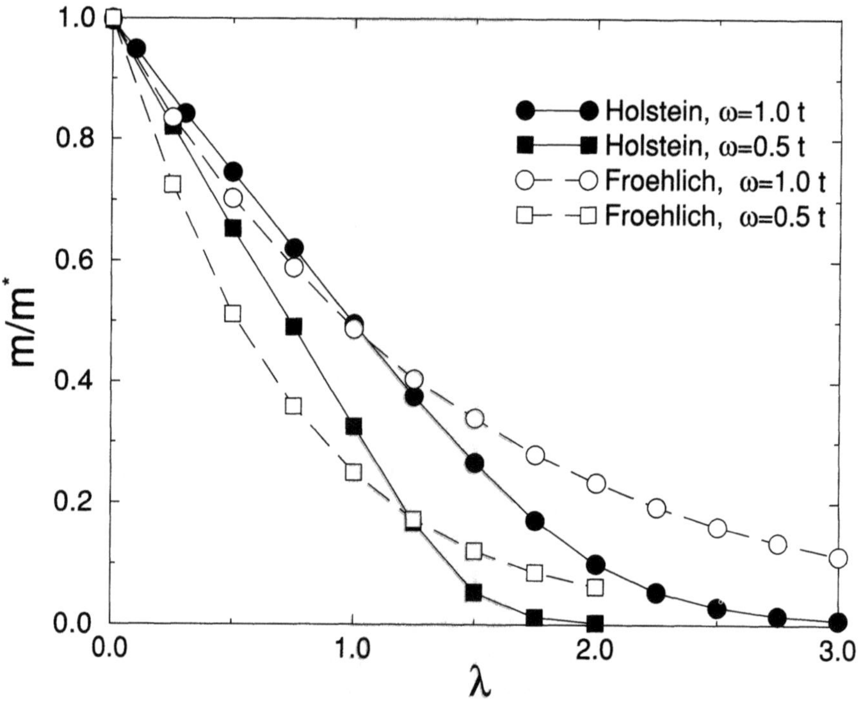

FIGURE 8.8. Inverse effective polaron mass in units of $1/m = 2t(a)a^2$.

coupling strength. It is not the case for the Holstein polaron. If the interaction is short ranged, the same analytical technique is applied only in the nonadiabatic regime $\omega/t(a) > 1$.

Another interesting point is that the size of SFP and the length, over which the distortion spreads, are *different*. In the strong-coupling limit, the polaron is almost localized on one site **m**. Hence, the size of its wavefunction is the atomic size. On the other hand, the ion displacements, proportional to the displacement force $f(\mathbf{m} - \mathbf{n})$, spread over a large distance. Their amplitude at a site **n** falls with the distance, as $|\mathbf{m} - \mathbf{n}|^{-3}$ in our one-dimensional model. The polaron cloud (i.e., lattice distortion) is more extended than the polaron itself. Such a polaron tunnels with a larger probability than the Holstein polaron due to a smaller *relative* lattice distortion around two neighboring sites. For a short-range e-ph interaction, the *entire* lattice deformation disappears at one site and then forms at its neighbor, when the polaron tunnels from site to site. Therefore $\gamma = 1$ and the polaron is very heavy already at $\lambda \approx 1$. To the contrary, if the interaction is long ranged, only a fraction of the total deformation changes every time the polaron tunnels from one site to its neighbor, and γ is smaller than 1. The fact that the Lang–Firsov transformation is rather accurate for the long-range interaction in a wide region of parameters allows us to generalize this result. Including all phonon branches in the three-dimensional lattice, we obtain $m^*_{\text{SFP}} = (m^*_{\text{SHP}})^{\gamma}$, where the masses are measured in units of the band mass and

$$\gamma = \frac{\sum_{\mathbf{q},\nu} \gamma^2(\mathbf{q},\nu)[1-\cos(\mathbf{q}\cdot\mathbf{m})]}{\sum_{\mathbf{q},\nu} \gamma^2(\mathbf{q},\nu)}. \tag{8.149}$$

For the Fröhlich interaction [i.e., $\gamma(\mathbf{q}) \propto 1/q$], we find $\gamma = 0.57$ in a cubic lattice and $\gamma = 0.255$ for the in-plane oxygen hole in cuprates [20]. A lighter mass of SFP compared with the nondispersive SHP is a generic feature of any dispersive e-ph interaction. As an example, a short-range interaction with dispersive acoustic phonons [$\gamma(\mathbf{q}) \propto 1/q^{1/2}$, $\omega_{\mathbf{q}} \propto q$] also leads to a lighter polaron in the strong-coupling regime compared with SHP. Actually, Holstein [10] pointed out in his original paper that the dispersion is a vital ingredient of the polaron theory.

8.4. ATTRACTIVE CORRELATIONS OF SMALL POLARONS

Applying the polaron canonical transformation to a generic "Fröhlich–Coulomb" Hamiltonian allows us to explicitly calculate the effective attraction of small polarons [61]. The Hamiltonian includes the infinite-range Coulomb, V_c, and e-ph interactions. The implicitly present infinite on-site repulsion (Hubbard U) prohibits double occupancy and removes the need to distinguish the fermionic spin. Introducing spinless fermion operators $c_{\mathbf{n}}$ and phonon operators $d_{\mathbf{m}\nu}$, the Hamiltonian is written as

$$\begin{aligned} H = {} & \sum_{\mathbf{n}\neq\mathbf{n}'} t(\mathbf{n}-\mathbf{n}')c^{\dagger}_{\mathbf{n}}c_{\mathbf{n}'} + \sum_{\mathbf{n}\neq\mathbf{n}'} V_c(\mathbf{n}-\mathbf{n}')c^{\dagger}_{\mathbf{n}}c_{\mathbf{n}}c^{\dagger}_{\mathbf{n}'}c_{\mathbf{n}'} \\ & + \omega \sum_{\mathbf{n}\neq\mathbf{m},\nu} g_{\nu}(\mathbf{m}-\mathbf{n})(\mathbf{e}_{\nu}\cdot\mathbf{e}_{\mathbf{m}-\mathbf{n}})c^{\dagger}_{\mathbf{n}}c_{\mathbf{n}}(d^{\dagger}_{\mathbf{m}\nu}+d_{\mathbf{m}\nu}) \\ & + \omega \sum_{\mathbf{m},\nu}\left(d^{\dagger}_{\mathbf{m}\nu}d_{\mathbf{m}\nu}+\frac{1}{2}\right). \end{aligned} \tag{8.150}$$

The e-ph term is written in real space (Section 8.3.4), which is more convenient in working with complex lattices, and the phonon mode index ν is introduced for generality.

This many-body Hamiltonian is of considerable complexity. However, we are interested in the limit of the strong e-ph interaction. In this case, the kinetic energy is a perturbation and the model can be grossly simplified using the canonical transformation in the Wannier representation for electrons and phonons:

$$S = \sum_{\mathbf{m}\neq\mathbf{n},\nu} g_{\nu}(\mathbf{m}-\mathbf{n})(\mathbf{e}_{\nu}\cdot\mathbf{e}_{\mathbf{m}-\mathbf{n}})c^{\dagger}_{\mathbf{n}}c_{\mathbf{n}}(d^{\dagger}_{\mathbf{m}\nu}-d_{\mathbf{m}\nu}).$$

The transformed Hamiltonian is

$$\tilde{H} = e^{-S} H e^{S} = \sum_{\mathbf{n}\neq\mathbf{n}'} \hat{\sigma}_{\mathbf{nn}'} c_{\mathbf{n}}^{\dagger} c_{\mathbf{n}'} + \omega_0 \sum_{\mathbf{m}\alpha} \left(d_{\mathbf{m}\nu}^{\dagger} d_{\mathbf{m}\nu} + \frac{1}{2} \right) \tag{8.151}$$
$$+ \sum_{\mathbf{n}\neq\mathbf{n}'} v(\mathbf{n}-\mathbf{n}') c_{\mathbf{n}}^{\dagger} c_{\mathbf{n}} c_{\mathbf{n}'}^{\dagger} c_{\mathbf{n}'} - E_p \sum_{\mathbf{n}} c_{\mathbf{n}}^{\dagger} c_{\mathbf{n}}.$$

The last term describes the energy gained by polarons due to e-ph interaction. E_p is the familiar polaron level shift

$$E_p = \omega \sum_{\mathbf{m}\nu} g_{\nu}^2(\mathbf{m}-\mathbf{n})(\mathbf{e}_{\nu} \cdot \mathbf{e}_{\mathbf{m}-\mathbf{n}})^2, \tag{8.152}$$

which is independent of $\mathbf{n}$. The third term on the right-hand side is the polaron–polaron interaction:

$$v(\mathbf{n}-\mathbf{n}') = V_c(\mathbf{n}-\mathbf{n}') - V_{ph}(\mathbf{n}-\mathbf{n}'), \tag{8.153}$$

where

$$V_{ph}(\mathbf{n}-\mathbf{n}') = 2\omega \sum_{\mathbf{m},\nu} g_{\nu}(\mathbf{m}-\mathbf{n}) g_{\nu}(\mathbf{m}-\mathbf{n}')$$
$$\times (\mathbf{e}_{\nu} \cdot \mathbf{e}_{\mathbf{m}-\mathbf{n}})(\mathbf{e}_{\nu} \cdot \mathbf{e}_{\mathbf{m}-\mathbf{n}'}).$$

The phonon-induced interaction $V_{\rm ph}$ is due to displacements of common ions caused by two electrons. Finally, the transformed hopping operator $\hat{\sigma}_{\mathbf{nn}'}$ is given by

$$\hat{\sigma}_{\mathbf{nn}'} = t(\mathbf{n}-\mathbf{n}') \exp\left[\sum_{\mathbf{m},\nu} [g_{\nu}(\mathbf{m}-\mathbf{n})(\mathbf{e}_{\nu} \cdot \mathbf{e}_{\mathbf{m}-\mathbf{n}}) \right.$$
$$\left. - g_{\nu}(\mathbf{m}-\mathbf{n}')(\mathbf{e}_{\nu} \cdot \mathbf{e}_{\mathbf{m}-\mathbf{n}'})](d_{\mathbf{m}\alpha}^{\dagger} - d_{\mathbf{m}\alpha}) \right]. \tag{8.154}$$

This term is a perturbation at large λ. It is absent in an isolated MQD (see Section 8.6.1), so that the canonical transformation solves the MQD problem exactly for any number of electrons. In a crystal, the term allows for the *bipolaron* tunneling and high-temperature superconductivity [30]. In crystal structures like perovskites, the tunnneling appears already in the first order in $t(\mathbf{n})$, so that $\hat{\sigma}_{\mathbf{nn}'}$ can be averaged over phonons. The result is

$$\tilde{t}(\mathbf{n}-\mathbf{n}') \equiv \langle\langle \hat{\sigma}_{\mathbf{nn}'} \rangle\rangle_{\rm ph} = t(\mathbf{n}-\mathbf{n}') \exp[-g^2(\mathbf{n}-\mathbf{n}')], \tag{8.155}$$

where

$$g^2(\mathbf{n}-\mathbf{n}') = \sum_{\mathbf{m},\nu} g_\nu(\mathbf{m}-\mathbf{n})(\mathbf{e}_\nu \cdot \mathbf{e}_{\mathbf{m}-\mathbf{n}}) \times \left[g_\nu(\mathbf{m}-\mathbf{n})(\mathbf{e}_\nu \cdot \mathbf{e}_{\mathbf{m}-\mathbf{n}}) - g_\nu(\mathbf{m}-\mathbf{n}')(\mathbf{e}_\nu \cdot \mathbf{e}_{\mathbf{m}-\mathbf{n}'})\right].$$

By comparing Eqs. (8.152),(8.153), and (8.155), the mass renormalization exponent can be expressed via E_p and V_{ph} as follows:

$$g^2(\mathbf{n}-\mathbf{n}') = \frac{1}{\omega_0}\left[E_p - \frac{1}{2}V_{\text{ph}}(\mathbf{n}-\mathbf{n}')\right]. \qquad (8.156)$$

When V_{ph} is larger than V_c, the full interaction becomes negative and small polarons form lattice bipolarons. The real-space representation allows us to elaborate more physics behind the lattice sums in Eqs. (8.152) and (8.153). If a carrier (electron or hole) acts on an ion with a force $\mathbf{f}$, it displaces the ion by some vector $\mathbf{x} = \mathbf{f}/k$. Here, k is the ion's force constant. The total energy of the carrier–ion pair is $-\mathbf{f}^2/2k$. This is precisely the summand in Eq. (8.152) expressed via dimensionless coupling constants. Now consider two carriers interacting with the *same* ion; see Fig. 8.9. The ion displacement is $\mathbf{x} = (\mathbf{f}_1 + \mathbf{f}_2)/k$ and the energy is $-\mathbf{f}_1^2/(2k) - \mathbf{f}_2^2/(2k) - (\mathbf{f}_1 \cdot \mathbf{f}_2)/k$. Here, the last term should be interpreted as an ion-mediated interaction between the two carriers. It depends on the scalar product of $\mathbf{f}_1$ and $\mathbf{f}_2$ and, consequently, on the relative positions of the carriers with respect

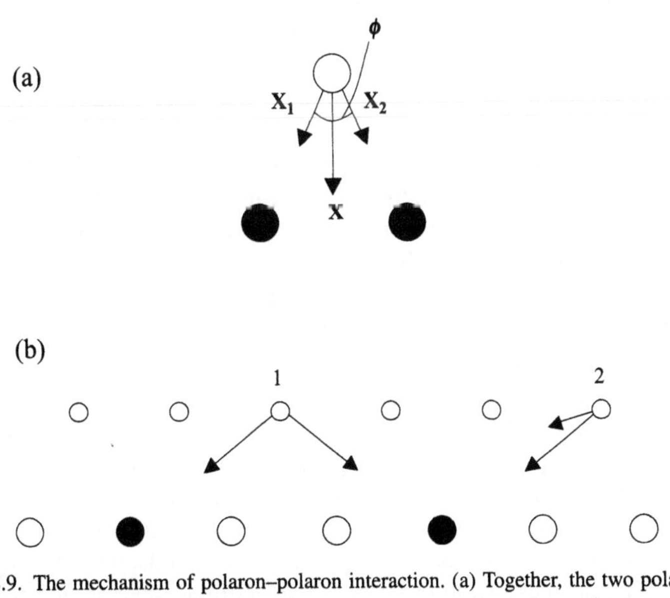

FIGURE 8.9. The mechanism of polaron–polaron interaction. (a) Together, the two polarons (solid circles) deform the lattice more effectively than separately. An effective attraction occurs when the angle ϕ is less than $\pi/2$. (b) A mixed situation. Ion 1 results in repulsion between two polarons, whereas ion 2 results in attraction.

to the ion. If the ion is an isotropic harmonic oscillator, as we assume here, then the following simple rule applies. If the angle ϕ between $\mathbf{f}_1$ and $\mathbf{f}_2$ is less than $\pi/2$, the polaron–polaron interaction will be attractive; otherwise it will be repulsive. In general, some ions will generate attraction and some will generate repulsion between polarons; see Fig. 8.9b.

The overall sign and magnitude of the interaction is given by the lattice sum in Eq. (8.153), the evaluation of which is elementary. One should also note that according to Eq. (8.156), an attractive interaction reduces the polaron mass (and, consequently, the bipolaron mass), whereas a repulsive interaction enhances the mass.

8.5. MOLECULAR SWITCHING: NEGATIVE-U HUBBARD MODEL

A few experimental studies [46, 58] provided evidence for various molecular switching effects, where the I–V characteristics show two branches with high and low current for the same voltage. This remarkable phenomenon can result from a conformational transformation of certain molecules containing a "moving part" like a bypyridinium ring, which changes its position if the voltage is sufficiently high. Such a transformation necessarily involves a large displacement of many atoms so that this *ionic* switching is rather slow, perhaps operating on a millisecond scale. Switching has also been observed for simple molecules (organic molecular films), and in some cases, it is strongly dependent on the choice of contacts and substrates [58]. In this and the last subsection, we overviewed the theory of a fast switching due to attractive polaron–polaron correlations in molecules, which could be the basis of future oscillators, amplifiers and other important circuit elements.

Repulsive electron correlations cause the "Coulomb blockade" in the I–V characteristics of QDs [62]. However, they cannot cause any switching. We show here that a negative Hubbard U of any origin can provide an intrinsic nonretarded current switching of MQDs, one mechanism that can produce a negative U in molecular systems is a strong e-ph (vibronic) interaction, as discussed in the previous sub section. If the tunneling time is comparable to or larger than the characteristic phonon times, a polaron is formed inside the molecular wire [54]. There is a wide range of bulk molecular conductors with polaronic carriers. Since the formation of polarons in polyacetylene (PA) was theoretically discussed [63], they were detected optically in PA [64], in conjugated polymers such as polyphenylene, polypyrrole, polythiophene, polyphenylene sulfide [65], Cs-doped biphenyl [66], N-doped bithiophene [67], polyphenylenevinylene (PPV)-based light-emitting diodes [68], and in other molecular systems. In contrast to bare electrons, polarons attract each other at short distances of the order of the interatomic spacing and form small bipolarons (Section 8.4). Bipolarons can strongly affect the transport properties of long molecular wires [69]. Even when bipolarons are not formed in MQDs because of the short lifetime of the carrier inside the molecule, the attractive correlations between carriers still remain.

8.5.1. *Steady Current Through MQDs*

We employ the Landauer-type expression for a steady current through a region of interacting electrons, derived by Meir and Wingreen [70] as

$$I(V) = -\frac{e}{\pi}\int_{-\infty}^{\infty} d\omega\,[f_1(\omega) - f_2(\omega)]\,\mathrm{Im}\,\mathrm{Tr}\left[\hat{\Gamma}(\omega)\hat{G}^R(\omega)\right], \tag{8.157}$$

where $f_{1(2)}(\omega) = \left\{\exp[(\omega + \Delta \mp \frac{1}{2}eV)/T] + 1\right\}^{-1}$, T is the temperature, and Δ is the position of the lowest unoccupied molecular level with respect to the chemical potential. $\hat{\Gamma}(\omega)$ depends on the density of states (DOS) in the leads and on the hopping integrals connecting one-particle states in the left (1) and the right (2) leads with the states in MQDs (see Fig. 8.10).

This formula includes, by means of the Fourier transform of the molecular retarded Green's function (GF), $\hat{G}^R(\omega)$, the e-ph and Coulomb interactions inside the MQDs and coupling to the leads. Since the leads are metallic, electron–electron and e-ph interactions in the leads and interactions of electrons in the leads with electrons and phonons in the MQDs can be neglected. We are interested in the tunneling near the conventional threshold, $eV = 2\Delta$ (see Fig. 8.10) within a voltage range about an effective attractive potential $|U|$.

The attractive energy is the difference of two large interactions: the Coulomb repulsion and the phonon mediated attraction, of the order of 1 eV each. Hence, $|U|$ is of the order of a few tens of electron volts. We neglect the energy dependence of $\hat{\Gamma}(\omega) \approx \Gamma$ on this scale and assume that the coupling to the leads is weak, $\Gamma \ll |U|$. In this case, $\hat{G}^R(\omega)$ does not depend on the leads. Moreover, we assume that there is a complete set of one-particle molecular states $|\mu\rangle$, where $\hat{G}^R(\omega)$ is diagonal. With these assumptions, we can reduce Eq. (8.157) to

$$I(V) = I_0\int_{-\infty}^{\infty} d\omega\,[f_1(\omega) - f_2(\omega)]\,\rho(\omega), \tag{8.158}$$

allowing for a transparent analysis of essential physics of the switching phenomenon. Here, $I_0 = e\Gamma$ and the molecular DOS, $\rho(\omega)$, is given by

$$\rho(\omega) = -\frac{1}{\pi}\sum_{\mu}\mathrm{Im}\,\hat{G}^R_{\mu}(\omega), \tag{8.159}$$

where $\hat{G}^R_{\mu}(\omega)$ is the Fourier transform of $\hat{G}^R_{\mu}(t) = -i\theta(t)\left\langle\left\{c_{\mu}(t), c^{\dagger}_{\mu}\right\}\right\rangle$, $\{\cdots,\cdots\}$ is the anticommutator, $c_{\mu}(t) = e^{iHt}c_{\mu}e^{-iHt}$, and $\theta(t) = 1$ for $t > 0$ and zero otherwise.

8.5.2. MQD Green's Function and Rate Equation

First, we calculate $\rho(\omega)$ exactly in the framework of the negative Hubbard-U Hamiltonian, which includes an attractive electron–electron interaction U in the molecular eigenstate ε_μ coupled with the *left* and *right* leads by the hopping integrals $t_{\alpha k\mu}$:

$$H = \sum_{\mu} \varepsilon_{\mu} \hat{n}_{\mu} + \frac{1}{2} U \sum_{\mu \neq \mu'} \hat{n}_{\mu} \hat{n}_{\mu'} + \sum_{k,\alpha} \xi_{\alpha k} a_{\alpha k}^{\dagger} a_{\alpha k} + \sum_{k,\mu,\alpha} (t_{\alpha k\mu} a_{\alpha k}^{\dagger} c_{\mu} + \text{h.c.}). \tag{8.160}$$

Here, $a_{\alpha k}$ and c_μ are the annihilation operators in the left ($\alpha = 1$) and right ($\alpha = 2$) leads and in the molecule, respectively, $\hat{n}_\mu = c_\mu^\dagger c_\mu$, $\xi_{\alpha k}$ is the energy dispersion in the leads, and $U < 0$. A similar Hamiltonian was applied to glassy semicon-

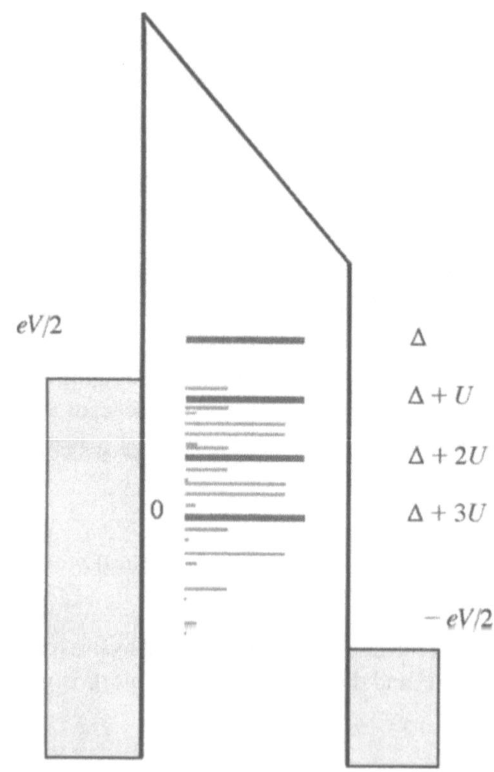

FIGURE 8.10. Schematic of the energy levels and phonon side-bands for MQDs under bias voltage V ($eV/2\Delta = 0.75$) with the coupling constant $\gamma^2 = 11/13$. The level is assumed to be fourfold degenerate ($d = 4$) with energies $\Delta + rU, r = 0, \ldots, (d-1)$ (thick bars). Thin bars show the vibron side-bands with the size of the bar proportional to the weight of the particular contribution in the density of states (see text) in the case of one vibron with frequency $\omega_0/\Delta = 0.2$ at $T = 0$. Only the bands in the energy window ($eV/2$, $-eV/2$) (shown) will contribute to current at zero temperature.

ductors [21], and high-T_c superconductors [30, 71], including doped fullerenes [72].

We use the perturbation theory with respect to the hopping integrals, neglecting any contribution to the current other than $t_{\alpha k\mu}^2$, but keeping all orders of the negative Hubbard U. Terms of higher order in $t_{\alpha k\mu}$ cannot change the gross I–V features for any voltage except the narrow transition regime from one branch to another. Applying the equations of motion for the Heisenberg operators $c_\mu(t)$, $\hat{n}_\mu(t)$, and $a_{\alpha k}(t)$, we obtain an *infinite* set of coupled equations for the molecular GFs as

$$i\frac{dG_\mu^{(N)}(t)}{dt} = \delta(t)\sum_{\mu_1\neq\mu_2\neq\cdots\neq\mu}\prod_{i=1}^{N-1} n_{\mu_i}(0) + [\varepsilon_\mu + (N-1)U]G_\mu^{(N)}(t) + UG_\mu^{(N+1)}(t), \qquad (8.161)$$

where $n_\mu(t) = \langle c_\mu^\dagger(t)c_\mu(t)\rangle$ is the expectation number of electrons on the molecular level, and the N-particle retarded GF is defined as

$$G_\mu^{(N)}(t) = -i\Theta(t)\sum_{\mu_1\neq\mu_2\neq\cdots\neq\mu}\left\langle\left\{c_\mu(t)\prod_{i=1}^{N-1}\hat{n}_{\mu_i}(t), c_\mu^\dagger\right\}\right\rangle \qquad (8.162)$$

for $2 \leqslant N < \infty$ and $G_\mu^{(1)}(t) \equiv \hat{G}_\mu^R(t)$.

For the sake of analytical transparency, we solve this system for a molecule having one d-fold degenerate energy level with $\varepsilon_\mu = 0$. In this case, the set is finite, and it can be solved using the Fourier transforms. The one-particle GF is found as

$$G_\mu^{(1)}(\omega) = \sum_{r=0}^{d-1}\frac{Z_r(n)}{\omega - rU + i\delta}, \qquad (8.163)$$

where $\delta = +0$, $n = n_\mu(0)$, and

$$Z_r(n) = \frac{(d-1)!}{r!(d-1-r)!}n^r(1-n)^{d-1-r}.$$

This is an exact solution with respect to correlations that satisfy all sum rules.

The electron density $n_\mu(t)$ obeys the rate equation, which is obtained by using the equations of motion:

$$\frac{dn_\mu(t)}{dt} = 2\sum_{\alpha,k} t_{\alpha k\mu}\,\mathrm{Im}\,A_{\alpha k\mu}^{(1)}(t), \qquad (8.164)$$

where

$$A_{\alpha k\mu}^{(N)}(t) = \sum_{\mu_1\neq\mu_2\neq\cdots\neq\mu}\left\langle c_\mu^\dagger(t)\prod_{i=1}^{N-1}\hat{n}_{\mu_i}(t)a_{\alpha k}(t)\right\rangle. \qquad (8.165)$$

These correlation functions should be calculated up to the first order with respect to the hopping integrals $t_{\alpha k\mu}$. In this order, they satisfy the infinite set of coupled equations

$$i\frac{dA^{(N)}_{\alpha k\mu}(t)}{dt} = t_{\alpha k\mu}[n_\mu(t) - f(\xi_{\alpha k})] \sum_{\mu_2 \neq \cdots \neq \mu} \prod_{i=1}^{N-1} n_{\mu_i}(t) + UA^{(N+1)}_{\alpha k\mu}(t) + [\xi_{\alpha k} - (N-1)U]A^{(N)}_{\alpha k\mu}(t). \quad (8.166)$$

When we have a finite number of molecular states, the set is finite. One readily solves the set in the stationary case, when $n_{\mu_i}(t)$ and $A^{(N)}_{\alpha k\mu}(t)$ do not depend on time. For the d-fold degenerate energy level, the one-particle correlation function is found to be

$$A^{(1)}_{\alpha k\mu} = [n - f(\xi_{\alpha k})]t_{\alpha k\mu} \sum_{r=0}^{d-1} \frac{Z_r(n)}{\xi_{\alpha k} - rU + i\delta}. \quad (8.167)$$

Substituting it into Eq. (8.166), we obtain the stationary rate equation for the electron density on the molecule:

$$\sum_\alpha \sum_{r=0}^{d-1} [n - f_\alpha(rU)]Z_r(n) = 0. \quad (8.168)$$

8.5.3. *Switching Effect*

Now using Eqs. (8.158), (8.159), and (8.163), the current is found to be

$$j = \sum_{r=0}^{d-1} [f_1(rU) - f_2(rU)]Z_r(n), \quad (8.169)$$

where $j = I/(dI_0)$. Let us consider the twofold, fourfold, and sixfold degenerate molecular levels. For $d = 2$, the kinetic equation is linear in n and there is only one solution:

$$n = \frac{\sum_\alpha f_\alpha(0)}{2 + \sum_\alpha [f_\alpha(0) - f_\alpha(U)]}. \quad (8.170)$$

The current through a twofold degenerate molecular dot is found to be

$$j = 2\frac{f_1(0)[1 - f_2(U)] - f_2(0)[1 - f_1(U)]}{2 + \sum_\alpha [f_\alpha(0) - f_\alpha(U)]}. \quad (8.171)$$

There is no current bistability in this case. Moreover, if the temperature is low ($T \ll \Delta, |U|$), there is practically no effect of correlations on the current,

$j \approx V\Theta(e|V| - 2\Delta)/|V|$. Remarkably, fourfold or higher degenerate negative-U dots reveal a switching effect. In this case, the kinetic equation is nonlinear, allowing for a few solutions. If $eV < 2(\Delta - |U|)$, the only physically allowed solution of Eq. (8.168) for $d = 4$ and $d = 6$ at zero temperature is $n = 0$. If $2(\Delta - |U|) < eV < 2\Delta$ and $T = 0$, the rate equation is reduced to

$$2n = 1 - (1 - n)^{d-1}. \tag{8.172}$$

For $d = 4$, it has *two* physical roots: $n = 0$ and $n = (3 - 5^{1/2})/2 \approx 0.38$. In this voltage range, $f_1(0) = f_2(rU) = 0$, but $f_1(U) = f_1(2U) = f_1(3U) = 1$ at $T = 0$. Using the sum rule $\sum_{r=0}^{d-1} Z_r(n) = 1$ and the rate equation (8.168), the current is simplified in this voltage range to $j = 2n$. Hence, we obtain two stationary states of the molecule with low (zero at $T = 0$) and high current, $I \approx 0.76I_0$ for the same voltage in the range $2(\Delta - |U|) < eV < 2\Delta$. For $d = 6$, the kinetic equation has two physical roots in this voltage range: $n = 0$ and $n \approx 0.48$, which corresponds to $I = 0$ and $I \approx 0.96I_0$, respectively. Above the standard threshold, $eV > 2\Delta$, where $f_1(rU) = 1$ and $f_2(rU) = 0$, there is only one solution: $n = 0.5$ with the current $I = I_0$.

One can better understand the origin of the switching phenomenon by taking the limit $d \gg 1$. The physical roots to Eq. (8.172) are $n = 0$ and $n = 0.5$ in this limit, with the current $I = 0$ and $I = I_0$, respectively. This is precisely the solution of the problem in the mean-field approximation (MFA), which is a reasonable approximation for $d \gg 1$. Indeed, using MFA, one replaces the exact two-body interaction in the Hamiltonian for a mean-field potential as $\frac{1}{2}U\sum_{\mu\neq\mu'}\hat{n}_\mu\hat{n}_{\mu'} \approx U\sum_{\mu\neq\mu'}\hat{n}_\mu n_{\mu'} - \frac{1}{2}U\sum_{\mu\neq\mu'}n_\mu n_{\mu'}$. Then the MFA DOS is given by $\rho_\mu(\omega) = \delta[\omega - U(d-1)n]$. Using the Fermi–Dirac golden rule, the rate equation for n becomes

$$\frac{dn}{dt} = -2\Gamma n + \Gamma\sum_\alpha f_\alpha[nU(d-1)]. \tag{8.173}$$

For $T = 0$, there are two stationary solutions of Eq. (8.173): $n = 0$ and $n = 0.5$ in the voltage range $2(\Delta - |\tilde{U}|) < eV < 2\Delta$; there is only one solution: $n = 0.5$ for $eV > 2\Delta$, where $\tilde{U} = U(d-1)/2$. The MFA current is found to be

$$j = f_1(2n\tilde{U}) - f_2(2n\tilde{U}). \tag{8.174}$$

Combining this equation and the rate equation (8.173) with $dn/dt = 0$, we obtain the I–V characteristic equation as

$$\frac{|\tilde{U}|}{\Delta}(1 - R) = 1 - \frac{T}{\Delta}\ln\left[\frac{(1+R)\sinh(eV/2T)}{j} - \cosh\left(\frac{eV}{2T}\right)\right],$$

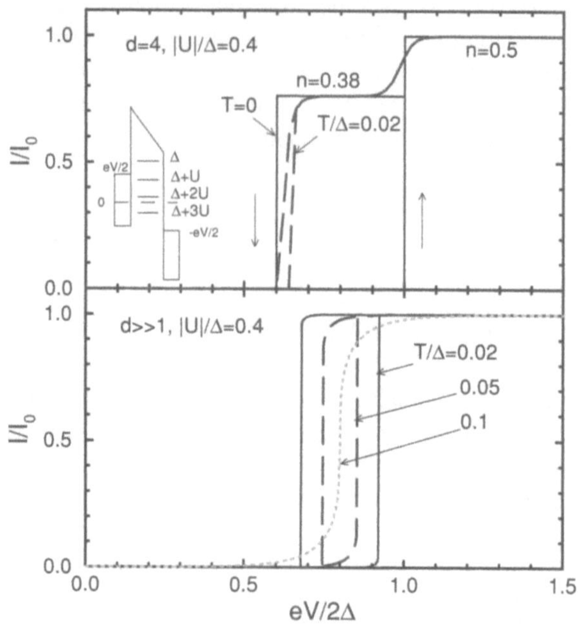

FIGURE 8.11. The I–V hysteresis loop in the degenerate negative-U model of a MQD for two temperatures.

where

$$R = \left\{ \left[1 - j \coth\left(\frac{eV}{2T} \right) \right]^{1/2} - \frac{j^2}{\sinh^2(eV/2T)} \right\}^{1/2}. \tag{8.175}$$

The I–V curves are shown in Fig. 8.11 for different temperatures and $|\tilde{U}| = 0.9\Delta$. Interestingly, the temperature narrows the voltage range of the hysteresis loop, but the transition from the low (high)-current branch to the high (low)-current branch remains discontinuous.

Let us examine the stability of each branch in the framework of the MFA rate equation. Introducing small fluctuations of the electron density as $n(t) = n + \delta n \exp(\gamma t)$ and linearizing Eq. (8.173) with respect to δn, we find the increment γ:

$$\gamma = -2\Gamma + \frac{|\tilde{U}|\Gamma}{2T} \cosh^{-2}\left(\frac{\Delta - 2n|\tilde{U}| - eV/2}{2T} \right) + \frac{|\tilde{U}|\Gamma}{2T} \cosh^{-2}\left(\frac{\Delta - 2n|\tilde{U}| + eV/2}{2T} \right). \tag{8.176}$$

One can see from this equation that at temperatures $T \ll |\tilde{U}|$, the low-current branch ($n = 0$) loses its stability at the threshold $V_2 = 2\Delta/e$, whereas the high-current branch loses its stability at $V_1 = 2(\Delta - |\tilde{U}|)/e$. In the voltage range $V_1 < V < V_2$, both branches are stable: $\gamma \approx -2\Gamma < 0$.

Finally, let us analyze the effect of a splitting of the degenerate molecular level on the bistability. The degeneracy could be removed because of Jahn–Teller distortions and/or the coupling with the leads. We assume that $d \gg 1$ levels are evenly distributed in a band of a width W. Then the MFA rate equation is modified as

$$\frac{dn_\mu}{dt} = -2\Gamma n_\mu + \Gamma \sum_\alpha f_\alpha(\varepsilon_\mu + NU), \tag{8.177}$$

where $N = \sum_{\mu'} n_{\mu'}$. For $T = 0$ in the stationary regime $(dn_\mu/dt = 0)$, it has two solutions: $N = 0$ and $N = d/2$, in the voltage range $V_1 + W < V < V_2$ with the current $j = 0$ and $j = 1$, respectively. We conclude that the level splitting $W < |U|$ leads to a narrowing of the voltage range of the bistability similar to the temperature narrowing shown in Fig. 8.11. A parameter-free estimate of the negative Hubbard U in some oxides yields $|U|$ about a few tens of an electron volts [73].

8.6. POLARONIC SWITCHING

Let us now apply the polaron formalism to MQDs [59]. Here, bipolarons might not exist because of a finite lifetime of electrons on a molecule connected with the leads; nevertheless, the attractive correlations could strongly modify the I–V characteristics. The appropriate molecular Hamiltonian, including the Coulomb repulsion, U^C, and the electron–vibron interaction is now

$$\begin{aligned} H = {} & \sum_\mu \varepsilon_\mu \hat{n}_\mu + \frac{1}{2} \sum_{\mu \neq \mu'} U^C_{\mu\mu'} \hat{n}_\mu \hat{n}_{\mu'} \\ & + \sum_{\mu,q} \hat{n}_\mu \omega_q (\gamma_{\mu q} d_q + \text{h.c.}) + \sum_q \omega_q \left(d_q^\dagger d_q + \frac{1}{2} \right). \end{aligned} \tag{8.178}$$

Here, d_q annihilates phonons, ω_q is the phonon (vibron) frequency, and $\gamma_{\mu q}$ are the e-ph coupling constant (q enumerates the vibron modes). This Hamiltonian conserves the occupation numbers of molecular states $\hat{n}_\mu$. Hence, it is compatible with Eq. (8.158).

8.6.1. MQD Density of States: Correlation and Phonon Side Bands

We apply the canonical polaron unitary transformation e^S, as in Section 8.3.5, integrating out phonons. The electron and phonon operators are transformed as

$$\tilde{c}_\mu = c_\mu X_\mu \tag{8.179}$$

and

$$\tilde{d}_q = d_q - \sum_{\mu} \hat{n}_\mu \gamma^*_{\mu q}, \tag{8.180}$$

respectively. Here,

$$X_\mu = \exp\left[\sum_q \gamma_{\mu q} d_q - \text{h.c.}\right].$$

The Lang–Firsov canonical transformation shifts ions to new equilibrium positions with no effect on the phonon frequencies. The diagonalization is exact in MQDs:

$$\tilde{H} = \sum_i \tilde{\varepsilon}_\mu \hat{n}_\mu + \sum_q \omega_q \left(d_q^\dagger d_q + \frac{1}{2}\right) + \frac{1}{2} \sum_{\mu \neq \mu'} U_{\mu\mu'} \hat{n}_\mu \hat{n}_{\mu'}, \tag{8.181}$$

where

$$U_{\mu\mu'} \equiv U^C_{\mu\mu'} - 2 \sum_q \gamma^*_{\mu q} \gamma_{\mu' q} \omega_q \tag{8.182}$$

is the interaction of polarons comprising their interaction via molecular deformations (vibrons) and nonvibron (e.g., Coulomb repulsion) $U^C_{\mu\mu'}$. To simplify the discussion, we will assume that the Coulomb integrals do not depend on the orbital index (i.e., $U_{\mu\mu'} = U$).

The molecular energy levels are shifted by the polaron level shift due to a deformation well created by a polaron:

$$\tilde{\varepsilon}_\mu = \varepsilon_\mu - \sum_q |\gamma_{\mu q}|^2 \omega_q. \tag{8.183}$$

Applying the same transformation in the retarded GF, we obtain

$$\begin{aligned} G^R_\mu(t) &= -i\theta(t) \langle \{ c_\mu(t) X_\mu(t),\ c^\dagger_\mu X^\dagger_\mu \} \rangle \\ &= -i\theta(t) [\langle c_\mu(t) c^\dagger_\mu \rangle \langle X_\mu(t) X^\dagger_\mu \rangle \\ &\quad + \langle c^\dagger_\mu c_\mu(t) \rangle \langle X^\dagger_\mu X_\mu(t) \rangle], \end{aligned} \tag{8.184}$$

where now electron and phonon operators are averaged over the quantum state of the transformed Hamiltonian $\tilde{H}$. There is no coupling between polarons and vibrons in the transformed Hamiltonian, so that

$$\langle X_\mu(t) X^\dagger_\mu \rangle = \exp\left[\sum_q \frac{|\gamma_{\mu q}|^2}{\sinh(\beta\omega_q/2)} \left[\cos\left(\omega t + i\frac{\beta\omega_q}{2}\right) - \cosh\frac{\beta\omega_q}{2}\right]\right], \tag{8.185}$$

where $\beta = 1/T$ and $\langle X_\mu^\dagger X_\mu(t)\rangle = \langle X_\mu(t) X_\mu^\dagger\rangle^*$.

Next, we introduce the N-particle GFs, which will necessarily appear in the equations of motion for $\langle c_\mu(t) c_\mu^\dagger\rangle$, as

$$G_\mu^{(N,+)}(t) \equiv -i\theta(t) \sum_{\mu_1 \neq \mu_2 \neq \cdots \neq \mu} \left\langle c_\mu(t) c_\mu^\dagger \prod_{i=1}^{N-1} \hat{n}_{\mu_i} \right\rangle \tag{8.186}$$

and

$$G_\mu^{(N,-)}(t) \equiv -i\theta(t) \sum_{\mu_1 \neq \mu_2 \neq \cdots \neq \mu} \left\langle c_\mu^\dagger c_\mu(t) \prod_{i=1}^{N-1} \hat{n}_{\mu_i} \right\rangle. \tag{8.187}$$

Then, using the equation of motion for the Heisenberg polaron operators, we derive the following equations for the N-particle GFs:

$$i\frac{dG_\mu^{(N,+)}(t)}{dt} = \delta(t)(1 - n_\mu) \sum_{\mu_1 \neq \mu_2 \neq \cdots \neq \mu} \prod_{i=1}^{N-1} n_{\mu_i} + [\tilde{\varepsilon}_\mu + (N-1)U] G_\mu^{(N,+)}(t) + U G_\mu^{(N+1,+)}(t) \tag{8.188}$$

and

$$i\frac{dG_\mu^{(N,-)}(t)}{dt} = \delta(t) n_\mu \sum_{\mu_1 \neq \mu_2 \neq \cdots \neq \mu} \prod_{i=1}^{N-1} n_{\mu_i} + [\tilde{\varepsilon}_\mu + (N-1)U] G_\mu^{(N,-)}(t) + U G_\mu^{(N+1,-)}(t), \tag{8.189}$$

where $n_\mu = \langle c_\mu^\dagger c_\mu\rangle$ is the expectation number of electrons on the molecular level μ.

We readily solve this set of coupled equations for MQDs with one d-fold degenerate energy level and with the e-ph coupling, $\gamma_{\mu q} = \gamma_q$, which does not break the degeneracy. Assuming that $n_\mu = n$, Fourier transformation of the set yields for $N = 1$

$$G_\mu^{(1,+)}(\omega) = (1 - n) \sum_{r=0}^{d-1} \frac{Z_r(n)}{\omega - rU + i\delta}, \tag{8.190}$$

$$G_\mu^{(1,-)}(\omega) = n \sum_{r=0}^{d-1} \frac{Z_r(n)}{\omega - rU + i\delta}, \tag{8.191}$$

where $\delta = +0$ and $Z_r(n)$ is the same as in Eq. (8.163).

In approximation, where we retain a coupling to a single mode with the characteristic frequency ω_0 and $\gamma_q \equiv \gamma$, the molecular DOS is readily found as an imaginary part of the Fourier transform of Eq. (8.184) using Eqs. (8.185), (8.190), and (8.191):

$$\rho(\omega) = \mathcal{Z} d \sum_{r=0}^{d-1} Z_r(n) \sum_{l=0}^{\infty} I_l(\xi) \\ \times \Big[e^{\beta\omega_0 l/2} \left[(1-n)\delta(\omega - rU - l\omega_0) + n\delta(\omega - rU + l\omega_0) \right] \\ + (1-\delta_{l0}) e^{-\beta\omega_0 l/2} [n\delta(\omega - rU - l\omega_0) \\ + (1-n)\delta(\omega - rU + l\omega_0)] \Big], \tag{8.192}$$

where

$$\mathcal{Z} = \exp\left[-\sum_{\mathbf{q}} |\gamma_q|^2 \coth\frac{\beta\omega_q}{2} \right], \tag{8.193}$$

$\xi = |\gamma|^2/\sinh(\beta\omega_0/2)$, $I_l(\xi)$ is the modified Bessel function, and δ_{lk} is the Kronecker symbol. The important feature of the DOS, Eq. (8.192), is its nonlinear dependence on the occupation number n, which leads to the switching effect and hysteresis in the I–V characteristics for $d > 2$, as will be shown next. It contains full information about all possible correlation and inelastic effects in transport—in particular, all of the vibron-assisted tunneling processes and vibron side bands—and describes the renormalization of hopping to the leads.

8.6.2. Nonlinear Rate Equation and Switching

Generally, the electron density n_μ obeys an infinite set of rate equations for many-particle GFs that can be derived in the framework of a tunneling Hamiltonian, including correlations as in Section 8.5. In the case of MQDs only weakly coupled with leads, one can apply the Fermi–Dirac golden rule to obtain an equation for n. Equating incoming and outgoing numbers of electrons in MQDs per unit time, we obtain the self-consistent equation for the level occupation n:

$$(1-n)\int_{-\infty}^{\infty} d\omega \{\Gamma_1 f_1(\omega) + \Gamma_2 f_2(\omega)\}\, \rho(\omega) \\ - n\int_{-\infty}^{\infty} d\omega \{\Gamma_1[1-f_1(\omega)] + \Gamma_2[1-f_2(\omega)]\}\, \rho(\omega) = 0, \tag{8.194}$$

where $\Gamma_{1(2)}$ are the transition rates from left (right) leads to MQDs. Taking into account that $\int_{-\infty}^{\infty} \rho(\omega) = d$, Eq. (8.194) for the symmetric leads, $\Gamma_1 = \Gamma_2$, reduces to

$$2nd = \int d\omega\, \rho(\omega)(f_1 + f_2), \tag{8.195}$$

which automatically satisfies $0 \leq n \leq 1$. Explicitly, the self-consistent equation for the occupation number is

$$n = \frac{1}{2}\sum_{r=0}^{d-1} Z_r(n)[na_r + (1-n)b_r], \tag{8.196}$$

where

$$a_r = \mathcal{Z}\sum_{l=0}^{\infty} I_l(\xi)\left(e^{\beta\omega_0 l/2}[f_1(rU - l\omega_0) + f_2(rU - l\omega_0)]\right.$$
$$\left. + (1-\delta_{l0})e^{-\beta\omega_0 l/2}[f_1(rU + l\omega_0) + f_2(rU + l\omega_0)]\right), \tag{8.197}$$

$$b_r = \mathcal{Z}\sum_{l=0}^{\infty} I_l(\xi)\left(e^{\beta\omega_0 l/2}[f_1(rU + l\omega_0) + f_2(rU + l\omega_0)]\right.$$
$$\left. + (1-\delta_{l0})e^{-\beta\omega_0 l/2}[f_1(rU - l\omega_0) + f_2(rU - l\omega_0)]\right). \tag{8.198}$$

The current is expressed as

$$j \equiv \frac{I(V)}{dI_0} = \sum_{r=0}^{d-1} Z_r(n)[na'_r + (1-n)b'_r], \tag{8.199}$$

where

$$a'_r = \mathcal{Z}\sum_{l=0}^{\infty} I_l(\xi)\left(e^{\beta\omega_0 l/2}[f_1(rU - l\omega_0) - f_2(rU - l\omega_0)]\right.$$
$$\left. + (1-\delta_{l0})e^{-\beta\omega_0 l/2}[f_1(rU + l\omega_0) - f_2(rU + l\omega_0)]\right), \tag{8.200}$$

$$b'_r = \mathcal{Z}\sum_{l=0}^{\infty} I_l(\xi)\left(e^{\beta\omega_0 l/2}[f_1(rU + l\omega_0) - f_2(rU + l\omega_0)]\right.$$
$$\left. + (1-\delta_{l0})e^{-\beta\omega_0 l/2}[f_1(rU - l\omega_0) - f_2(rU - l\omega_0)]\right). \tag{8.201}$$

There is only one physical ($0 < n < 0.5$) solution of the rate equation (8.196) and no switching for nondegenerate, $d = 1$, and double-degenerate, $d = 2$, MQDs. However, the switchingappears for $d > 2$. For example, for $d = 4$ the rate equation is of the fourth power in n:

$$\begin{aligned} 2n = {} & (1-n)^3[na_0 + (1-n)b_0] \\ & + 3n(1-n)^2[na_1 + (1-n)b_1] \\ & + 3n^2(1-n)[na_2 + (1-n)b_2] \\ & + n^3[na_3 + (1-n)b_3]. \end{aligned} \tag{8.202}$$

Different from the nondegenerate or double-degenerate MQDs, the rate equation (8.202) for $d = 4$ has two stable physical roots in a certain voltage range and the I–V characteristics show a hysteretic behavior. Ermakov [52] also calculated the I–V curves of the fourfold degenerate dot with the Coulomb and e-ph interactions and found a switching effect in the numerical I–V curves. However, he obtained an unphysical population of each molecular state, $n = 1$. Moreover, the exact theory [59] showed that the averaging procedure over phonons in the transformed Hamiltonian used in Ref. 52 cannot be applied in MQDs since it misses the vibron side bands (see Fig. 8.10).

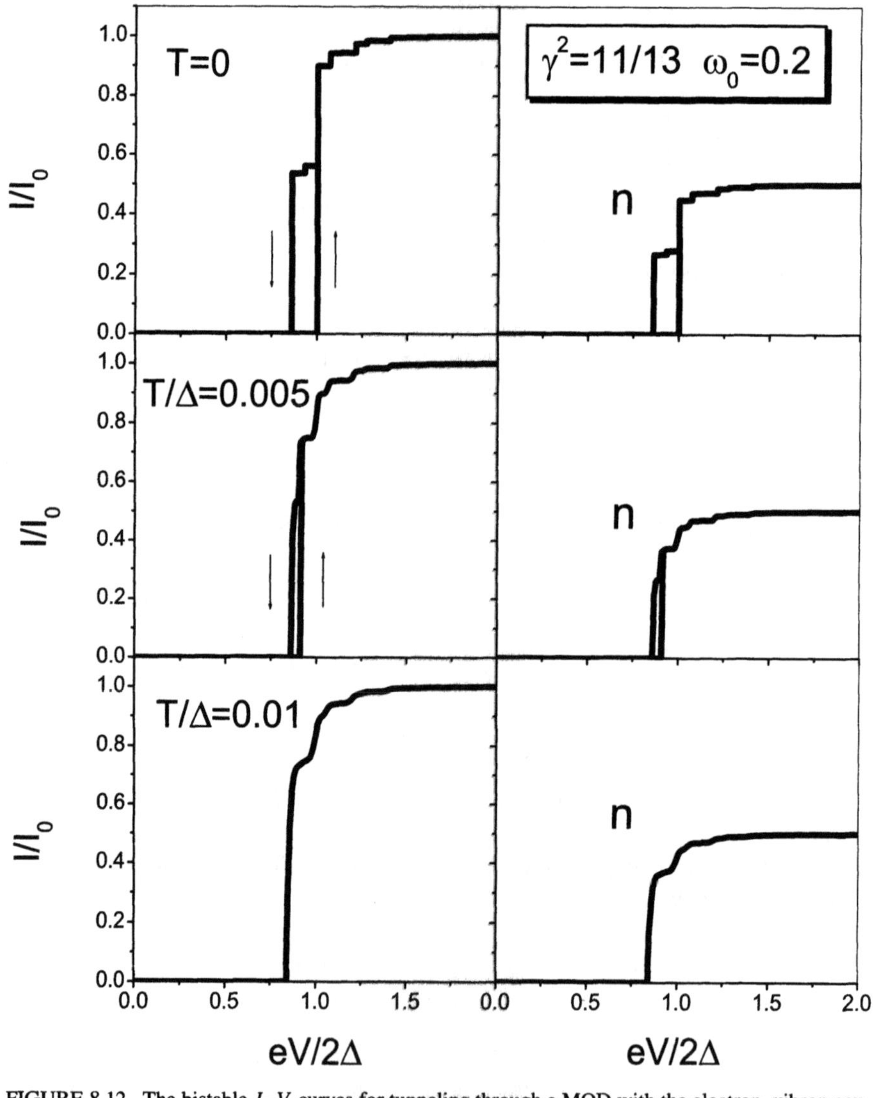

FIGURE 8.12. The bistable I–V curves for tunneling through a MQD with the electron–vibron coupling constant $\gamma^2 = 11/13$ and $\omega_0/\Delta = 0.2$. The up arrows show that the current picks up at some voltage when it is biased and then drops at a lower voltage when the bias is being reduced. The bias dependence of current basically repeats the shape of the level occupation n (right column). Steps on the curve correspond to the changing population of the phonon side bands, which are shown in Fig. 8.10.

Our numerical results for $\omega_0 = 0.2$ (in units of Δ, as all the energies in the problem) and $U^C = 0$ for the coupling constant $\gamma^2 = 11/13$ are shown in Fig. 8.12. This case formally corresponds to a negative Hubbard $U = -2\gamma^2\omega_0 \approx -0.4$ of Section 8.5. (We selected those values of γ^2 to avoid accidental commensurability of the correlated levels separated by U and the phonon side bands.) The threshold for the onset of bistability appears at a voltage bias $eV/2\Delta = 0.86$ for $\gamma^2 = 11/13$ and $\omega_0 = 0.2$. The steps on the I–V curve (Fig. 8.8) are generated by the phonon side bands originating from correlated levels in the dot with the energies Δ, $\Delta + U$, $\ldots$, $\Delta + (d-1)U$. Since ω_0 is not generally commensurate with U, we obtain a quite irregular picture of the steps in I–V curves. The bistability region shrinks with temperature.

8.7. CONCLUSION

The integrated circuit was invented in 1959 and Jack Kilby won the Nobel Prize in Physics for this invention. Since that date, the number of transistors that can be fabricated onto a single chip has been doubling about every 18 months: an observation commonly known as Moore's law. This exponential process has taken the world from a crude chip with a single transistor to integrated circuits with 100 million active components in only 40 years. At the same time, the amount of useful computational work that comes out of an integrated circuit for each unit of electrical power put into it has also increased by roughly 100 million. How much longer can this exponential growth continue? Ever since it was first proposed, there has been a great deal of discussion about when Moore's law would reach a limiting plateau. There is a growing belief that the progress of the silicon technology will reach its physical, engineering, and economic limits in about 10 years. The fact that silicon is approaching the upper limit, however, does not mean that progress in computing will slow down. What will take us beyond 2010 are new technologies that are being pursued in university and corporate laboratories around the world. In particular, at HP Labs, in conjunction with UCLA, and in a few other laboratories, molecular switching devices and systems that will assemble themselves through molecular recognition are designed and being investigated.

There are broad categories of molecular switching behavior (i.e., electrically controlled switching molecules, electromechanically switched molecules, photo-switched molecules, electrochemically switched molecules, etc.). In particular, the switching phenomenon can result from a conformational transformation of certain molecules containing a "moving part," like a bypyridinium ring, which changes its position if the voltage is sufficiently high, or from the interaction of metallic leads with the molecules. Such transformations necessarily involve a large displacement of many atoms so that this *ionic* switching is rather slow, perhaps operating on a millisecond scale. Reversible switching was also observed in some simple molecules. Any molecular devices that exhibit intrinsic switching could be the basis of future active elements of molecular electronics. Molecular self-assembly is widely used to construct well-defined molecular nanostructures.

These nanostructures are formed spontaneously by molecules that assemble under the influence of intermolecular forces. Molecular electronics require switchable devices as well as the interfaces to exchange information with the outside world. Furthermore, these assemblies must be controllable, reversible, and readable at the molecular level.

Consequently, the actual mechanisms of molecular switching and transport through molecular nanowires are of the highest current experimental and theoretical value [74]. The further progress will depend on finding molecules and understanding *intrinsic* mechanisms for their reversible switching from a low- to a high-current state. Here, we have reviewed the multipolaron theory of a novel *current controlled* switching mechanism. The degenerate MQD with strong electron–vibron coupling shows a hysteretic volatile memory if the degeneracy of the molecular level is larger than 2 ($d > 2$). The hysteretic behavior strongly depends on the electron–vibron coupling and characteristic vibron frequencies. The current bistability vanishes above some critical temperature. The origin of the bistability is illustrated in Fig. 8.13. When the current flows through the MQD, the highest occupied molecular orbital–lowest unoccupied molecular orbital (HOMO–LUMO) gap is renormalized down to a lower value due to attractive correlations, so the current-off voltage V_1 turns out to be smaller than the current-on voltage V_2.

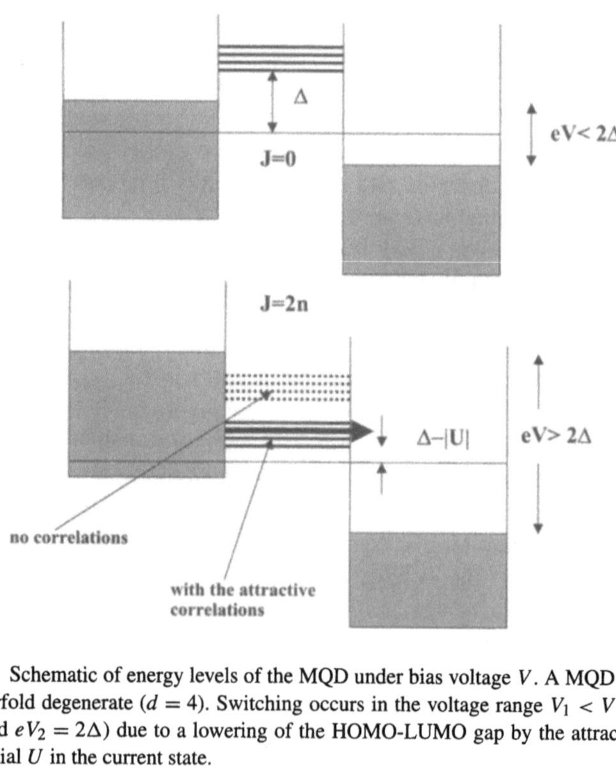

FIGURE 8.13. Schematic of energy levels of the MQD under bias voltage V. A MQD is assumed to be (quasi) fourfold degenerate ($d = 4$). Switching occurs in the voltage range $V_1 < V < V_2$ ($eV_1 = 2(\Delta - |U|)$ and $eV_2 = 2\Delta$) due to a lowering of the HOMO-LUMO gap by the attractive electron–electron potential U in the current state.

Among potential candidates for the negative–U QD are a single C_{60} molecule ($d = 6$), where the electron–vibronic coupling proved to be particularly strong [48], or other carbon nanostructures, including short nanotubes ($d \gg 1$) connected to metal electrodes. There should be no retardation of the switching on the timescale above the inverse vibron (phonon) frequency, which is $10^{-14}s$ or less in carbon-based compounds. It would be very interesting to look for an experimental realization of the model, possibly in a system containing a certain conjugated central part, which exhibits the attractive correlations of carriers with large degeneracy $d > 2$.

Acknowledgment. This work was supported by the Leverhulme Trust (UK) and by EPSRC (UK) (Grant No. EP/C518365/1).

REFERENCES

1. L.D. Landau, *Physikalische Zeitschrift der Sowjetunion* **3**, 664 (1933).
2. S.I. Pekar, Autolocalization of the electron in a dielectric inertially polarizing medium, *Zh. Eksp. Teor. Fiz.* **16**, 335 (1946).
3. H. Froehlich, Electrons in lattice fields, *Adv. Phys.* **3**, 325 (1954).
4. R.P. Feynman, Slow electrons in a polar crystal, *Phys. Rev.* **97**, 660 (1955).
5. E.I. Rashba, Theory of strong interactions of electron excitations with lattice vibrations in molecular crystals, *Optika i spektroskopia* **2**, 75 (1957).
6. J.T. Devreese, Polarons in *Encyclopedia of Applied Physics*, vol. 14, VCH New York, 1996, p. 383.
7. S.V. Tyablikov, Electron energy spectrum in a polar crystal, *Zh. Eksp. Teor. Fiz.* **23**, 381 (1952).
8. J. Yamashita and T. Kurosawa, On electronic current in NiO, *J. Phys.Chem. Solids* **5**, 34 (1958).
9. G.L. Sewell, Electrons in polar crystals, *Phil. Mag.* **3**, 1361 (1958).
10. T. Holstein, Studies of polaron motion, *Ann. Phys.* **8**, 325, 343 (1959); L. Friedman and T. Holstein, Studies of polaron motion 3: the hall mobility of the small polaron, *Ann. Phys.* **21** 494 (1963); D. Emin and T. Holstein, Studies of small-polaron motion 4: adiabatic theory of hall effect, Ann. Phys. **53**, 439 (1969).
11. I.G. Lang and Yu.A. Firsov, Kinetic theory of semiconductors with low mobility, *Sov. Phys. JETP* **16**, 1301 (1963).
12. D.M. Eagles, Optical absorption in ionic crystals involving small polarons, optical absorption in ionic crystals involving small polarons, *Phys. Rev.* **130**, 1381 (1963).
13. Yu.A. Firsov (ed.) *Polarons*, (Moscow: Nauka), 1975; H. Boettger and V.V. Bryksin, *Hopping Conduction in Solids*, Academie-Verlag, Berlin, 1985; A.M. Stoneham, Small polarons and polaron transitions, *J. Chem. Soc. Faraday II* 85, 505 (1989); G.D. Mahan, *Many Particle Physics* , Plenum, New York, 1990; A.L. Shluger and A.M. Stoneham, Small polarons in real crystals—concepts and problems, *J. Phys. Condens. Matter* **1**, 3049 (1993); A.S. Alexandrov and N.F. Mott, *Polarons and Bipolarons*, World Scientific, Singapore, 1995; N. Itoh and A.M. Stoneham, *Materials Modification by Electronic Excitation* Cambridge University Press, Cambridge, 2001.
14. V.L. Vinetskii and M.S. Giterman, On the theory of the interaction of excess charges in ionic crystals, *Sov. Phys. JETP* **6**, 560 (1958).
15. S.G. Suprun and B.Y. Moizhes, Electron correlation effect in pekar bipolaron formation, *Sov. Phys. Solid State* **24**, 903 (1982).

16. J. Adamowski, Formation of frohlich bipolarons, *hys. Rev. B* **39**, 3649 (1989).
17. F. Bassani, M. Geddo, G. Iadonisi, and D. Ninno, Variational calculations of bipolaron binding-energies, *Phys. Rev. B***43**, 5296 (1991).
18. G. Verbist, F.M. Peters, and J.T. Devreese, Large bipolarons in 2 and 3 dimensions, *Phys. Rev. B* **43**, 2712 (1991).
19. K.A. Müller, Jahn-Teller bipolarons and their condensation, *Physica Scripta T* **102** 39 (2002), and references therein.
20. A.S. Alexandrov, Bipolaron anisotropic flat bands, Hall mobility edge, and metal-semiconductor duality of overdoped high-T-c oxides, *Phys. Rev. B* **53**, 2863 (1996).
21. P.W. Anderson, Model for electronic-structure of amorphous-semiconductors, *Phys. Rev. Lett.* **34**, 953 (1975); R.A. Street and N.F. Mott, States in gap in glassy semiconductors, *Phys. Rev. Lett.* **35**, 1293 (1975).
22. B.K. Chakraverty, Possibility of insulator to superconductor phase-transition, *J. Phys. Lett. (Paris)* **40**, L-99, (1979).
23. A.S. Alexandrov and J. Ranninger, Theory of bipolarons and bipolaronic bands, *Phys. Rev. B* **23**, 1796 (1981); A.S. Aleksandrov (Alexandrov), Bipolarons in narrow-zone crystals, *Russ. J. Phys. Chem.* **57**, 167 (1983).
24. S. Aubry, High-Tc superconductivity with polarons and bipolarons: an approach from the insulating state, in *Polarons and Bipolarons in High-Tc Superconductors and Related Materials*, edited by E.K.H. Salje, A.S. Alexandrov, and W.Y. Liang , Cambridge University Press, Cambridge, 1995, p. 271; A.R. Bishop and M. Salkola, Polarons in Peierls-Hubbard models, in *Polarons and Bipolarons in High-Tc Superconductors and Related Materials*, edited by E.K.H. Salje, A.S. Alexandrov, and W.Y. Liang , Cambridge University Press, Cambridge, 1995, p. 353.
25. F. Marsiglio, Pairing in the holstein model in the dilute limit, *Physica C* **244**, 21 (1995).
26. Y. Takada and T. Higuchi, Vertex function for the coupling of an electron with intramolecular phonons—exact results in the antiadiabatic limit, *Phys. Rev. B* **52** 12720 (1995).
27. A.J. Millis, P.B. Littlewood, and B.I. Shraiman, Double exchange alone does not explain the resistivity of La1-xSrxMnO3, *Phys. Rev. Lett.* **74**, 5144 (1995).
28. C. Baesens C and R.S. MacKay, Finite coherence length for equilibrium states of generalized adiabatic Holstein models, *J. Math. Phys.* **38**, 2104 (1997).
29. H. Fehske, J. Loos, and G. Wellein, Spectral properties of the 2D Holstein polaron, *Z. Phys. B* **104**, 619 (1997).
30. A.S. Alexandrov, *Theory of superconductivity: from weak to strong coupling*, IoP Publishing, Bristol, 2003.
31. T. Gerisch, R. Münzner, and A. Rieckers, Canonical versus grand-canonical free energies and phase diagrams of a bipolaronic superconductor model, *J. Statisti. Phys.* **93**, 1021 (1998).
32. T. Frank and M. Wagner, Contrasting unitary transformations for the standard bipolaron model, *Phys. Rev. B* **60**, 3252 (1999).
33. P. Benedetti and R. Zeyher, Holstein model in infinite dimensions at half-filling, *Phys. Rev. B* **58**, 14320 (1998).
34. A.S. Alexandrov, V.V. Kabanov, and D.K. Ray, From electron to small polaron—an exact cluster solution, *Phys. Rev. B* **49**, 9915 (1994).
35. H. Fehske, J. Loos, and G. Wellein, Spectral properties of the 2D Holstein polaron, *Zeitschrift fur Physik B: Cond. Mat.* **104**, 619 (1997).
36. A.H. Romero, D.W. Brown, and K. Lindenberg, Converging toward a practical solution of the Holstein molecular crystal model, *J. Chem. Phys.* **109**, 6504 (1998).
37. A. LaMagna and R. Pucci, Variational study of the discrete Holstein model, *Phys. Rev. B* **53**, 8449 (1996).
38. J. Bonca, S.A. Trugman, and I. Batistic, Holstein polaron, *Phys. Rev. B***60**, 1633 (1999).
39. L. Proville and S. Aubry, Small bipolarons in the 2-dimensional Holstein-Hubbard model:The adiabatic limit, *Eur. Phys. J. B* **11**, 41 (1999).

40. S.A. Trugman, J. Bonca, and L.C. Ku, Statics and dynamics of coupled electron-phonon systems, *Int. J. Mod. Phys. B* 15, 2707 (2001), and references therein.
41. P.E. Kornilovitch, Continuous-Time Quantum Monte Carlo Algorithm for the Lattice Polaron, *Phys. Rev. Lett.* **81**, 5382 (1998).
42. N.V. Prokof'ev and B.N. Svistunov, Polaron Problem by Diagrammatic Quantum Monte Carlo, *Phys. Rev. Lett.* **81**, 2514 (1998) .
43. J.M. Lehn, Perspectives in supramolecular chemistry—from molecular recognition towards molecular information-processing and self-organization, *Angew. Chem. Int. Ed.* Engl. **29**, 1304 (1990).
44. J.M. Tour, Molecular electronics: Synthesis and testing of components, *Acc. Chem. Res.* **33**, 791 (2000).
45. A. Aviram and M. Ratner (eds.), *Molecular Electronics: Science and Technology*, New York Academy of Science, New York, 1998.
46. C.P. Collier, E.W. Wong, M. Belohradsky, F.M. Raymo, J.F. Stoddart, P.J. Kuekes, R.S. Williams, and J.R. Heath, Electronically configurable molecular-based logic gates, *Science* **285**, 391 (1999); D.I. Gittins, D. Bethell, D.J. Schiffrin, and R.J. Nichols, A nanometre-scale electronic switch consisting of a metal cluster and redox-addressable groups, *Nature (London)* **408**, 67 (2000).
47. N.B. Zhitenev, H. Meng, and Z. Bao, Conductance of small molecular junctions, *Phys. Rev. Lett.* **88**, 226801 (2002).
48. J. Park, A.N. Pasupathy, J.I. Goldsmith, C. Chang, Y. Yaish, J.R. Retta, M. Rinkoski, J.P. Sethna, H.D. Abruña, P.L. McEuen, and D.C. Ralph, Coulomb blockade and the Kondo effect in single-atom transistors, *Nature (London)* **417** 722 (2002).
49. L.I. Glazman and R.I. Shekhter, Inelastic resonance tunneling of electrons through a potential barrier, *Zh. Eksp. Teor. Fiz.* **94**,292 (1987); N.S. Wingreen, K.W. Jacobsen, and J.W. Wilkins, Inelastic-scattering in resonant tunneling, *Phys. Rev. B* **40**, 11834 (1989).
50. Xi Li, Hao Chen, and Shi-xun Zhou, Conductance of a quantum dot with a Hubbard interaction in the presence of a boson field, *Phys. Rev. B* **52**, 12202 (1995).
51. K. Kang, Transport through an interacting quantum dot coupled to two superconducting leads, *Phys. Rev. B* **57**, 11891 (1998).
52. V.N. Ermakov, Resonant electron tunneling through double-degenerate local state with account of strong electron-phonon interaction, *Physica E* **8**, 99 (2000).
53. M. Di Ventra1, S.-G. Kim, S. T. Pantelides, and N.D. Lang, Temperature Effects on the Transport Properties of Molecules, *Phys. Rev. Lett.* **86**, 288 (2001).
54. N. Ness, S.A. Shevlin, and A.J. Fisher, Coherent electron-phonon coupling and polaronlike transport in molecular wires, *Phys. Rev. B* **63**, 125422 (2001).
55. U. Lundin and R.H. McKenzie, Temperature dependence of polaronic transport through single molecules and quantum dots, *Phys. Rev. B* **66**, 075303 (2002).
56. A.S. Alexandrov, I.K. Yanson, and J. Demsar (eds.), *Molecular Nanowires and Other Quantum Objects*, Kluwer Academic, Amsterdam, 2004.
57. A.S. Alexandrov, A.M. Bratkovsky, and R.S. Williams, Bistable tunneling current through a molecular quantum dot, *Phys. Rev. B* **67**, 075301 (2003).
58. D. Stewart, Y. Chen, and R.S. Williams, unpublished data.
59. A.S. Alexandrov and A.M. Bratkovsky, Memory effect in a molecular quantum dot with strong electron-vibron interaction, *Phys. Rev. B* **67**, 235312 (2003).
60. A.S. Alexandrov and P.E. Kornilovich, Mobile Small Polaron, *Phys. Rev. Lett.* **82**, 807 (1999).
61. A.S. Alexandrov and P.E. Kornilovitch, The Frohlich-Coulomb model of high-temperature superconductivity and charge segregation in the cuprates, *J. Phys.: Condens. Matter* **14**, 5337 (2002).
62. N.S. Wingreen and Y. Meir, Anderson model out of equilibrium: Noncrossing-approximation approach to transport through a quantum dot, *Phys. Rev. B* **49**, 11040 (1994) and references therein.

63. W.P. Su and J.R. Schrieffer, Soliton dynamics in polyacetylene, *Proc. Natl. Acad. Sci.* **77**, 5626 (1980).
64. A. Feldblum, J.H. Kaufman, S. Etemad, and A.J. Heeger, Opto-electrochemical spectroscopy of trans-(CH)x, *Phys. Rev. B* **26**, 815 (1982).
65. R.R. Chance, J.L. Bredas, and R. Silbey, Bipolaron transport in doped conjugated polymers, *Phys. Rev. B* **29**, 4491 (1984).
66. M.G. Ramsey, D. Steinmuller, and F.P. Netzer, Explicit evidence for bipolaron formation: Cs-doped biphenyl, *Phys. Rev. B* **42**, 5902 (1990).
67. D. Steinmuller, M.G. Ramsey, and F.P. Netzer, Polaron and bipolaronlike states in n-doped bithiophene, *Phys. Rev. B* **47**, 13323 (1993).
68. L.S. Swanson, J. Shinar, A.R. Brown, D.D.C. Bradley, R.H. Friend, P.L. Burn, A. Kraft, and A.B. Holmes, Electroluminescence-detected, conductivity-detected, and photoconductivity-detected magnetic-resonance study of poly(p-phenylenevinylene)-based light-emitting-diodes, *Synth. Metals* **55**, 241 (1993).
69. A.S. Alexandrov, A.M. Bratkovsky, and P.E. Kornilovitch, Two-electron elastic tunneling in low-dimensional conductors, *Phys. Rev. B* **65**, 155209 (2002).
70. Y. Meir and N.S. Wingreen, Landauer formula for the current through an interacting electron region, *Phys. Rev. Lett.* **68**, 2512 (1992).
71. R. Micnas, J. Ranninger, and S. Robaszkiewicz, Superconductivity in narrow-band systems with local nonretarded attractive interactions, *Rev. Mod. Phys.* **62**, 113 (1990), and references therein.
72. A.S. Alexandrov and V.V. Kabanov, Theory of superconducting Tc of doped fullerenes, *Phys. Rev. B* **54**, 3655 (1996).
73. A.S. Alexandrov and A.M. Bratkovsky, The essential interactions in oxides and spectral weight transfer in doped manganites, *J. Phys.- Condens. Mat.* **11**, L531 (1999).
74. J.R. Heath, J.F. Stoddart, and R.S. Williams, More on molecular electronics, *Science* **303**, 1136 (2004).

Index

D

MIX
Papier aus verantwortungsvollen Quellen
Paper from responsible sources
FSC® C105338

If you have any concerns about our products,
you can contact us on
ProductSafety@springernature.com

In case Publisher is established outside the EU,
the EU authorized representative is:
Springer Nature Customer Service Center GmbH
Europaplatz 3, 69115 Heidelberg, Germany

Printed by Libri Plureos GmbH
in Hamburg, Germany